Lecture Notes in Bioinformatics 7036

Edited by S. Istrail, P. Pevzner, and M. Waterman

Subseries of Lecture Notes in Computer Science

Marco Loog   Lodewyk Wessels
Marcel J.T. Reinders   Dick de Ridder (Eds.)

Pattern Recognition in Bioinformatics

6th IAPR International Conference, PRIB 2011
Delft, The Netherlands, November 2-4, 2011
Proceedings

 Springer

Series Editors

Sorin Istrail, Brown University, Providence, RI, USA
Pavel Pevzner, University of California, San Diego, CA, USA
Michael Waterman, University of Southern California, Los Angeles, CA, USA

Volume Editors

Marco Loog
Marcel J.T. Reinders
Dick de Ridder
Delft University of Technology
Pattern Recognition & Bioinformatics Group
Mekelweg 4, 2628 CD Delft, The Netherlands
E-mail: {m.loog, m.j.t.reinders, d.deridder}@tudelft.nl

Lodewyk Wessels
Netherlands Cancer Institute
Bioinformatics and Statistics
Plesmanlaan 121, 1066 CX Amsterdam, The Netherlands
E-mail: l.wessels@nki.nl

ISSN 0302-9743   e-ISSN 1611-3349
ISBN 978-3-642-24854-2   ISBN 978-3-642-24855-9 (eBook)
DOI 10.1007/978-3-642-24855-9
Springer Heidelberg Dordrecht London New York

Library of Congress Control Number: 2011939081

CR Subject Classification (1998): J.3, I.5, F.2.2, I.2, I.4, H.3.3, H.2.8

LNCS Sublibrary: SL 8 – Bioinformatics

Typesetting: Camera-ready by author, data conversion by Scientific Publishing Services, Chennai, India

Printed on acid-free paper

Springer is part of Springer Science+Business Media (www.springer.com)

Preface

In modern biology, high-throughput measurement devices allow life scientists to gather data at unprecedented rates. To make sense of these data, computational biologists and system biologists construct quantitative models, many of which depend on pattern recognition techniques. Application of these techniques is challenging due to the large volumes of data and background information, noisy measurements and target outputs, highly diverse data types etc. To address these bioinformatics challenges, research in pattern recognition has been forced to move beyond the simple application of classifiers to high-throughput data, with which the field started over a decade ago. In 2006, the Pattern Recognition in Bioinformatics (PRIB) meeting series was established, under the auspices of the International Association of Pattern Recognition (IAPR), to support the exchange of ideas on how pattern recognition can be developed and extended to support the life sciences.

The Sixth conference in this series was organized in Delft, The Netherlands, November 2–4, 2011. In total, 35 papers were submitted, of which 29 were selected for publication in these proceedings after peer review by the Program Committee. The contributions demonstrate the wide range of possible applications of pattern recognition in bioinformatics: novel algorithms to handle traditional pattern recognition problems such as (bi)clustering, classification and feature selection; applications of (novel) pattern recognition techniques to infer and analyze biological networks and studies on specific problems such as biological image analysis and the relation between sequence and structure. In addition to the 29 paper presentations, the conference saw invited presentations by Yves Moreau (Bioinformatics Research Group, Katholieke Universiteit Leuven, Belgium), Mark Girolami (Department of Statistical Science, University College London, UK), Pierre Baldi (Institute of Genomics and Bioinformatics, University of California, Irvine, USA), Alvis Brazma (European Bioinformatics Institute, Hinxton, UK) and Berend Snel (Department of Biology, Utrecht University, The Netherlands).

We would like to thank all participants in the conference, in particular the authors who contributed their work to this book. We are grateful for the support of IAPR's Technical Committee 20 and the PRIB Steering Committee. The help of the members of the Program Committee was invaluable in evaluating the submissions and defining the final program. EasyChair (http://www.easychair.org) was of great help in handling the submissions, reviews, decisions, and accepted papers. We would also like to thank the co-organizers (Jeroen de Ridder, Manuele Bicego, David Tax and Thomas Abeel) as well as members of the Pattern Recognition & Bioinformatics group (Delft University of Technology) for their assistance before and during the conference as well as the Netherlands Bioinformatics Centre and the EU PASCAL2 Network of Excellence for financial support. Last but not least, the conference could not have been organized without the invaluable help of Saskia Peters, who took care of administration and local arrangements.

August 2011

Marco Loog
Lodewyk Wessels
Marcel Reinders
Dick de Ridder

Organization

Program Committee

Jesús S. Aguilar-Ruiz	Pablo de Olavide University, Spain
Shandar Ahmad	National Institute of Biomedical Innovation, Japan
Tatsuya Akutsu	Kyoto University, Japan
Jaume Bacardit	University of Nottingham, UK
Rainer Breitling	University of Groningen, The Netherlands
Sebastian Böcker	Friedrich Schiller University Jena, Germany
Frederic Cazals	INRIA Sophia, France
Dick De Ridder	Delft University of Technology, The Netherlands
Tjeerd Dijkstra	Radboud University Nijmegen, The Netherlands
Federico Divina	Pablo de Olavide University, Spain
Bas E. Dutilh	CMBI / NCMLS / UMCN, The Netherlands
Richard Edwards	University of Southampton, UK
Maurizio Filippone	University College London, UK
Rosalba Giugno	University of Catania, Italy
Michael Habeck	MPIs for Biological Cybernetics and Developmental Biology, Germany
Jin-Kao Hao	University of Angers, France
Morihiro Hayashida	Kyoto University, Japan
Tom Heskes	Radboud University Nijmegen, The Netherlands
Pavol Jancura	Radboud University Nijmegen, The Netherlands
Zhenyu Jia	University of California, Irvine, USA
Giuseppe Jurman	Fondazione Bruno Kessler, Italy
Visakan Kadirkamanathan	The University of Sheffield, UK
Seyoung Kim	Carnegie Mellon University, USA
Walter Kosters	Leiden University, The Netherlands
Xuejun Liu	Nanjing University of Aeronautics and Astronautics, China
Stefano Lonardi	UC Riverside, USA
Elena Marchiori	Radboud University, The Netherlands
Francesco Masulli	University of Genoa, Italy
Jason Moore	Dartmouth College, USA
Alison Motsinger-Reif	North Carolina State University, USA

Table of Contents

Session 4: Biomarker Selection and Classification (2)

Session 5: Image Analysis (1)

Session 6: Biomarker Selection and Classification (3)

Session 7: Network Inference and Analysis (2)

Session 8: Sequence, Structure, and Interactions

Session 9: Image Analysis (2)

A New Framework for Co-clustering of Gene Expression Data[*]

Shuzhong Zhang[1],[**], Kun Wang[2], Bilian Chen[3], and Xiuzhen Huang[4],[***]

[1] Industrial and Systems Engineering Program, University of Minnesota,
Minneapolis, MN 55455, USA
zhangs@umn.edu
[2] Department of Computer Science, Arkansas State University,
Jonesboro, AR 72467, USA
kun.wang@smail.astate.edu
[3] Department of Systems Engineering and Engineering Management,
The Chinese University of Hong Kong, Shatin, Hong Kong
blchen@se.cuhk.edu.hk
[4] Department of Computer Science, Arkansas State University,
Jonesboro, AR 72467, USA
xhuang@astate.edu

Abstract. A new framework is proposed to study the co-clustering of gene expression data. This framework is based on a generic tensor optimization model and an optimization method termed *Maximum Block Improvement* (MBI) recently developed in [3]. Not only can this framework be applied for co-clustering gene expression data with genes expressed at different conditions represented in 2D matrices, but it can also be readily applied for co-clustering more complex high-dimensional gene expression data with genes expressed at different tissues, different development stages, different time points, different stimulations, etc. Moreover, the new framework is so flexible that it poses no difficulty at all to incorporate a variety of clustering quality measurements. In this paper, we demonstrate the effectiveness of this new approach by providing the details of one specific implementation of the algorithm, and presenting the experimental testing on microarray gene expression datasets. Our results show that the new algorithm is very efficient and it performs well for identifying patterns in gene expression datasets.

1 Introduction

Microarray and next-generation sequencing (also, high-throughput sequencing) technologies produce huge amount of datasets of genome-wide gene expression at

---

[*] This research is partially supported by NIH Grant # P20 RR-16460 from the IDeA Networks of Biomedical Research Excellence (INBRE) Program of the National Center for Research Resources.

[**] On leave from Department of Systems Engineering and Engineering Management, The Chinese University of Hong Kong, Shatin, Hong Kong zhang@se.cuhk.edu.hk

[***] Corresponding author.

M. Loog et al. (Eds.): PRIB 2011, LNBI 7036, pp. 1–12, 2011.
© Springer-Verlag Berlin Heidelberg 2011

different tissues, different development stages, different time points, and different stimulations. These datasets could significantly facilitate and benefit biological hypothesis testing and discovery. However, the availability of these gene expression datasets at the same time brings the challenge of how to transform the large amount of gene expression data to information meaningful for biologists and life-scientists. Especially this imposes increasing demands for efficient computational models and approaches for processing and analyzing these data.

Clustering, as an effective approach, is usually applied to partition gene expression data into groups, where each group aggregates genes with similar expression levels. All the classical clustering algorithms are focused on clustering genes into a number of groups based on their similar expression on all the considered conditions.

Cheng and Church [4] introduced the concept of co-cluster gene expression data and developed an effective measure of the co-clusters based on the mean square residue and a greedy node-deletion algorithm. Their algorithm could cluster genes and conditions simultaneously and thus could discover the similar expression of a certain group of genes on a certain group of conditions and vice versa. Later many different co-clustering algorithms were developed. For example, the authors in [5] formulated the objective functions based on minimizing two measures of squared residue that are similar to those used by Cheng and Church [4] and Hartigan [6]. Their iterative algorithm could directly minimize the squared residues and find $k * l$ co-clusters simultaneously as opposed to finding a single co-cluster at a time like Cheng and Church. Readers may refer to [11,7,5] for the ideas of other co-clustering algorithms.

In this paper we propose a new framework to study the co-clustering of gene expression data. This new framework is based on a generic tensor optimization model and a method termed *Maximum Block Improvement* (MBI). This framework not only can be used for co-clustering of gene expression data with genes expressed at different conditions (genes × conditions) represented in 2D matrices, but also it can be readily applied for co-clustering of gene expression data in 3D, 4D, 5D with genes expressed at different tissues, different development stages, different time points, different stimulations, and so on and so forth (e.g., genes×tissues×development stages×time points×stimulations) and even more complex high-dimensional matrices. Moreover, this framework is flexible enough to incorporate different objective functions. We demonstrate this new framework by providing the details of the algorithm for one model with one specific objective function under the framework, the implementation of the algorithm and the experimental testing on microarray gene expression datasets. Our algorithm turns out to be very efficient (which runs for only a few minutes on a regular PC for large gene expression datasets) and performs well for identifying patterns in microarray data sets compared with other approaches (refer to the section of experimental results).

The reminder of the paper is organized as follows. Section 2 presents the new generic co-clustering framework. Section 3 describes the algorithm for one specific 2D gene expression co-clustering model. Section 4 presents experimental testing results on gene expression datasets. Section 5 concludes the paper.

2   A New Generic Framework for Co-clustering

In this section we first present our model for the co-clustering problem based on tensor optimization and then give a generic algorithm for high-dimensional gene expression data co-clustering.

2.1   Background of Tensor Operations

Readers are referred to [9] for different tensor operations. We will need in the following the operation *mode product* between a tensor X and a matrix P. Suppose that $X \in \Re^{p_1 \times p_2 \times \cdots \times p_d}$ is an d-dimensional tensor and $P \in \Re^{p_i \times m}$ is a 2D matrix. Then, $X \times_i P$ is a tensor in $\Re^{p_1 \times p_2 \times \cdots \times p_{i-1} \times m \times p_{i+1} \times \cdots \times p_d}$, whose $(j_1, j_2, \cdots, j_{i-1}, j_i, j_{i+1}, \cdots, j_d)$-th component is defined by

$$(X \times_i P)_{j_1, j_2, \cdots, j_{i-1}, j_i, j_{i+1}, \cdots, j_d} = \sum_{\ell=1}^{p_i} X_{j_1, j_2, \cdots, j_{i-1}, \ell, j_{i+1}, \cdots, j_d} P_{\ell, j_i}.$$

The mode product is communicative, i.e.,

$$X \times_i P \times_j Q = X \times_j Q \times_i P.$$

2.2   The Optimization Model of the Co-clustering Problem

The co-clustering problem is described as follows. Suppose that $A \in \Re^{n_1 \times n_2 \times \cdots \times n_d}$ is a d-dimensional tensor. Let $I_j = \{1, 2, \cdots, n_j\}$ be the set of indices on the j-th dimension, $j = 1, 2, ..., d$. We wish to find a p_j-partition of the index set I_j, say $I_j = I_1^j \cup I_2^j \cup \cdots \cup I_{p_j}^j$, where $j = 1, 2, ..., d$, in such a way that each of the *sub-tensor* $A_{I_{i_1}^1 \times I_{i_2}^2 \times \cdots \times I_{i_d}^d}$ is as tightly packed up as possible, where $1 \leq i_j \leq n_j$ and $j = 1, 2, ..., d$.

Suppose that $X \in \Re^{p_1 \times p_2 \times \cdots \times p_d}$ is the tensor for the co-cluster values. Let $X_{j_1, j_2, \cdots, j_{i-1}, j_i, j_{i+1}, \cdots, j_d}$ be the value of the co-cluster $(j_1, j_2, \cdots, j_{i-1}, j_i, j_{i+1}, \cdots, j_d)$ with $1 \leq j_i \leq p_i$, $i = 1, 2, ..., d$.

Next, we define a row-to-column assignment matrix $Y^j \in \Re^{n_j \times p_j}$ for the indices for the j-th array of tensor A, with:

$$Y_{ik}^j = \begin{cases} 1, & \text{if } i \text{ is assigned to the } k\text{-th partition } I_k^j; \\ 0, & \text{otherwise.} \end{cases}$$

Then, we introduce a *proximity* measure $f(s) : \Re \to \Re_+$, with the property that $f(s) \geq 0$ for all $s \in \Re$ and $f(s) = 0$ if and only if $s = 0$. The co-clustering problem can be formulated as

$$(CC) \min \sum_{j_1=1}^{n_1} \sum_{j_2=1}^{n_2} \cdots \sum_{j_d=1}^{n_d} f\left(A_{j_1, j_2, \cdots, j_d} - (X \times_1 Y^1 \times_2 Y^2 \times_3 \cdots \times_d Y^d)_{j_1, j_2, \cdots, j_d}\right)$$

$$\text{s.t. } X \in \Re^{p_1 \times p_2 \times \cdots \times p_d}, Y^j \in \Re^{n_j \times p_j} \text{ is an assignment matrix, } j = 1, 2, ..., d.$$

A variety of proximity measures could be considered. For instance, if $f(s) = s^2$, then (CC) can be written as

(P_1) min $\left\| A - X \times_1 Y^1 \times_2 Y^2 \times_3 \cdots \times_d Y^d \right\|_F$
  s.t. $X \in \Re^{p_1 \times p_2 \times \cdots \times p_d}$, $Y^j \in \Re^{n_j \times p_j}$ is an assignment matrix, $j = 1, 2, ..., d$.

If $f(s) = |s|$ then (CC) can be written as

(P_2) min $\displaystyle\sum_{j_1=1}^{n_1} \sum_{j_2=1}^{n_2} \cdots \sum_{j_d=1}^{n_d} \left| A_{j_1, j_2, \cdots, j_d} - (X \times_1 Y^1 \times_2 Y^2 \times_3 \cdots \times_d Y^d)_{j_1, j_2, \cdots, j_d} \right|$
  s.t. $X \in \Re^{p_1 \times p_2 \times \cdots \times p_d}$, $Y^j \in \Re^{n_j \times p_j}$ is an assignment matrix, $j = 1, 2, ..., d$.

A third possible formulation can be

(P_3) min $\displaystyle\max_{1 \leq j_i \leq n_i;\, i=1,2,...,d} \left| A_{j_1, j_2, \cdots, j_d} - (X \times_1 Y^1 \times_2 Y^2 \times_3 \cdots \times_d Y^d)_{j_1, j_2, \cdots, j_d} \right|$
  s.t. $X \in \Re^{p_1 \times p_2 \times \cdots \times p_d}$, $Y^j \in \Re^{n_j \times p_j}$ is an assignment matrix, $j = 1, 2, ..., d$.

2.3   A Generic Algorithm for Co-clustering

In this section we provide an algorithm for the (CC) model of the co-clustering problem. The algorithm is based on a recent work in mathematical optimization ([3]), where the authors considered a generic optimization model in the form of

$$(G) \ \max \ f(x^1, x^2, \cdots, x^d)$$
$$\text{s.t.} \ \ x^i \in S^i \subseteq \Re^{n_i}, \ i = 1, 2, ..., d,$$

where $f : \Re^{n_1 + \cdots + n_d} \to \Re$ is a general continuous function, and S^i is a general set, $i = 1, 2, ..., d$. They proposed a new method termed Maximum Block Improvement (MBI) for solving the optimization problem (G).

Note that in [3], the authors proved that the Maximum Block Improvement Method method guarantees to converge to a stationary point:

Theorem 1. *([3]) If S^i is compact, $i = 1, 2, ..., d$, then any cluster point of the iterates $(x_k^1, x_k^2, \cdots, x_k^d)$, say $(x_*^1, x_*^2, \cdots, x_*^d)$, will be a stationary point for (G); i.e.,*

$$x_*^i = \arg \max_{x^i \in S^i} f(x_*^1, \cdots, x_*^{i-1}, x^i, x_*^{i+1}, \cdots, x_*^d), \ \textit{for } i = 1, 2, ..., d.$$

We can see that all our formulations, (P_1), (P_2) and (P_3), are in the format of (G), which are suitable for the application of the MBI method. Refer to Figure 1 for our generic algorithm for the co-clustering problem based on the MBI method.

The model contains the block variables X, Y^1, Y^2, ..., Y^d. For the fixed Y^j variables, $j = 1, 2, ..., d$, the search of X becomes:

– In the case of (P_1), the problem is a least square problem;
– In the case of (P_2) or (P_3), the problems are linear programming.

To appreciate the computational complexity of the models under consideration, we remark here that even if the X block variable is fixed, to search for the *two* joint optimal assignments of, say, Y^1 and Y^2, while all other Y's are fixed, is already NP-hard. (We omit the proof here due to the space limit).

Generic co-clustering algorithm

Input: $A \in \Re^{n_1 \times n_2 \times \cdots \times n_d}$ is an d-dimensional tensor. Parameters k_1, k_2 and k_d, are all positive integers, $0 < k_i \le n_i$, $1 \le i \le d$.
Output: $k_1 \times k_2 \times \cdots \times k_d$ co-clusters of A.

Main Variables:
A non-negative integer k as the loop counter;
A $k_1 \times k_2 \cdots \times k_d$-tensor X with each entry a real number as the artificial central point of one of the co-clusters;
A $n_i \times k_i$-matrix Y_i as the assignment matrix with $\{0, 1\}$ as the value of each entry, $1 \le i \le d$.

begin

  0 *(Initialization)*. $Y^0 = X$; Choose a feasible solution $(Y_0^0, Y_0^1, Y_0^2, \cdots, Y_0^d)$ and compute the initial objective value $v_0 := f(Y_0^0, Y_0^1, Y_0^2, \cdots, Y_0^d)$. Set the loop counter $k := 0$.

  1 *(Block Improvement)*. For each $i = 0, 1, 2, \ldots, d$, solve

$$(G_i) \ \max f(Y_k^0, Y_k^1, \cdots, Y_k^{i-1}, Y^i, Y_k^{i+1}, \cdots, Y_k^d)$$
$$\text{s.t.} \quad Y^i \in \Re^{n_j \times p_j} \text{ is an assignment matrix,}$$

and let

$$y_{k+1}^i := \arg \max f(Y_k^0, Y_k^1, \cdots, Y_k^{i-1}, Y^i, Y_k^{i+1}, \cdots, Y_k^d)$$
$$w_{k+1}^i := f(Y_k^0, Y_k^1, \cdots, Y_k^{i-1}, y_{k+1}^i, Y_k^{i+1}, \cdots, Y_k^d).$$

  2 *(Maximum Improvement)*. Let $w_{k+1} := \max_{1 \le i \le d} w_{k+1}^i$ and $i^* = \arg \max_{1 \le i \le d} w_{k+1}^i$. Let

$$Y_{k+1}^i := Y_k^i, \ \forall \ i \in \{0, 1, 2, \cdots, d\} \backslash \{i^*\}$$
$$Y_{k+1}^{i^*} := y_{k+1}^{i^*}$$
$$v_{k+1} : := w_{k+1}.$$

  3 *(Stopping Criterion)*. If $|v_{k+1} - v_k| < \epsilon$, go to Step 4. Otherwise, set $k := k + 1$, and go to Step 1.

  4 *(Outputting Co-clusters)*. According to the assignment matrices $Y_{k+1}^1, Y_{k+1}^2, \cdots, Y_{k+1}^d$, print the $k_1 \times k_2 \times \cdots \times k_d$ co-clusters of A.

end

Fig. 1. Algorithm Based on the MBI Method in [3]

3   Algorithm for Co-clustering 2D Matrix Data

We have implemented the algorithm for co-clustering gene expression data in 2D-matrices when the (P_1) formulation is used. Given a 2D-matrix A with m rows and n columns, which represents the gene expressions of m different genes under n different conditions. We apply our co-clustering algorithm to partition the genes and conditions at the same time to get $k_1 \times k_2$ submatrices, where k_1 is the number of partitions of the m genes and k_2 is the number of partitions of the n conditions. Refer to Figure 2 for the details of our algorithm.

4   Experimental Results

We use two microarray datasets to test our algorithm and make comparisons with other clustering and co-clustering methods. The first dataset is the gene expression of a yeast cell cycle dataset with 2884 genes and 17 conditions, where the expression values are in the range 0 to 595. The second dataset is the gene expression of a human B-cell lymphoma dataset with 4026 genes and 96 conditions, where the values are in the range -749 and 642. The detailed information about the datasets could be found in Cheng and Church [4], Tavazoie et al. [13] and Alizadeh et al. [2].

4.1   Implementation Details and Some Discussions

Our algorithm is implemented using C++. The experimental testing is performed on a regular PC (configuration: processor: Pentium dual-core CPU, T4200 @ 2.00GHz; memory: 3GB; operating system: 64-bit windows 7; compiler: Microsoft Visual C++ 2010). The figures are generated using MATLAB R2010a.

We tested our algorithm using different initial values of the three matrices X, Y_1 and Y_2 (refer to Figure 2). The setup of the initial values of the three matrices includes using random values for the three matrices, using subsets of values in A to initialize X, limiting the number of 1s to be one in each row of matrices Y_1 and Y_2, and using the values of the matrices Y_1 and Y_2 to calculate the values of the matrix X. We found out that the initial values of the three matrices will not significantly affect the convergence of our algorithm (refer to Figure 3 for the final objective function values and the running times over 50 runs for the yeast dataset to generate 30×3 co-clusters).

We also tested our algorithm for different numbers of partitions of the rows and the columns, that is, different values of k_1 and k_2. For example, when $k_1 = 30$ and $k_2 = 3$, our program generates the co-clusters of the yeast cell dataset in 40.252 seconds with the final objective function value -7386.75, and when $k_1 = 100$ and $k_2 = 5$, our program generates the co-clusters of the yeast cell dataset in 90.138 seconds with the final objective function value -6737.86. The running time of our algorithm is comparable to the running time of the algorithms developed in [5].

Algorithm for 2D-matrix co-clustering based on the P1 model

Input: A 2D-matrix A with m rows and n columns. Two parameters k_1 and k_2, where k_1 and k_2 are both positive integers.

Output: $(k_1 \times k_2)$ co-clusters of the matrix A, where k_1 is the number of partitions of the m rows and k_2 is the number of partitions of the n columns.

Main Variables:

A non-negative integer k as the loop counter;

A $k_1 \times k_2$ matrix X with each entry a real number as the artificial central point of one of the $k_1 * k_2$ co-clusters of the matrix A;

A $m \times k_1$ matrix Y_1 as the row assignment matrix with $\{0, 1\}$ as the value of each entry; and

A $n \times k_2$ matrix Y_2 as the column assignment matrix with $\{0, 1\}$ as the value of each entry.

begin

0 *(Initialization)*. Set the loop counter $k := 0$. Randomly set the initial values of the three matrices X^k, Y_1^k and Y_2^k and compute the initial objective value $v_0 := \max - \left\| A - X \times_1 Y_1 \times_2 Y_2 \right\|_F$.

1 *(Block Improvement)*.

1.1 Based on the values in matrices X^k and Y_1^k, get the optimal column assignment matrix Y_2' and compute the objective value $v_{Y_2} := \max - \left\| A - X^k \times_1 Y_1^k \times_2 Y_2' \right\|_F$;

1.2 Based on the values in matrices X^k and Y_2^k, get the optimal row assignment matrix Y_1' and compute the objective value $v_{Y_1} := \max - \left\| A - X^k \times_1 Y_1' \times_2 Y_2^k \right\|_F$;

1.3 Based on the values in matrices Y_1^k and Y_2^k, get the optimal matrix X' and compute the objective value $v_X := \max - \left\| A - X' \times_1 Y_1^k \times_2 Y_2^k \right\|_F$.

2 *(Maximum Improvement)*. $v_{k+1} := \max\{v_{Y_2}, v_{Y_1}, v_x\}$;
If $v_{k+1} = v_{Y_2}$ then update Y_2:
$X^{k+1} = X^k$, $Y_1^{k+1} = Y_1^k$, and $Y_2^{k+1} = Y_2'$;
If $v_{k+1} = v_{Y_1}$ then update Y_1:
$X^{k+1} = X^k$, $Y_1^{k+1} = Y_1'$, and $Y_2^{k+1} = Y_2^k$;
If $v_{k+1} = v_X$ then update X:
$X^{k+1} = X'$, $Y_1^{k+1} = Y_1^k$, and $Y_2^{k+1} = Y_2^k$;

3 *(Stopping Criterion)*. If $|v_{k+1} - v_k| < \epsilon$, go to Step 4. Otherwise, set $k := k + 1$, and go to Step 1.

4 *(Outputting Co-clusters)*. According to the assignment matrices Y_{k+1}^1, Y_{k+1}^2, print the $k_1 \times k_2$ co-clusters of A.

end

Fig. 2. Algorithm for 2D-matrix Co-clustering

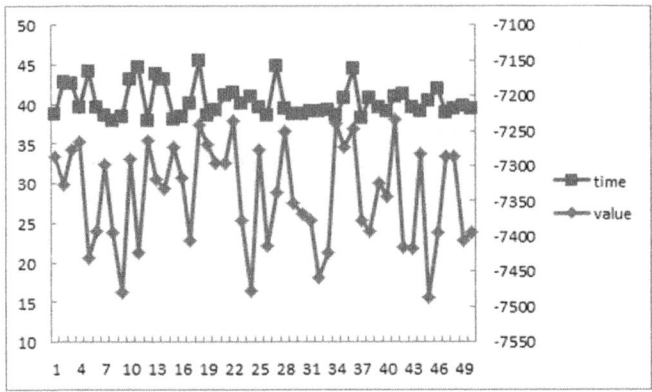

Fig. 3. The final objective function values (the right axis) and the running time (the left axis, in seconds) of 50 runs of our algorithm with random initial values of the three matrices X, Y_1 and Y_2 on the yeast dataset to generate 30×3 co-clusters

Refer to Figure 4 for the objective function value versus iteration of our algorithm on the yeast cell dataset and the human lymphoma dataset. The average initial and final objective function values over 20 runs for the yeast dataset to generate 30×3 co-clusters are -25818.1 and -7323.42. The average initial and final objective function values over 20 runs for the human lymphoma dataset to generate 150×7 co-clusters are -143958 and -119766. There are 100 iterations of our implemented algorithm. We can see that our algorithm converges rapidly.

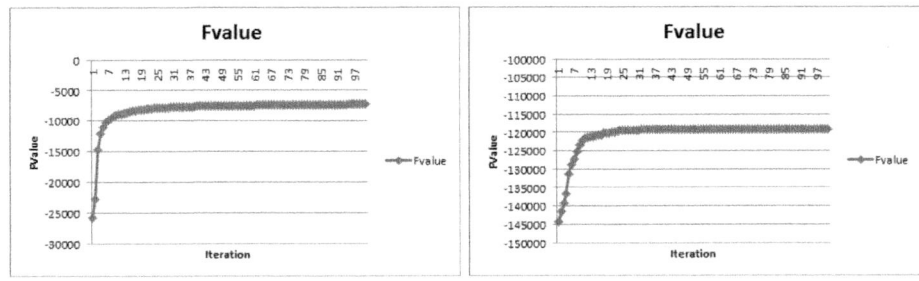

Fig. 4. The figure on the left shows the objective function value vs. iteration of our algorithm on the yeast dataset to generate 30×3 co-clusters. The figure on the right shows the objective function value vs. iteration of our algorithm on the human dataset to generate 150×7 co-clusters.

4.2   Testing Results Using Microarray Datasets

In the following we present some exemplary co-clusters identified by our algorithm. We compare the co-clusters with those identified by other approaches.

For all the figures presented here, the x-axis represents the different number of conditions and the y-axis represents the values of the gene expression level. (Due to the space limit, some detailed testing results and identified co-clusters are not shown here).

Figure 5 shows four co-clusters of the yeast cell dataset generated when the two parameters $k_1 = 20$ and $k_2 = 3$.

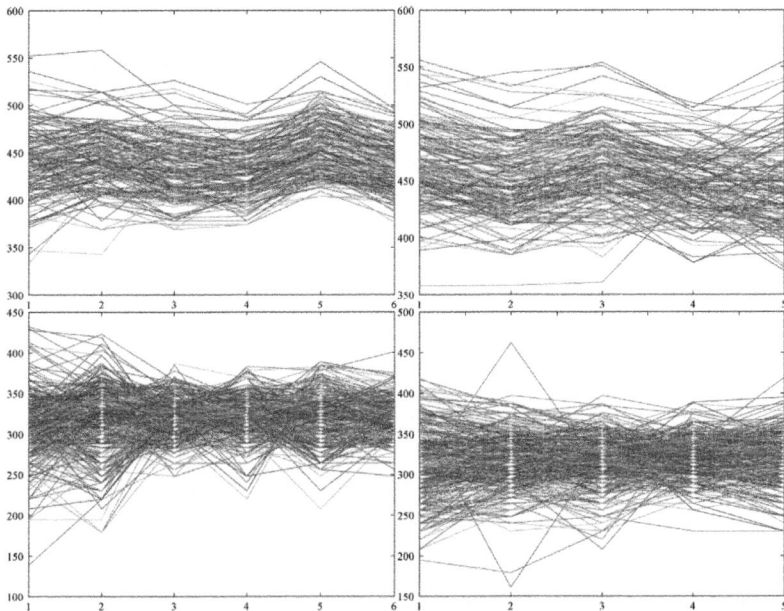

Fig. 5. Four co-clusters of the yeast cell dataset generated when the two parameters $k_1 = 20$ and $k_2 = 3$. The two co-clusters in the same row contain the same sets of genes but in two different sets of conditions, and the two co-clusters in the same column show two different groups of genes on the same set of conditions. Each of the four co-clusters from top-left to bottom-right has the following (number of genes, [list of conditions]) respectively (148, [condition 0, 1, 5, 8, 11, 12]), (148, [condition 2, 3, 4, 6, 7]), (292, [condition 0, 1, 5, 8, 11, 12]), and (292, [condition 2, 3, 4, 6, 7]).

We can see from the co-clusters shown in Figure 5, 6 and other generated co-clusters that our algorithm can effectively identify groups of genes and groups of conditions that exhibit similar expression patterns. It can discover the same subset of genes that have different expression levels over different subsets of conditions, and can also discover different subsets of genes that have different expression levels over the same subset of conditions.

The four co-clusters in Figure 5 are closely related to the clusters of Tavazoie *et al.* [13], where the classical k-means clustering algorithm was applied and the yeast cell cycle gene expression dataset was clustered into 30 clusters. The bottom two co-clusters are mainly related to their clusters 2, 3, 4, 6 and 18. The top two co-clusters are mainly related to their cluster 1. This shows that the same group of genes have different expression patterns over different subsets of conditions. This also shows that one or more than one co-clusters could correspond to one cluster of Tavazoie *et al.* [13].

We use the mean square residue score developed in [4] to evaluate the co-clusters generated by our algorithm. We identify 12 co-clusters with the best mean square residue scores of the yeast cell dataset when $k_1 = 30$ and $k_2 = 3$. The list of the scores are 168.05, 182.04, 215.69, 335.72, 365.01, 378.37, 408.98, 410.03, 413.08, 416.63, 420.37, and 421.49. All the 12 co-clusters have the mean square residue scores less than 450. They are meaningful co-clusters.

We conduct similar experimental testing on the human lymphoma dataset. Figure 6 shows four exemplary co-clusters of the dataset generated when the two parameters $k_1 = 150$ and $k_2 = 7$.

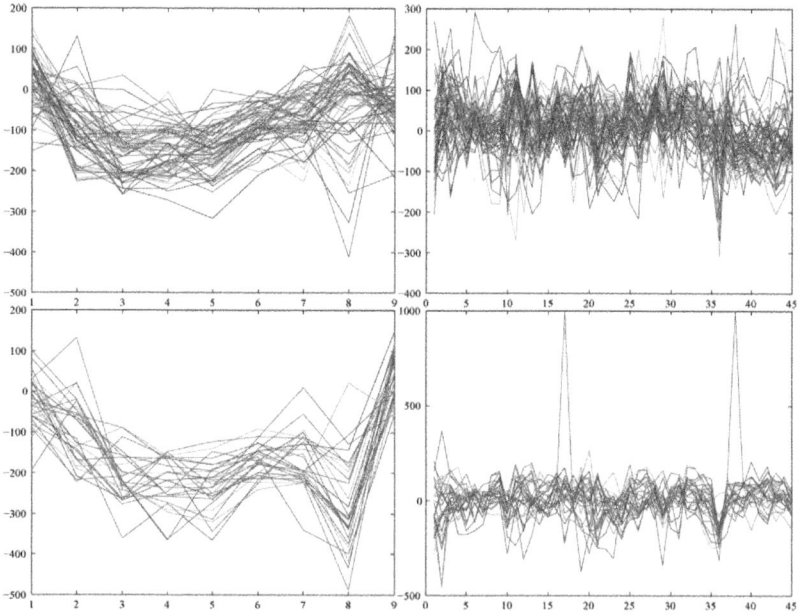

Fig. 6. Four co-clusters of human cell dataset generated when the two parameters $k_1 = 150$ and $k_2 = 7$. Note that two co-clusters in the same row contain the same sets of genes but in different sets of conditions, and the two co-clusters in the same column show two different groups of genes on the same set of conditions. Each of the four co-clusters has the following (number of genes, number of conditions): (57, 9), (57,45), (27,9), and (27,45).

4.3   Testing Using 3D Synthesis Dataset

We test our algorithm using the 3D synthetic dataset from [12] which has six files with each file containing 1,000 genes measured over 10 conditions with 6 time-points for each condition. The co-clusters in Figure 7 show clear coherent patterns of the 3D dataset.

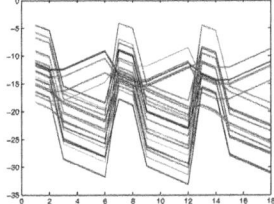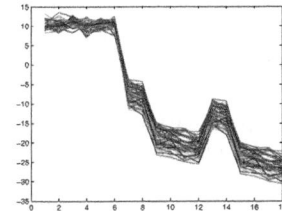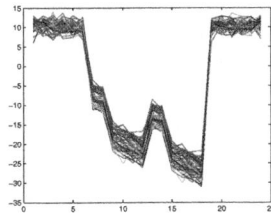

Fig. 7. Co-clusters of the 3D dataset generated when the three parameters $k_1 = 10$, $k_2 = 1$ and $k_3 = 3$. Each curve corresponds to the expression of one gene. The x-axis represents the different number of time points with every 6 time-points in one condition, while the y-axis represents the values of the gene expression level.

5   Summary and Future Works

We have developed a new framework for co-clustering gene expression data, which includes an optimization model and a generic algorithm for the co-clustering problem. We implemented and tested our algorithm on two 2D microarray datasets and one 3D synthesis dataset.

In the near future, we will extend our algorithm to handle gene expression datasets in high-dimensional tensors, such as genes expressed at different tissues, different development stages, different time points, different stimulations. We will study and test the co-clustering model for identifying scaling and shifting patterns [1] and overlapped co-clusters [4]. We are currently conducting the testing of our models for analyzing different microarray and next-generation sequencing datasets from real-life biological experiments. It will also be very useful to consider pre-processing experimental data such as removing trivial co-clusters as in [4], post-processing the identified co-clusters, and incorporating biological constraints into the co-clustering model. To extract meaningful information for biologists and life-scientists out of the vast experimental gene expression datasets is a hugely challenging task. The marvelous prospect of its success, however, may arguably justify the toil in the process.

References

1. Aguilar-Ruiz, J.S.: Shifting and scaling patterns from gene expression data. Bioinformatics 21(20), 3840–3845 (2005)
2. Alizadeh, A.A., Eisen, M.B., Davis, R.E., Ma, C., Lossos, I.S., Rosenwald, A., Boldrick, J.C., Sabet, H., Tran, T., Yu, X., Powell, J.I., Yang, L., Marti, G.E., Moore, T., Hudson Jr., J., Lu, L., Lewis, D.B., Tibshirani, R., Sherlock, G., Chan, W.C., Greiner, T.C., Weisenburger, D.D., Armitage, J.O., Warnke, R., Levy, R., Wilson, W., Grever, M.R., Byrd, J.C., Botstein, D., Brown, P.O., Staudt, L.M.: Distinct types of diffuse large B-cell lymphoma identified by gene expression profiling. Nature 403(6769), 503–511 (2000)
3. Chen, B., He, S., Li, Z., Zhang, S.: Maximum block improvement and polynomial optimization (submitted for publication, 2011)
4. Cheng, Y., Church, G.M.: Biclustering of expression data. In: Proc. Int. Conf. Intell. Syst. Mol. Biol., vol. 8, pp. 93–103 (2000)
5. Cho, H., Dhillon, I.S., Guan, Y., Sra, S.: Minimum sum-squared residue co-clustering of gene expression data. In: Proceedings of The fourth SIAM International Conference on Data Mining, pp. 114–125 (2004)
6. Hartigan, J.A.: Direct clustering of a data matrix. Journal of the American Statistical Association 67(337), 123–129 (1972)
7. Hochreiter, S., Bodenhofer, U., Heusel, M., Mayr, A., Mitterecker, A., Kasim, A., Khamiakova, T., Sanden, S.V., Lin, D., Talloen, W., Bijnens, L., Ghlmann, H.W.H., Shkedy, Z., Clevert, D.: FABIA: factor analysis for bicluster acquisition. Bioinformatics 26(12), 1520–1527 (2010)
8. Kilian, J., Whitehead, D., Horak, J., Wanke, D., Weinl, S., Batistic, O., D'Angelo, C., Bornberg-Bauer, E., Kudla, J., Harter, K.: The AtGenExpress global stress expression data set: protocols, evaluation and model data analysis of UV-B light, drought and cold stress responses. The Plant Journal 2, 347–363 (2007)
9. Kolda, T.G., Bader, B.W.: Tensor decompositions and applications. SIAM Review 51(3), 455–500 (2009)
10. Jegelka, S., Sra, S., Banerjee, A.: Approximation algorithms for tensor clustering. In: Gavaldà, R., Lugosi, G., Zeugmann, T., Zilles, S. (eds.) ALT 2009. LNCS, vol. 5809, pp. 368–383. Springer, Heidelberg (2009)
11. Madeira, S.C., Oliveira, A.L.: Biclustering algorithms for biological data analysis: a survey. IEEE/ACM Trans. Comput. Biology Bioinform. 1(1), 24–45 (2004)
12. Supper, J., Strauch, M., Wanke, D., Harter, K., Zell, A.: EDISA: extracting biclusters from multiple time-series of gene expression profiles. BMC Bioinformatics 8, 334–347 (2007)
13. Tavazoie, S., Hughes, J.D., Campbell, M.J., Cho, R.J., Church, G.: Systematic determination of genetic network architecture. Nat. Genet. 22(3), 281–285 (1999)

Biclustering of Expression Microarray Data Using Affinity Propagation

Alessandro Farinelli, Matteo Denitto, and Manuele Bicego

University of Verona, Department of Computer Science, Verona, Italy

Abstract. Biclustering, namely simultaneous clustering of genes and samples, represents a challenging and important research line in the expression microarray data analysis. In this paper, we investigate the use of Affinity Propagation, a popular clustering method, to perform biclustering. Specifically, we cast Affinity Propagation into the Couple Two Way Clustering scheme, which allows to use a clustering technique to perform biclustering. We extend the CTWC approach, adapting it to Affinity Propagation, by introducing a stability criterion and by devising an approach to automatically assemble couples of stable clusters into biclusters.

Empirical results, obtained in a synthetic benchmark for biclustering, show that our approach is extremely competitive with respect to the state of the art, achieving an accuracy of 91% in the worst case performance and 100% accuracy for all tested noise levels in the best case.

1 Introduction

The recent wide employment of microarray tools in molecular biology and genetics have produced an enormous amount of data, which has to be processed to infer knowledge. Due to the dimension and complexity of those data, automatic tools coming from the Pattern Recognition research area have been successfully employed. Among others, clear examples are tools aiding the microarray probe design, the image processing-based techniques for the quantification of the spots (segmentation spot/background, grid matching, noise suppression [6]) and methodologies for classification or clustering [18,24,25,8]. In this paper we focus on this last class of problems and in particular on the clustering issue. Within this context, a recent trend is represented by the study and development of *biclustering* methodologies, namely techniques able to simultaneously group genes and samples; a bicluster may be defined as a subset of genes that show similar activity patterns under a specific subset of samples [20]. This kind of analysis may have a clear biological impact in the microarray scenario, where a bicluster may be associated to a biological process that is active only in some samples and may involve only a subset of genes. Different approaches to biclustering expression microarray data have been presented in the literature in the past, each one characterized by different features, like computational complexity, effectiveness, interpretability, optimization criterion and others (for a review see [20,22]).

M. Loog et al. (Eds.): PRIB 2011, LNBI 7036, pp. 13–24, 2011.

It is worth noticing that, in many cases, successful methodologies have been obtained by adapting and tailoring advanced techniques developed in other fields of the Pattern Recognition research area. One clear example is represented by the topic models [14,5], initially designed for text mining and computer vision applications, and recently successfully applied in the microarray context [23,3,2]. Clearly, the peculiar context may lead to substantial changes in the model so to improve results [21].

This paper follows this promising direction, preliminary investigating the capabilities, in the expression microarray biclustering context, of a recent and powerful clustering technique, called Affinity Propagation (AP [10]). This technique is based on the idea of iteratively exchanging messages between data points until a proper set of representatives (called exemplars) are found. Such exemplars identify clusters (which are all points represented by a given exemplar). The efficacy of this algorithm (in terms of clustering accuracy) and its efficiency (due to its really fast learning algorithm) have been shown in many different application scenarios, including image analysis, gene detection and document analysis [10]. Moreover, AP seems to be a promising approach for the microarray scenario for two main reasons: first, AP does not need to know the number of clusters beforehand and in a microarray clustering problem it is often difficult to estimate a priori the number of groups, especially for the gene part; second, and more important, it is known that under some assumptions (e.g. sparse input data) this technique is very efficient for large scale problems such as microarray data which involves thousands of genes. Actually, in recent years some papers appeared in the literature with the aim of studying the application of this algorithm in the expression microarray field, in its basic version or in some more tailored ones [1,19,16,7]. Nevertheless, all these papers deal with the clustering problem, whereas the biclustering problem has not been addressed yet.

In this paper, we propose an approach to use AP for biclustering based on a biclustering scheme called Coupled Two-Way Clustering (CTWC) [11], which iteratively performs samples and genes clustering using the supeparamagnetic clustering (SPC) algorithm, maintaining only clusters that satisfies a stability criterion (directly provided by the SPC clustering approach). In its original formulation CTWC does not provide an explicit representation of the obtained biclusters (which remain implicitly defined). Nevertheless, an automatic mechanism able to explicitly list the components of a bicluster is crucial for validation purposes (to the best of our knowledge, no biological validation tools deal with implicit or probabilistic memberships). To this end, in this paper we also proposed an automatic reassembling strategy, which may in principle be applied also to the original approach proposed in [11].

In more details this paper makes the following contribution to the state of the art:

— Proposes the use of AP for biclustering. We cast AP into the CTWC biclustering scheme and propose a different stability criterion inspired from the bootstrapping, which is more general than the stability criterion used by CTWC and is well suited for the AP approach.

– Extends CTWC by devising an automatic reassembling strategy for the biclusters. The strategy takes as input the clusters obtained on rows and columns and returns only the couples of clusters that forms a bicluster in the input data.
– Empirically tests the biclustering approach on a literature benchmark, using the synthetic data and protocols described in [22]. While a biological validation is clearly important to assess the practical significance of the approach, a synthetic validation permits to quantitatively compare the approach to other state of the art methods. In our experimental evaluation, we show that the proposed approach is very competitive with the literature, encouraging further investigations on the use of AP clustering algorithm in the microarray biclustering scenario.

The remainder of the paper is organized as follows: Sect. 2 will introduce the Affinity Propagation clustering algorithm, whereas the proposed biclustering scheme is presented in Sect. 3. The experimental evaluation is detailed in Sect. 4; finally, in Sect. 5 conclusions are drawn and future perspectives are envisaged.

2 Background: Affinity Propagation

Affinity Propagation (AP) is a well known clustering technique recently proposed by Frey and Dueck [10] and successfully used in many different clustering contexts.

The main idea behind AP is to perform clustering by finding a set of exemplar points that best represent the whole data set. This is carried out by viewing the input data as a network where each data point is a node, and selecting the exemplars by iteratively passing messages among the nodes. Messages convey the affinity that each point has for choosing another data point as its exemplar and the process stops when good exemplars emerge or after a predefined number of iterations.

In more details, AP takes as input a similarity matrix, where each entry $s(i, j)$ defines how much point j is suited to be an exemplar for i. The similarity is clearly domain dependent and can be any measure of affinity between points, in particular it does not need to be a metric. A peculiar characteristic of AP, when compared to other popular exemplar based clustering techniques such as k-means, is that it does not require to specify a priori the number of clusters to be formed. Such number is automatically computed by the algorithm, and it is influenced by the values $s(i, i)$, given as an input, which represents the *preference* for point k of being itself an exemplar. In particular, the higher the preferences the larger the number of clusters that will be formed and vice versa. While tuning preferences is an important issue to have more accurate clustering, usually all preferences are set to a common value that depends on the input similarity matrix and a common choice is the median of such matrix [10].

Given the similarity matrix, AP finds the exemplars by iteratively exchanging messages between points. In particular there are two types of messages that data points exchange: responsibility and availability messages. Intuitively, a responsibility message sent from point i to point j represents how much j is well suited to be a representative for i while an availability message from i to j indicates how much i is well suited to be a representative of j. Both availability and responsibility are updated by accumulating information coming from neighboring data points and considering the preferences.

More in details, at each iteration, messages are updated according to the following equations [10]:

$$r_{i \to j} = s(i,j) - \max_{k \neq j}\{a_{k \to i} + s(i,k)\} \tag{1}$$

$$a_{i \to j} = \begin{cases} \min\{0, r(i,i) + \sum_{k \neq j} max\{0, r_{k \to i}\}\} & \text{if } i \neq j \\ \sum_{k \neq i} max\{0, r_{k \to i}\} & \text{otherwise} \end{cases} \tag{2}$$

where $r_{i \to j}$ and $a_{i \to j}$ represent respectively a responsibility and an availability message from data point i to j. At the beginning all availabilities are set to zero. The responsibility update is obtained by combining the similarity between point i and point j with the maximum similarity of point i and all other points and their availability of being a representative for point i. Intuitively, if the availability of a point k becomes negative over the iterations, its contribution of being a representative will also decrease. The availability update adds to the self responsibility ($r(i,i)$) the positive responsibilities message from other point, which represent information gathered by other points about how good point i would be as an exemplar. The update for self availability consider all incoming positive responsibilities from other points.

The exemplar for each data point i is the data point j that maximizes the sum of $a_{j \to i} + r_{i \to j}$. Exemplars can be computed at every iteration and the message update process converges when for a given amount of iterations the exemplars do not change, or it can be stopped after a predetermined amount of iterations.

Notice that the message update rules reported above can be derived by representing the clustering problem with a factor graph [17] and then by running the max-sum algorithm [4] to solve it. We refer the interested reader to [12] for further details on this.

As previously mentioned, a key element of the success of AP is the ability to efficiently cluster large amount of sparse data. This is possible because messages need not be exchanged among points that can not (or are very unlikely to) be part of the same cluster (i.e. points that has extremely low similarities). In fact, AP can leverage the sparsity of the input data by exchanging messages only among the relevant subsets of data point pairs, thus dramatically speeding up the clustering process.

Being a clustering technique AP can not directly perform biclustering analysis, therefore, in the next section, we present our approach to use AP for biclustering analysis of microarray data.

3   The Proposed Approach

The approach we propose in this paper is built on a scheme called Coupled Two-Way Clustering (CTWC) [11]. The basic idea of this scheme is to use a clustering algorithm to independently cluster samples and genes and select clusters that meet some stability criteria; sample clusters are then coupled with gene clusters effectively forming sub-matrices of the input data; then the CTWC method is run again on each of these sub-matrices in an iterative fashion. This process is repeated until no new stable clusters are formed.

In [11], the superparamagnetic clustering approach was used to perform row and column clustering. Nevertheless, as the authors claim in [11], any clustering method can be used to cluster samples and genes. Here we propose the use of Affinity Propagation and extend the CTWC scheme to automatically assemble genes and samples clusters into an explicit bicluster representation.

More in details, our proposed approach can be described as follows:

1. Given an n by m input matrix we independently cluster rows and columns by using Affinity Propagation. To do this each row (column) is considered as an m(n) dimensional data point. Affinity Propagation takes in input a similarity matrix between each pair of the data points and the preferences, it then clusters the data points automatically detecting the best number of clusters as described in Section 2.
2. We maintain a subset of the clusters returned by Affinity Propagation by selecting only those that meet a stability criteria inspired by the bootstrap method (see Section 3.1).
3. Following the CTWC scheme we couple *all* stable clusters together and iterate the process on all the obtained sub-matrices. Notice that when coupling the clusters together we consider also stable clusters that were formed in previous iterations of the approach.
4. We assemble clusters in biclusters by coupling stable clusters on rows and columns and testing whether each cluster couple forms a bicluster in the input data (see 3.2).

The above steps are iterated until no new stable cluster is formed.

Algorithm 1 reports the pseudo-code of our approach: in particular the algorithm takes in input an n by m matrix that represents the expression microarray data and returns a set of biclusters. The queue Q represents the sub-matrices that have to be analyzed. It is used to control the algorithm iterations and it is initialized with the input data matrix (line 1). The two sets of clusters, rows S_r and columns S_c, represents stable clusters and are initialized with a single cluster each, which includes all rows and columns of the input matrix (lines 2 and 3). The while loop describes the algorithm iterations and stops when no elements are present in the queue (lines 4 to 11). At every iteration, a sub-matrix $currA$ is extracted from the queue and affinity propagation is used to find stable clusters on rows (line 6) and columns (line 7). In particular, the stableAP method runs affinity propagation on a set of multidimensional points and selects stable clusters, according to the stability criteria described in section 3.1. Relevant data

---

Algorithm 1. ap-ctwc

Require: A : and n by m input data matrix
Ensure: B : a set of biclusters
1: $Q \leftarrow A$
2: $S_r \leftarrow \{(1, \ldots, n)\}$
3: $S_c \leftarrow \{(1, \ldots, m)\}$
4: **while** Q is not empty **do**
5:     $currA \leftarrow pop(Q)$
6:     $\{r_1, \ldots, r_s\} \leftarrow stableAP(rows(currA))$
7:     $\{c_1, \ldots, c_t\} \leftarrow stableAP(col(currA))$
8:     $S_r \leftarrow S_r \cup \{r_1, \ldots, r_s\}$
9:     $S_c \leftarrow S_c \cup \{c_1, \ldots, c_t\}$
10:    $Q \leftarrow push(Q, allNewCouples(S_r, S_c))$
11: **end while**
12: **return** $B \leftarrow assembleBiclusters(S_r, S_c)$

---

structures are then updated. Specifically, the new stable clusters are added to the set of row and columns clusters (lines 8 and 9), and clusters are coupled and pushed in the queue (line 10). Notice that the $allNewCouples(S_r, S_c)$ function returns all couples of clusters involving new clusters, i.e. all couples $< r_i, c_j >$ where $r_i \in \{r_1, \ldots, r_s\}$ and $c_j \in S_c \cup \{c_1, \ldots, c_t\}$ and vice versa. Finally, the set of row and column clusters are assembled in biclusters (line 12) as explained in section 3.2.

A final postprocessing has been carried out in order to present data in a more meaningful way. First, we clean the set of biclusters by removing smaller size biclusters contained in a more stable bicluster, since usually the aim is to find maximal biclusters which are maximally coherent. Second, we order the biclusters according to size (from bigger to smaller) and, in case of ties, we order based on stability, where the stability of a bicluster is defined as the sum of the stability of the row and column clusters that forms the bicluster. The rationale behind the ordering is that researchers are usually interested in clusters of bigger size because they usually yield more knowledge about the interconnections among gene behaviors across experiments.

3.1   Stability Criterion

Stability or robustness of clusters is a well known but yet unsolved problem in the clustering field [15]. In this paper we adopt a very simple approach, starting from the consideration that AP injects noise in the clustering process to avoid ties. The approach we propose is to perform N different clusterings, measuring the stability of a cluster as the fraction of times where it is found by the algorithm. Clearly, if a cluster is stable, small perturbations of the input data won't affect its detection. This is the same rationale under the bootstrap method used in the phylogeny domain [9], where noise is added to the data and a final consensus tree is devised (stability of a group is exactly the fraction of times where such group appears).

3.2   Assembling Biclusters

Given the two sets of row and column clusters, biclusters are assembled by considering all possible couples of row-columns clusters in such lists and by selecting only the couples of clusters that can mutually generate each other. More in details, consider a specific couple $b_{ij} =< r_i, c_j >$, we consider b_{ij} as a valid bicluster if and only if the clusters obtained by clustering the sub-matrix $A[r_i, :]$ along the columns contains c_j and vice versa. Here, $A[r_i, :]$ represents the sub-matrix of the input data obtained by considering only the subset of rows contained in the cluster r_i and all the columns. This condition effectively avoids that couples of clusters that do not form a biclusters in the input matrix are assembled together.

To further clarify this point, Figure 1 reports an example of a couple of clusters that do form a bicluster in the input data. Specifically Figure 1(a) shows the input data and the row and column clusters that we want to test. Let's consider the column submatrix (Figure 1(b)), clustering along the rows we obtain three clusters that include the input row cluster. The same happens when we consider the row submatrix of Figure 1(c), hence we can conclude that the input cluster couple does form a bicluster of the input data, as Figure 1(d) shows.

On the other hand, Figure 2 reports an example of a couple that does not form a bicluster. Here, we consider the same input data matrix as before but a different cluster couple as Figure 2(a) shows. Running the test for this couple we can see that by considering the row submatrix we do obtain the input column cluster (2(b)). However, when we consider the column submatrix we can not find the input row cluster (2(c)). Therefore this cluster couple would not pass our test and in fact this is not a bicluster of the input data as Figure 2(d) shows.

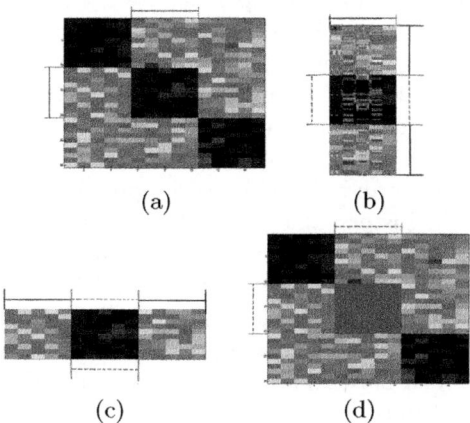

(a)   (b)

(c)   (d)

Fig. 1. Couple of clusters that form a bicluster: (a) Input data and the couple of clusters to test. (b) Submatrix considering the column cluster. (c) Submatrix considering the row cluster. (d) Bicluster formed by the couple of clusters.

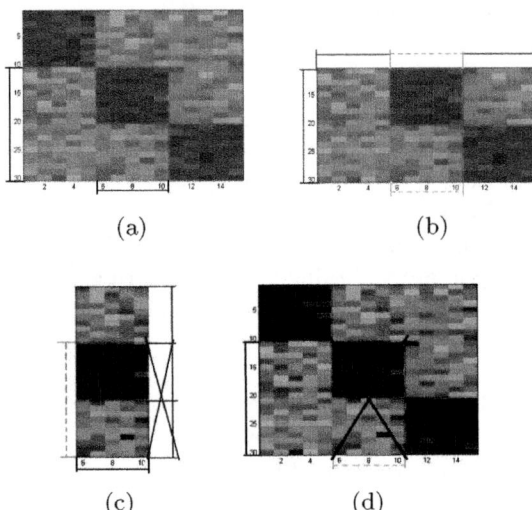

(a)   (b)

(c)   (d)

Fig. 2. Couple of clusters that do not form a bicluster: (a) Input data and the couple of clusters to test. (b) Submatrix considering the row cluster. (c) Submatrix considering the column cluster. (d) Submatrix formed of row and columns clusters, which is not a bicluster of the input data.

Having described our approach the next session will report and discuss results obtained in the empirical evaluation of our method.

4   Experimental Evaluation

The methodology proposed in this paper has been tested in a synthetic benchmark ([22]), which includes synthetic expression matrices, perturbed with different schemes[1]. In this setting, biclusters represent transcription modules; these modules are defined by (i) a set G of genes regulated by a set of common transcription factors, and (ii) a set C of conditions in which these transcription factors are active. In the original paper two scenarios are considered, one with non overlapping biclusters and one with overlapping biclusters. Here we consider only non overlapped biclusters, since the CTWC scheme does not permit to extract overlapped biclusters. In the experiments, 10 non-overlapping transcription modules, each extending over 10 genes and 5 conditions, emerge. Each gene is regulated by exactly one transcription factor and in each condition only one transcription factor is active. The corresponding datasets contain 10 implanted biclusters and have been used to study the effects of noise on the performance of the biclustering methods.

The accuracy of the biclustering has been assessed with the so-called *Gene Match Score* [22], which reflects the similarity of the biclusters obtained by an

---

[1] All datasets may be downloaded from: www.tik.ee.ethz.ch/sop/bimax

algorithm and the original known biclustering (it varies between 0 and 1, the higher the better the accuracy), for all details on the datasets and the evaluation protocol please refer to [22].

Even if our proposed approach is also able to extract biclusters with an average expression value close to zero, in this setting we remove such groups, in order to adapt our results to the synthetic evaluation of [22]. The proposed approach has been tested using as similairty the negative Euclidean distance and the Pearson coefficient, the latter being a very common choice in the microarray analysis scenario. Results were qualitatively and quantitatively really similar, thus here we report only those with the negative Euclidean distance. As for the stability threshold, after a preliminary evaluation we have found that a 50% value represents a good compromise between quality of clusters and level of details. Concerning the affinity propagation clustering algorithm, it is well known that setting the preferences may be crucial in order to obtain a proper clusterization [10]. Even if some sophisticated solutions have been proposed (e.g. [26]) the most widely used approach is to set all the preferences as the median value of the similarity matrix [10]. Here we slightly enlarge the scope of this rule, by also setting as preferences the median +/- the Median Absolute Deviation (MAD) (which represents a robust estimator of the standard deviation [13]), defined, for a given population $\{x_i\}$ with median med, as

$$MAD = median_i\{|x_i - med|\} \qquad (3)$$

The results are reported in Fig. 3, for the different initialization of the preference values. Following [22], we report both *bicluster relevance* (i.e., to what extent the generated biclusters represent true biclusters), and *module recovery* (i.e., how well true biclusters are recovered).

Reported results support two main conclusions: i) the method performs extremely well on this dataset, with a worst case performance that still provides 91% accuracy. By comparing these results with those published in [22,2], we can

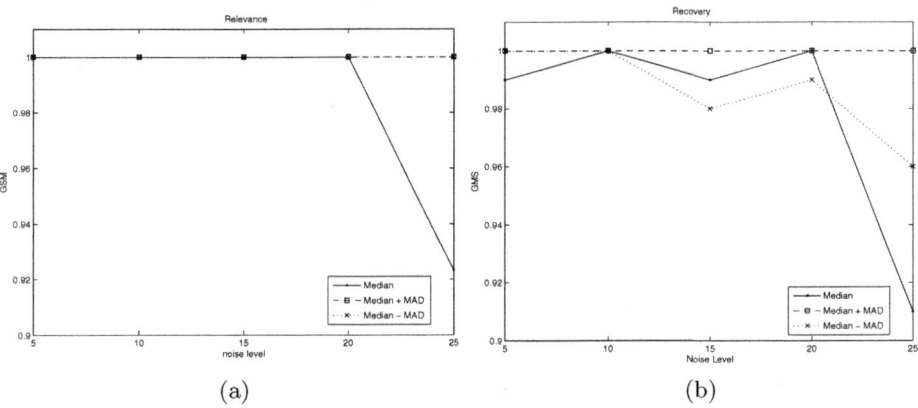

Fig. 3. Results on the synthetic dataset: (a) bicluster relevance, (b) module recovery

observe that our approach is very competitive with respect to the state of the art. ii) while in principle the initialization of preference values does make a difference in the AP method, the approach is not very sensitive to this parameter, reaching very good performances for preference values within the tested range (especially in the first four conditions).

Moreover, it is worth noticing that the different performance of the three preference settings confirm the intuition that higher preference values lead to more clusters in AP. In fact, the reason why the $median - MAD$ has worst average performance, in terms of $recovery$, is due to the fact that AP forms less clusters thus failing to detect some of the biclusters that are present in the data set when the level of noise increases. On the other hand, by setting higher preferences AP finds more clusters thus resulting in a GMS that is more robust to the increasing noise level while maintaining a very good level of $relevance$.

As a further analysis, we tried to understand which kind of errors are produced by the approach. In particular, in Fig. 4 we reported the biclusters extracted

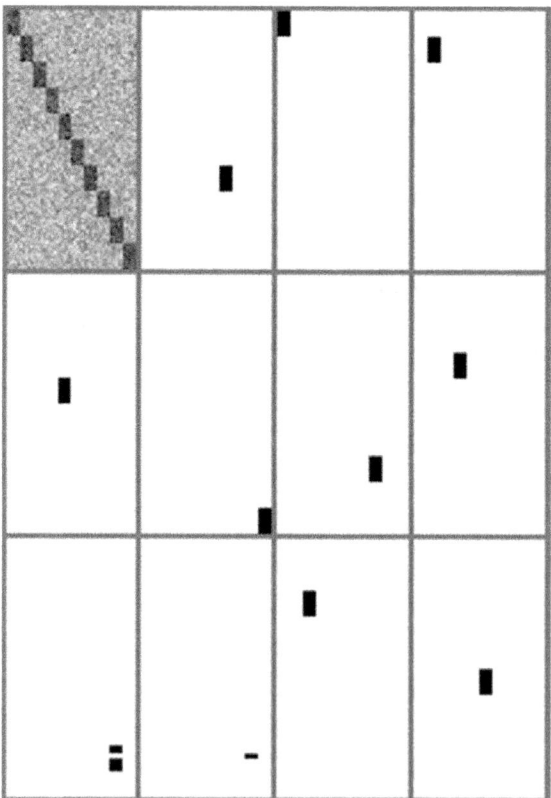

Fig. 4. Example of the result of the algorithm: the original expression matrix is shown in the top left corner, in the remaining boxes the obtained biclusters are displayed

by the proposed approach in one of the run of the algorithm, within the last condition.

It is clear that almost all the biclusters have been obtained, there is just one which has been divided in two. Therefore, the elements are correctly grouped together (the algorithm does not group together expressions which are not supposed to be together), but oversegmentation occurs. This issue may be possibly faced by selecting the preferences in the AP clustering module in a more careful way.

5   Conclusions and Future Works

In this paper we propose a method to use Affinity Propagation [10] (a recently proposed, promising clustering techniques), to perform biclustering of expression microarray data. Our method builds on the CTWC biclustering scheme [11] and extends it in two main directions: i) we propose a stability criterion, inspired from the bootstrap method, which is suited for AP, and more general that the one used in the original version of CTWC. ii) we propose a method to automatically assemble couples of stable clusters into biclusters.

We empirically evaluated our approach in a synthetic benchmark [22], and results show that our method is very competitive with respect to state of the art.

Future work in this area includes two main research directions: i) testing the approach on real biological data sets to assess the practical significance of the approach, ii) investigate extensions of the approach to deal with overlapping biclusters.

References

1. Bay, A., Granitto, P.: Clustering gene expression data with a penalized graph-based metric. BMC Bioinformatics 12 (2011)
2. Bicego, M., Lovato, P., Ferrarini, A., Delledonne, M.: Biclustering of expression microarray data with topic models. In: Proceedings of the International Conference on Pattern Recognition, pp. 2728–2731 (2010)
3. Bicego, M., Lovato, P., Oliboni, B., Perina, A.: Expression microarray classification using topic models. In: ACM Symposium on Applied Computing (Bioinformatics and Computational Biology track) (2010)
4. Bishop, C.: Pattern Recognition and Machine Learning. Springer, Heidelberg (2006)
5. Blei, D., Ng, A., Jordan, M.: Latent Dirichlet allocation. Journal of Machine Learning Research 3, 993–1022 (2003)
6. Brändle, N., Bischof, H., Lapp, H.: Robust DNA microarray image analysis. Machine Vision and Applications 15, 11–28 (2003)
7. Chiu, T.Y., Hsu, T.C., Wang, J.S.: Ap-based consensus clustering for gene expression time series. In: Proc. Int. Conf. on Pattern Recognition, pp. 2512–2515 (2010)
8. de Souto, M., Costa, I., de Araujo, D., Ludermir, T., Schliep, A.: Clustering cancer gene expression data: A comparative study. BMC Bioinformatics 9 (2008)

9. Felsenstein, J.: Confidence limits on phylogenies: an approach using the bootstrap. Evolution 39, 783–791 (1985)
10. Frey, B., Dueck, D.: Clustering by passing messages between data points. Science 315, 972–976 (2007)
11. Getz, G., Levine, E., Domany, E.: Coupled two-way clustering analysis of gene microarray data. Proc. Natl. Acad. Sci. USA 97(22), 12079–12084 (2000)
12. Givoni, I., Frey, B.: A binary variable model for affinity propagation. Neural Computation 21(6), 1589–1600 (2009)
13. Hampel, F., Rousseeuw, P., Ronchetti, E., Stahel, W.: Robust Statistics: the Approach Based on Influence Functions. John Wiley & Sons (1986)
14. Hofmann, T.: Unsupervised learning by probabilistic latent semantic analysis. Machine Learning 42(1-2), 177–196 (2001)
15. Jain, A., Dubes, R.: Algorithms for clustering data. Prentice-Hall (1988)
16. Kiddle, S., Windram, O., McHattie, S., Mead, A., Beynon, J., Buchanan-Wollaston, V., Denby, K., Mukherjee, S.: Temporal clustering by affinity propagation reveals transcriptional modules in arabidopsis thaliana. Bioinformatics 26(3), 355–362 (2010)
17. Kschischang, F., Frey, B., Loeliger, H.A.: Factor graphs and the sum-product algorithm. IEEE Transactions on Information Theory 47(2), 498–519 (2001)
18. Lee, J.W., Lee, J.B., Park, M., Song, S.: An extensive comparison of recent classification tools applied to microarray data. Computational Statistics & Data Analysis 48(4), 869–885 (2005)
19. Leone, M., Weigt, S., Weigt, M.: Clustering by soft-constraint affinity propagation: applications to gene-expression data. Bioinformatics 23(20), 2708–2715 (2007)
20. Madeira, S., Oliveira, A.: Biclustering algorithms for biological data analysis: a survey. IEEE Trans. on Computational Biology and Bioinformatics 1, 24–44 (2004)
21. Perina, A., Lovato, P., Murino, V., Bicego, M.: Biologically-aware Latent Dirichlet Allocation (BaLDA) for the Classification of Expression Microarray. In: Dijkstra, T.M.H., Tsivtsivadze, E., Marchiori, E., Heskes, T. (eds.) PRIB 2010. LNCS, vol. 6282, pp. 230–241. Springer, Heidelberg (2010)
22. Prelic, A., Bleuler, S., Zimmermann, P., Wille, A., Buhlmann, P., Gruissem, W., Hennig, L., Thiele, L., Zitzler, E.: A systematic comparison and evaluation of biclustering methods for gene expression data. Bioinformatics 22(9), 1122–1129 (2006)
23. Rogers, S., Girolami, M., Campbell, C., Breitling, R.: The latent process decomposition of cdna microarray data sets. IEEE/ACM Transactions on Computational Biology and Bioinformatics 2(2), 143–156 (2005)
24. Statnikov, A., Aliferis, C., Tsamardinos, I., Hardin, D., Levy, S.: A comprehensive evaluation of multicategory classification methods for microarray gene expression cancer diagnosis. Bioinformatics 21(5), 631–643 (2005)
25. Valafar, F.: Pattern recognition techniques in microarray data analysis: A survey. Annals of the New York Academy of Sciences 980, 41–64 (2002)
26. Zhang, X., Wu, F., Zhuang, Y.: Clustering by evidence accumulation on affinity propagation. In: Proc. Int. Conf. on Pattern Recognition, pp. 1–4 (2008)

A Two-Way Bayesian Mixture Model for Clustering in Metagenomics

Shruthi Prabhakara and Raj Acharya

Pennsylvania State University, University Park, PA, 16801
{shruthi,acharya}@psu.edu

Abstract. We present a new and efficient Bayesian mixture model based on Poisson and Multinomial distributions for clustering metagenomic reads by their species of origin. We use the relative abundance of different words along a genome to distinguish reads from different species. The distribution of word counts within a genome is accurately represented by a Poisson distribution. The Multinomial mixture model is derived as a standardized Poisson mixture model. The Bayesian network efficiently encodes the conditional dependencies between word counts in a DNA due to overlaps and hence is most consistent with the data. We present a two-way mixture model that captures the high dimensionality and sparsity associated with the data. Our method can cluster reads as short as 50 bps with accuracy over 80%. The Bayesian mixture models clearly outperform their Naive Bayes counterparts on datasets of varying abundances, divergences and read lengths. Our method attains comparable accuracy to that of state-of-art Scimm and converges at least 5 times faster than Scimm for all the cases tested. The reduced time taken, by our method, to obtain accurate results is highly significant and justifies the use of our proposed method to evaluate large metagenome datasets.

Keywords: Clustering, Mixture Modeling, Metagenomics.

1 Introduction

Metagenomics is defined as the study of genomic content of microbial communities in their natural environments, bypassing the need for isolation and laboratory cultivation of individual species[6]. It has shown tremendous potential to discover and study the vast majority of species that are resistant to cultivation and sequencing by traditional methods. Unlike single genome sequencing, assembly of a metagenome is intractable and is by large, an unsolved mystery.

A crucial step in metagenomics that is not required in single genome assembly, is binning the reads belonging to a species i.e. the need to associate the reads with its source organism. Clustering methods aim to identify the species present in the sample, classify the sequences by their species of origin and quantify the abundance of each of these species. The efficacy of clustering methods depends on the number of reads in the dataset, the read length and relative abundances of source genomes in the microbial community.

M. Loog et al. (Eds.): PRIB 2011, LNBI 7036, pp. 25–36, 2011.

2   Related Work

Current approaches to metagenome clustering can be mainly classified into similarity-based and composition-based methods. The similarity-based approaches align reads to close phylogenetic neighbors and hence depend on the availability of closely related genomes in existing databases[7,11]. As most of the extant databases are highly biased in their representation of true diversity, such methods fail to find homologs for reads derived from novel species. On the other hand, composition-based methods rely on the intrinsic features of the reads such as oligomer/word distributions[15,3,5,13,12], codon usage preference[1] and GC composition[2] to ascertain the origin of the reads. The underlying basis is that the distribution of words in a DNA is specific to each species and undergoes only slight variations along the genome.

Most of the existing clustering methods are supervised and depend on the availability of reference data for training[15,3,19,5]. A metagenome may however, contain reads from unexplored phyla which cannot be labeled into one of the existing classes. Most clustering methods until now have been relatively inaccurate in classifying short reads. The poor performance on short reads can be attributed to the high dimensionality and sparsity associated with the data[5].

LikelyBin is an unsupervised method that clusters metagenomic sequences via a Monte Carlo Markov Chain approach[13]. Scimm is a recently developed state-of-art model-based approach where interpolated Markov models represent clusters and optimization is performed using a variant of the k-means algorithm (initialized by LikelyBin and CompostBin)[12]. Later, we compare the performance of our proposed method with Scimm.

Mixture models have become popular tools for analyzing biological sequence data. The dominant patterns in the data are captured by its component distributions. Most mixture models assume an underlying normal distribution[19]. However, the distribution of word counts within a genome vary according to a Poisson distribution[17,18]. The Poisson distribution is adequately approximated by a normal distribution for short words with high count. However, when the count is low, a Poisson distribution more accurately represents the data[24]. Figure 1 illustrates the distribution of dimers and pentamers across reads sampled from the genome of Haemophilus Influenzae. Therefore, the problem of clustering metagenomic reads where distribution of each word varies according to a Poisson distribution can be cast as a multivariate mixture of Poissons.

Bayesian and Naive Bayes Models: Bayesian networks have been an active area of research[10]. Bayesian networks can efficiently represent complex probability distributions. It encodes the joint probability distribution of a set of n variables, $\{X_1, X_2, ..., X_n\}$ as a directed acyclic graph and a set of conditional probability distributions (CPDs). The set of parents of X_i are denoted by Pa_i. Each X_i is conditionally dependent only on its parents Pa_i. The joint probability distribution is given by,

$$p(X_1, X_2, ..., X_n|\Theta) = \prod_{i=1}^{n} p(X_i|Pa_i, \Theta) \tag{1}$$

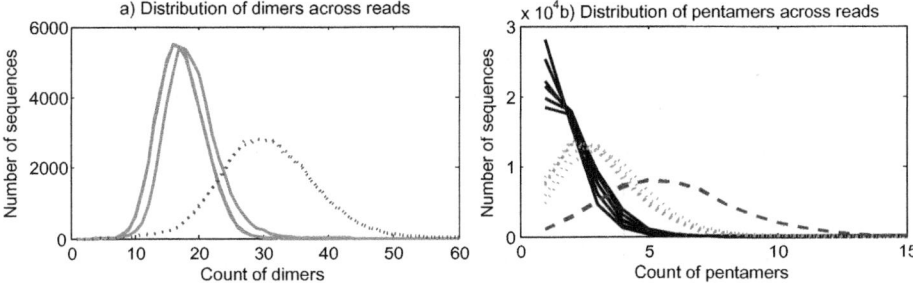

Fig. 1. Distribution of dimers and pentamers across 50,000 reads sampled from the genome of Haemophilus Influenzae(Only a few distributions are shown). a) Distribution of dimers tends to Gaussian and is approximated by a Poisson, two groups can be observed. b) Distribution of pentamers tends to Poisson, three groups are observed.

Typically, even if the sequence of bases in a DNA are independently and identically distributed, distribution of word counts are not independent due to overlaps. Hence, Bayesian networks are ideal for representing sequence data. Though, in practice, methods for exact inference of the structures in Bayesian networks are often computationally expensive. An alternative to Bayesian networks is the Naive Bayes method that assumes independence between the variables. It takes time linear in the number of components. The joint probability distribution is given by,

$$p(X_1, X_2, ..., X_n|\Theta) = \prod_{i=1}^{n} p(X_i|\Theta) \tag{2}$$

Naive Bayes is the simplest Bayesian network that does not represent any variable dependencies. In [20], we described a Naive Bayes mixture of Poissons model to cluster the metagenome reads. We implicitly assume that the variables within a class are independent. The motivation in this paper is to overcome the bottleneck of Naive Bayes by taking into account the conditional dependencies between the word counts within the reads. We focus on developing a tractable Bayesian network for a mixture of Poisson and Multinomial distributions.

3   Methods

We are given a metagenome, $\mathbf{X} = \{\mathbf{x_1}, \mathbf{x_2}, ..., \mathbf{x_N}\}$, containing N reads from M species. Let α_m be the proportion of species m in the dataset, with $\sum_{m=1}^{M} \alpha_m = 1$. We assume that $\mathbf{X}$ is observed and is governed by some density function $p(\mathbf{X}|\Theta)$ with parameter Θ. Our goal is to cluster the reads by their species of origin, based on the frequency of words that appear in the reads. For every species m, we want to determine α_m, its proportion in the dataset, and Θ, the parameter governing the distribution of words within the reads. Let $\mathbf{Y} = \{y_1, y_2, ..., y_N\}$, be the cluster labels. We assume that $y_i = m$ for $m \in 1, ...M$, if the ith read belongs to the mth species. Also, $p(y_i = m) = \alpha_m$. Cluster label $\mathbf{Y}$ is unknown. We call $(\mathbf{X}, \mathbf{Y})$, the complete dataset.

We use Bayesian networks to represent the conditional dependencies between words. Let read $\mathbf{x}$ be of length n , $\mathbf{x} = (c_1 c_2 ... c_n)$, where each $c_k \in (A, C, T, G)$. We assume that the probability of the read is determined by a set of $p = 4^l$ probabilities corresponding to words of length l.

$$p(\mathbf{x}|\Theta) = p(c_1 c_2 ... c_l) \prod_{k=l+1}^{n} p(c_k|c_{k-l}...c_{k-1}, \Theta) = \prod_{k=1}^{n} p(c_k|pa_k, \Theta) \qquad (3)$$

In a read, any given nucleotide c_k can be preceded by its parents pa_k in the read, where $c_k \in (A, C, T, G)$ and $pa_k \in \{pa_k^1, pa_k^2, .., pa_k^p\}$ denote different word configurations of parents. In the next section, we will formulate the Bayesian mixture of Poissons from first principles. In section 3.2, we present the two-way Bayesian mixture of Poissons that uses "word grouping" to handle high-dimensionality and sparsity associated with the metagenome. In section 3.3, we briefly introduce the Bayesian mixture of Multinomials as standardized Bayesian mixture of Poissons and the corresponding two-way Bayesian mixture of Multinomials.

3.1 Bayesian Mixture of Poissons

We represent each read $\mathbf{x_i}$ by a $4 \times p$ matrix $\mathbb{N}_i = \{N_i(c_k|pa_j) : j = 1, ..., p\}$ and $c_k \in (A, C, T, G)$, where $N_i(c_k|pa_j)$ is the count of the number of occurrences of parent word pa_j followed by nucleotide c_k in read $\mathbf{x_i}$. The distribution of words within the reads of a species follow the parameters of a Poisson distribution, $\Theta = (\lambda_1, \lambda_2, ..., \lambda_m)$, where $\lambda_m = ((\lambda_{m,c_k|pa_j})_{\forall c_k})_{\forall pa_j}$, i.e., each local species distribution is a collection of Poisson distributions, one for every configuration pa_j of parents and c_k, and has the same parameters across reads of a species.

$$\Theta = \{\lambda_m : \forall m \in 1, .., M\}$$
$$\lambda_m = \{\lambda_{m,c_k|pa_j} : \forall c_k \in (A, C, T, G) \text{ and } \forall pa_j \in \{pa_j^1, pa_j^2, .., pa_j^p\}\} \qquad (4)$$

Therefore, the likelihood of the data will be,

$$p(\mathbf{x_i}|y_i = m) = p(\mathbf{x_i}|\lambda_m) = \prod_{pa_j} \prod_{c_k} p(N_i(c_k|pa_j)|\lambda_{m,c_k|pa_j})$$

$$= \prod_{pa_j} \prod_{c_k} \frac{\lambda_{m,c_k|pa_j}^{N_i(c_k|pa_j)} e^{-\lambda_{m,c_k|pa_j}}}{N_i(c_k|pa_j)!} \qquad (5)$$

EM Algorithm: We use Expectation-Maximization (EM) algorithm to infer the parameters [8]. In the expectation step, use the current parameter estimate $\Theta^{(i-1)}$ to find the posterior probability $p(y_i = m|\mathbf{x_i}, \Theta^{(i-1)})$ or $q_{i,m}$.

$$q_{i,m} \propto \alpha_m \cdot \prod_{pa_j} \prod_{c_k} p(N_i(c_k|pa_j)|\lambda_{m,c_k|pa_j}) \text{ subject to } \sum_{m=1}^{M} q_{i,m} = 1$$

In the maximization step, determine the expectation of the complete-data log likelihood.

$$Q(\Theta, \Theta^{(i-1)}) = \sum_{m=1}^{M} \sum_{i=1}^{N} q_{i,m} \left(\log(\alpha_m) \right.$$

$$\left. + \sum_{pa_j} \sum_{c_k} (N_i(c_k|pa_j) \log \lambda_{m,c_k|pa_j} - \lambda_{m,c_k|pa_j}) \right) \qquad (6)$$

subject to the constraint, $\sum_{m=1}^{M} \alpha_m = 1$. The maximum likelihood estimates for the Bayesian mixture of Poissons are,

$$\alpha_m = \frac{\sum_{i=1}^{N} q_{i,m}}{N} \; , \; \lambda_{m,c_k|pa_j} = \frac{\sum_{i=1}^{N} q_{i,m}.N_i(c_k|pa_j)}{\sum_{i=1}^{N} q_{i,m}} \qquad (7)$$

To initialize the EM algorithm, we randomly assign each read to a cluster m. The posterior probability $q_{i,m}$ is set to 1, if read i is assigned to cluster m and 0 otherwise. We then proceed with the M-step. Each iteration is guaranteed to increase the log-likelihood and the algorithm is guaranteed to converge to a local maximum of the likelihood function.

3.2   Two-Way Bayesian Mixture of Poissons

Higher order words are known to be more discriminative than shorter ones[21]. However, with the increase in the length of words, the length of the read vector grows exponentially (e.g, for $l = 10, 4^l \approx 10^6$). Moreover, many words will tend to similar distributions and hence, can be clustered together into a "word group". The feature matrix becomes high-dimensional and sparse. Hence, the model may fail to predict the true distribution of different components. Therefore, dimension reduction becomes necessary before estimating the components in the model. However, reduction of the number of words using feature selection cannot be too aggressive, otherwise the clustering accuracy will suffer.

In this paper, we handle the above challenge by "word grouping". This idea was first explored by Li *et al.* for a Naive Bayes mixture of Poisson distributions [14]. They called such a model a two-way mixture model, reflecting the observation that the mixture clusters induce a partition of the reads as well as of words. The cluster means are regularized by dividing the words into groups and constraining the parameters for the words within the same group to be identical. The grouping of the words is not pre-determined, but optimized as part of the model estimation. This implies that for every group, only one statistic for all the words in this group is needed to cluster reads. For instance, in Figure 1, the distributions of pentamers falls into three distinct groups. Thus, words following similar distributions can be clustered together into a "word group". Note that we make a distinction on the use of "cluster" for binning of reads within the same species and "group" for binning of words within a cluster. For simplicity, we assume that all clusters have the same number of word groups.

In the Bayesian mixture of Poissons, we group the set of Poisson parameters corresponding to each parent into its Poisson vector. Therefore, we have p different Poisson vectors corresponding to p configurations of parents. We divide the parents into groups and constrain the Poisson vector distributions corresponding to parents within the same group to have identical parameters. Let L be the number of groups within each cluster. Let $c(m, pa_j) \in 1, 2, ..., L$ denote the group that pa_j belongs to in class m. All parents in group l, have Poisson parameter $\lambda_{m,c_k|l}$. Let the number of parents in group l of class m be η_{ml}.

$$p(\mathbf{x_i}|\lambda_{\mathbf{m}}) = \prod_{pa_j} \prod_{c_k} \frac{(\lambda_{m,c_k|l}^{N_i(c_k|pa_j)} e^{-\lambda_{m,c_k|l}})}{N_i(c_k|pa_j)!} \quad \text{where } c(m, pa_j) = l \qquad (8)$$

Now, we can perform clustering using no more than order of ML dimensions. Word grouping leads to dimension reduction in this precise sense. We can derive an EM algorithm similar to the one outlined above to estimate the parameters.

$$\alpha_m = \frac{\sum_{i=1}^{N} q_{i,m}}{N} \; , \; \lambda_{m,c_k|pa_j} = \frac{\sum_{i=1}^{N} q_{i,m} \cdot \sum_{pa_j \in l} N_i(c_k|pa_j)}{\eta_{ml} \sum_{i=1}^{N} q_{i,m}} \qquad (9)$$

Once $\theta_{m,c_k|pa_j}^{(t+1)}$ is fixed, the word cluster index $c^{(t+1))}(m, j)$ can be found by doing a linear search over all components:

$$c(m, pa_j) = \arg \max_{l} \sum_{i=1} q_{i,m} \sum_{c_k} (x_{ij} \log(\lambda_{m,c_k|l}) - \lambda_{m,c_k|l})) \qquad (10)$$

3.3   Bayesian Mixture of Multinomials

Theorem: If $(X_1, X_2, .., X_p)$ are independent Poisson variables with parameters, $\lambda_1, \lambda_2, .., \lambda_p$ respectively, then the conditional distribution of $(X_1, X_2, .., X_p)$ given that $X_1 + X_2 + ... + X_p = n$ is multinomial with parameters λ_j/λ, where $\lambda = \sum \lambda_j$, i.e. $Mult(n, \pi)$, where $\pi = (\lambda_1/\lambda, \lambda_2/\lambda, ..., \lambda_p/\lambda)$[9].

The above theorem implies that the unconditional distribution $(X_1, X_2, ..., X_p)$ can be factored into a product of two distributions: a Poisson for the overall total, and a multinomial distribution of X, $X \sim Mult(n, \pi)$. Therefore, the likelihood based inferences about π are the same whether we regard $X_1, X_2, .., X_p$ as sampled from p independent Poissons or from a single multinomial. Here, n refers to the length of the reads and our interest lies in the proportion of words in the reads. Any estimates, tests, inferences about the proportions will be the same whether we regard n as random or fixed.

We can now derive the Bayesian mixture of Multinomials as standardized Bayesian mixture of Poissons. We assume that the distribution of words within the reads of a species is governed by the parameters of a multinomial distribution $\mathbf{\Theta} = (\theta_1, \theta_2, ..., \theta_\mathbf{m})$. Let $P_m(c_k|pa_j) = \theta_{m,c_k|pa_j}$. The sum of CPDs is well-defined, $\sum_{c_k \in (A,C,T,G)} \theta_{m,c_k|pa_j} = 1 \; \forall m$ and $\forall pa_j$. Each local species distribution is a collection of multinomial distributions, one for each configuration of pa_j. $\theta_\mathbf{m} = \{((\theta_{m,c_k|pa_j})_{\forall c_k})_{\forall pa_j}\}$. Therefore, within every species m, for each configuration pa_j of parents, we get an independent multinomial problem, $Mult(\theta_{\mathbf{m,c}|\mathbf{pa_j}}) = (\theta_{m,c_k|pa_j})_{\forall c_k}$, that has the same parameters across reads of a species.

$$\Theta = \{\theta_{\mathbf{m}} : \forall m \in 1, .., M\}$$
$$\theta_{\mathbf{m}} = \{\theta_{\mathbf{m,c}|\mathbf{pa_j}} : \forall pa_j \in \{pa_j^1, pa_j^2, .., pa_j^p\}\}$$
$$Mult(\theta_{\mathbf{m,c}|\mathbf{pa_j}}) = \theta_{\mathbf{m,c}|\mathbf{pa_j}} = \{\theta_{m,c_k|pa_j} : \forall c_k \in (A, C, T, G)\} \quad (11)$$

Therefore, the likelihood of the data will be,

$$p(\mathbf{x_i}|y_i = m) = p(\mathbf{x_i}|\theta_{\mathbf{m}}) = \prod_{pa_j} \prod_{c_k \in (A,C,T,G)} \theta_{m,c_k|pa_j}^{N_i(c_k|pa_j)} \quad (12)$$

The EM algorithm for Bayesian mixture of Multinomials and its corresponding two-way reduction can be derived similarly and we do not discuss it further.

4   Results

Datasets: Metagenomics being a relatively new field, lacks standard datasets for the purpose of testing clustering algorithms. As the "true solution" for sequence data generated from most metagenomic studies is still unknown, we focus on synthetic datasets for benchmarking. We use Metasim to simulate synthetic metagenomes[11]. It takes as input the sequencing technology to be used, a set of known genomes, length of the reads and a profile that determines the relative abundance of each genome in the dataset. We generated over 450 datasets with read lengths between 50 and 1000 bps and various abundance ratios.

The algorithms were implemented in Matlab. The space and time complexity scale linearly with the number of reads and species and quadratically with the number of dimensions in the search space. Our methods converged for all the cases we tested and was robust to the choice of initial conditions.

In order assess the robustness of our method, we ranked the 2-species datasets by a measure of intergenomic difference between sequences f and g, called the average dinucleotide relative abundance [4].

$$\delta^*(f, g) = \frac{1}{16} \sum_{X,Y} |\rho_{XY}^*(f) - \rho_{XY}^*(g)|$$

where $\rho_{XY}^*(f) = \dfrac{f_{XY}^*}{f_Y^* f_Y^*}$ and f_X^* denotes the frequency of X in f.   (13)

The δ^* values ranges from 34 to 340. In general, lower δ^* values correspond to "closely related species" and higher values correspond to "distant species". We use F-measure to calculate the clustering accuracy. F-measure is a combination of precision and recall. Precision represents the fraction of reads within the cluster that belong to the same species. And recall is the extent to which the reads within a cluster belong to the same species. In order to obtain global performance statistics, we combined the F-measure by weighting each cluster by the number of reads in the cluster.

The number of species in each dataset is supplied as an input. Determining the number of clusters from a statistical perspective is a difficult problem[22]. Maximum likelihood favors more complex models leading to over-fitting and hence is unable to address this issue. Previously, 16s/18s rDNA have been used for phylotyping and assessing species diversity using a rare-fraction curve. Most methods rely on heuristics to guide the choices of clusters. Determining species diversity is still an active area of research and we do not address it in this paper.

The Bayesian method overcomes the bottleneck of Naive Bayes methods in their assumption of independence between the words in a genome. In Figure 2, we compare the performance of our Bayesian methods with its Naive Bayes counterparts over 450 datasets with δ^* values ranging from 34 to 340. We observe a positive correlation between δ^* and the accuracy of our methods, as also noted in [13]. The Bayesian method outperforms the Naive Bayes in all instances. A Bayesian model regards the word counts as being multinomially distributed and hence captures the correlation between words counts. However, Naive Bayes methods was found to be on an average twice as faster than the corresponding Bayesian methods. Increased accuracy on datasets with short reads in yet another consequence of Bayesian networks (Figure 2). Though the classification accuracy is correlated to the read length, the drop in accuracy (bounded by 5%) with the decrease in read length from 1000 bps to 50 bps is hardly significant.

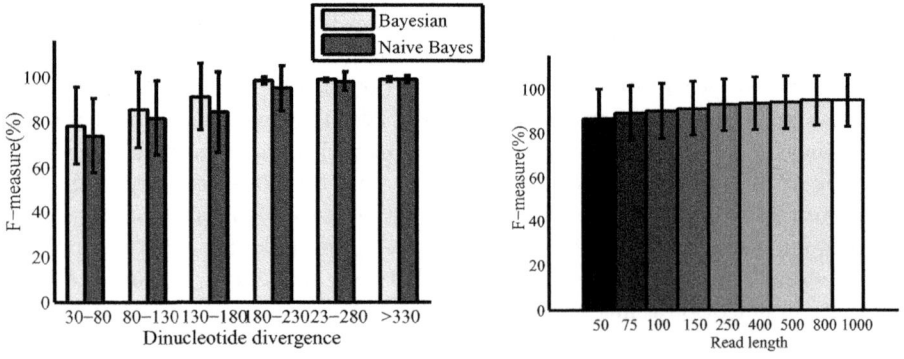

Fig. 2. Bayesian methods account for word overlaps. a) Comparison of performance of Bayesian mixture of Poissons and Multinomials with their Naive Bayes counterparts. We used a word length of 4. b) Effect of read length on clustering accuracy.

We systematically evaluated the robustness of our method to changes in the coverage ratio between species representative of various intergenomic differences. Binning results for 20 sets of simulated metagenomes with two species each is summarized in Figure 3.We varied the coverage ratio from 10:1 to 1:10 in stages, for the two species. We note that the accuracy drops at extreme coverages, when the fractional content of the species reduces to less than 10%.

Next, we analyzed the applicability of Bayesian mixture of Poissons in binning reads from low complexity communities, containing 2-6 species (Figure 3). The results were averaged over 50 datasets of varying divergences. Given that the multi-species dataset may contain reads from species with little intergenomic differences, there was a slight degradation in performance with the increase in number of species. This is in agreement with the results on variation with coverage, since the total coverage of each species is much lower in a multi-species dataset.

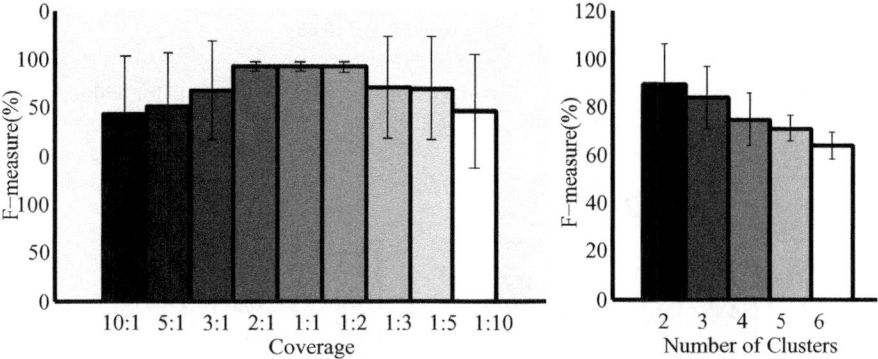

Fig. 3. Performance of Bayesian Poisson mixture model with varying coverage and different number of species (Word length of 4)

The above two observations are true of most composition based methods. Thus, even though composition based methods are suitable for binning reads from novel species, these methods by themselves are not sufficient for metagenomes containing a large number of species. For these methods to be practical on real metagenomes, we need to combine them with other similarity-based and abundance-based methods. In [12], the authors integrate the proposed composition based Scimm with similarity based methods to demonstrate increased performance on a medium-complexity simulated dataset. In our previous paper[20], we combined the Naive Bayes method with an abundance based method to characterize the Acid Mine Drainage dataset [23]. These results indicate that our method is suitable for binning reads belonging to dominant species, and that binning relatively rare species in a multi-species dataset may require modifications to the present Bayesian formulation.

In general, the discriminative power of the models increases with the length of words, despite the increasing space complexity. In our experiments, when we increased the word length from 2 to 7, initially the accuracy increases with word length. For word lengths beyond five, the accuracy begins to drop. This is because the feature matrix becomes high-dimensional and sparse. Hence, the model fails to predict the true feature distribution of different components. This necessitates dimension reduction before estimating the components in the model.

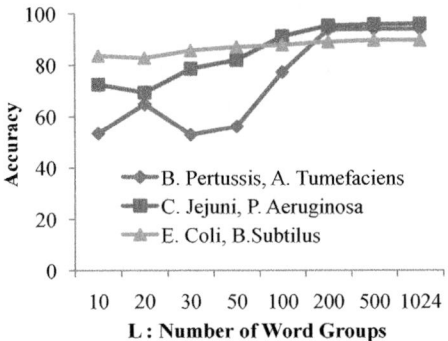

Fig. 4. Performance of Two-way Bayesian Poisson mixture model for values of word groups, L, varying from 10 to 1024. A word length of 5 is used.

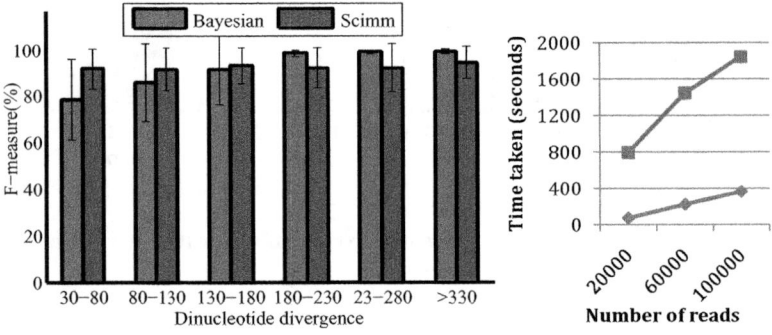

Fig. 5. Comparison of accuracy and time taken by Bayesian mixture of Poissons with Scimm. δ^* vary from 60 to 300. Read length of 200 bps and word length of 4.

In this paper, we perform "word grouping" to handle the above challenge. We propose a two-way mixture model where the mixture clusters induce a partition of the reads as well as of words. We used word length of 5 and varied the number of word groups from 10 to 1024 in stages (Figure 4). Performance stabilizes close to its optimal value at $L = 100$. This implies that the data can be classified using no more than ML dimensions, a significant reduction from the original number of dimensions. That is, the characteristic vectors are of a much lower dimension. Note that it is difficult to know a priori, the exact value of L that yields the best clustering. However, among the values we tested, lower values of L provided a higher accuracy.

Finally, in Figure 5, we compare the accuracy of our proposed multi-dimensional Bayesian model with state-of-art unsupervised composition-based method Scimm[12] averaged over 450 datasets. We used a read length of 100 bps and word length of 3. As the number of dimensions is relatively small, our method performs well without word grouping too. For δ^* values of 180 and above, our method performs better than Scimm. Though for δ^* values below 90,

Scimm does better than our method. However, our method converges at least 5 times faster than Scimm. The reduced time taken by our method to achieve comparable results justifies its use for clustering large metagenome datasets.

5 Conclusion

In this paper, we proposed a multivariate Bayesian methods based on Poisson and Multinomial mixture model to cluster the reads in a metagenome by their species of origin. This work demonstrates the use of statistically based mixture models for analysis of metagenome datasets by suitable choices of probability distributions. The Poisson and Multinomial models can effectively cluster the reads when the word counts are very low. Bayesian networks are used to represent the conditional dependencies between the words. We examined the sensitivity of the method to the number of species, abundance profile and length of reads within the dataset. Much work needs to be done to validate the usefulness of these model for real metagenome datasets. Our method is an unsupervised method that does not require any training data. However, we still need to specify the number of species for the algorithm. A future direction for our work is to overcome this limitation. Our framework complements the existing similarity-based and abundance-based methods and hence, can be combined with such methods to obtain a better performance. We intend to develop such hybrid methods in the future that can tackle the problem of classifying sequences in complex metagenomic communities. The methods have been tested on metagenomes, but can be adapted for use with a variety of discrete sequence data such as document clustering data, web-logs, purchase history or stock market data among others.

References

1. Bailly-Bechet, M., Danchin, A., Iqbal, M., Marsili, M., Vergassola, M.: Codon Usage Domains over Bacterial Chromosomes. PLoS Comput. Biol. 2(4), e37+ (2006)
2. Bentley, S.D., Parkhill, J.: Comparative genomic structure of prokaryotes. Annual Review of Genetics 38(1), 771–791 (2004)
3. Brady, A., Salzberg, S.L.: Phymm and PhymmBL: metagenomic phylogenetic classification with interpolated Markov models. Nature Methods 6(9), 673–676 (2009)
4. Campbell, A., Mrázek, J., Karlin, S.: Genome signature comparisons among prokaryote, plasmid, and mitochondrial DNA. Proceedings of the National Academy of Sciences of the United States of America 96(16), 9184–9189 (1999)
5. Chatterji, S., Yamazaki, I., Bai, Z., Eisen, J.: CompostBin: A DNA composition-based algorithm for binning environmental shotgun reads. ArXiv e-prints, 708 (August 2007)
6. Chen, K., Pachter, L.: Bioinformatics for whole-genome shotgun sequencing of microbial communities. PLoS Comput. Biol. 1(2), e24 (2005)
7. Dalevi, D., Ivanova, N.N., Mavromatis, K., Hooper, S.D., Szeto, E., Hugenholtz, P., Kyrpides, N.C., Markowitz, V.M.: Annotation of metagenome short reads using proxygenes. Bioinformatics 24(16), i7–i13 (2008)

8. Dempster, A.P., Laird, N.M., Rubin, D.B.: Maximum likelihood from incomplete data via the em algorithm. Journal of the Royal Statistical Society. Series B (Methodological) 39(1), 1–38 (1977)
9. Feller, W.: An Introduction to Probability Theory and Its Applications, vol. 1. Wiley (1968)
10. Heckerman, D.: A tutorial on learning with bayesian networks. Technical report, Learning in Graphical Models (1995)
11. Huson, D.H., Auch, A.F., Qi, J., Schuster, S.C.: MEGAN analysis of metagenomic data. Genome research 17(3), 377–386 (2007)
12. Kelley, D., Salzberg, S.: Clustering metagenomic sequences with interpolated markov models. BMC Bioinformatics 11(1), 544 (2010)
13. Kislyuk, A., Bhatnagar, S., Dushoff, J., Weitz, J.S.: Unsupervised statistical clustering of environmental shotgun sequences. BMC Bioinformatics 10(1), 316+ (2009)
14. Li, J., Zha, H.: Two-way poisson mixture models for simultaneous document classification and word clustering. Comput. Stat. Data Anal. 50, 163–180 (2006)
15. McHardy, A.C.C., Martín, H.G.G., Tsirigos, A., Hugenholtz, P., Rigoutsos, I.: Accurate phylogenetic classification of variable-length DNA fragments. Nature Methods 4(1), 63–72 (2007)
16. Rapp, M.S., Giovannoni, S.J.: The uncultured microbial majority. Annual Review of Microbiology 57(1), 369–394 (2003)
17. Reinert, G., Schbath, S., Waterman, M.S.: Probabilistic and Statistical Properties of Words: An Overview. Journal of Computational Biology 7(1-2), 1–46 (2000)
18. Robin, S., Rodolphe, F., Schbath, S.: DNA, Words and Models: Statistics of Exceptional Words. Cambridge University Press (2005)
19. Rosen, G., Garbarine, E., Caseiro, D., Polikar, R., Sokhansanj, B.: Metagenome fragment classification using n-mer frequency profiles
20. Shruthi Prabhakara, R.A.: A two-way multi-dimensional mixture model for clustering metagenomic sequences. In: ACM BCB (2011)
21. Teeling, H., Meyerdierks, A., Bauer, M., Amann, R., Glöckner, F.O.: Application of tetranucleotide frequencies for the assignment of genomic fragments. Environmental Microbiology 6(9), 938–947 (2004)
22. Tibshirani, R., Walther, G.: Cluster Validation by Prediction Strength. Journal of Computational & Graphical Statistics 14(3), 511–528 (2005)
23. Tyson, G.W., Chapman, J., Hugenholtz, P., Allen, E.E., Ram, R.J., Richardson, P.M., Solovyev, V.V., Rubin, E.M., Rokhsar, D.S., Banfield, J.F.: Community structure and metabolism through reconstruction of microbial genomes from the environment. Nature 428(6978), 37–43 (2004)
24. Willse, A., Tyler, B.: Poisson and multinomial mixture models for multivariate sims image segmentation. Analytical Chemistry 74(24), 6314–6322 (2002)

CRiSPy-CUDA: Computing Species Richness in 16S rRNA Pyrosequencing Datasets with CUDA

Zejun Zheng[1], Thuy-Diem Nguyen[1], and Bertil Schmidt[2]

[1] School of Computer Engineering, Nanyang Technological University, Singapore
{zjzheng,thuy1}@ntu.edu.sg
[2] Institut für Informatik, Johannes Gutenberg University Mainz, Germany
{bertil.schmidt}@uni-mainz.de

Abstract. Pyrosequencing technologies are frequently used for sequencing the 16S rRNA marker gene for metagenomic studies of microbial communities. Computing a pairwise genetic distance matrix from the produced reads is an important but highly time consuming task. In this paper, we present a parallelized tool (called CRiSPy) for scalable pairwise genetic distance matrix computation and clustering that is based on the processing pipeline of the popular ESPRIT software package. To achieve high computational efficiency, we have designed massively parallel CUDA algorithms for pairwise k-mer distance and pairwise genetic distance computation. We have also implemented a memory-efficient sparse matrix clustering program to process the distance matrix. On a single-GPU, CRiSPy achieves speedups of around two orders of magnitude compared to the sequential ESPRIT program for both the time-consuming pairwise genetic distance module and the whole processing pipeline, thus making CRiSPy particularly suitable for high-throughput microbial studies.

Keywords: Metagenomics, Pyrosequencing, Alignment, CUDA, MPI.

1 Introduction

Pyrosequencing technologies are frequently used for microbial community studies based on sequencing of hyper-variable regions of the 16S rRNA marker gene. Examples include profiling of microbial communities in seawater [1] and human gut [2]. The produced datasets contain reads of average length between 200 and 600 base-pairs. Typical dataset sizes range between a few tens of thousand up to around a million reads. Computational analysis of these datasets can be classified into two approaches: taxonomy-dependent and taxonomy-independent [3,4]. The taxonomy-dependent approach compares the input data against a reference database to assign each read to an organism based on the reported matches. The drawback of this approach is that existing databases are incomplete since the vast majority of microbes are still unknown.

The taxonomy-independent approach performs a hierarchical clustering and then bins the reads into OTUs (Operational Taxonomic Units) based on a distance threshold. Clustering is typically computed on a pairwise genetic distance

M. Loog et al. (Eds.): PRIB 2011, LNBI 7036, pp. 37–49, 2011.
© Springer-Verlag Berlin Heidelberg 2011

matrix derived from an all-against-all read comparison. The advantage of this approach is its ability to characterize novel microbes. However, the all-against-all comparison is highly compute-intensive. Furthermore, due to advances in pyrosequencing technologies, the availability and size of input read datasets is increasing rapidly. Thus, finding fast and scalable solutions is of high importance to research in this area.

Existing methods for the taxonomy-independent approach can be classified into four categories:

1. multiple sequence alignment (MSA), e.g. MUSCLE [5]
2. profile-based multiple sequence alignment (PMSA), e.g. RDP-aligner [6]
3. pairwise global alignment (PGA), e.g. ESPRIT [7]
4. greedy hierarchical clustering (GHC), e.g. UCLUST [8]

A number of recent performance evaluations [9,10] have shown that both the MSA and PMSA approaches often lead to less accurate genetic distance matrix values than the PGA approach. The GHC approach is faster than the other approaches but produced clusters which are generally of lower quality [10].The main drawback of the PGA approach is its high computational complexity. For an input dataset containing n reads of average length l, the time complexity of all optimal global pairwise alignments is $O(n^2 l^2)$. Thus, the application of PGA for metagenomic studies has so far been limited to a relatively small dataset size.

In this paper, we present an approach to extend the application of PGA to bigger datasets. We address the scalability problem by designing a fast solution for large-scale pairwise genetic distance matrix computations using commonly available massively parallel graphics hardware (GPUs) with the CUDA (Compute Unified Device Architecture) programming language. Recent works on using CUDA for fast biological sequences analysis [11,12] have motivated the use of CUDA for pairwise distance matrix computation. The presented tool, called CRiSPy, is based on the popular ESPRIT software [7], which is used by more than 150 major research institutes worldwide. Compared to sequential ESPRIT, CRiSPy provides up to two orders of magnitude speedup for both the pairwise alignment computation and the complete processing pipeline on a single-GPU. We achieve this acceleration by designing efficient massively parallel algorithms of the pairwise k-mer distance and genetic distance matrix computation and by implementing a more scalable hierarchical clustering module.

2   PGA Approach and ESPRIT

We consider an input dataset $R = \{R_1, \ldots, R_n\}$ consisting of n reads over the DNA alphabet $\Sigma = \{A, C, G, T\}$. Let the length of R_i be denoted as l_i and the average length of all reads be l. The PGA approach consists of three steps:

1. Computation of a symmetric matrix D of size $n \times n$, where $D_{i,j}$ is the genetic distance between two reads R_i and R_j
2. Hierarchical clustering of D
3. Using the dendrogram, group reads into non-overlapping OTUs at each given distance level d (e.g. output five OTU grouping for $d = 0.02, 0.03, 0.05, 0.1, 0.2$)

In the PGA approach, the genetic distance $D_{i,j}$ of the two reads R_i and R_j of length l_i and l_j is usually defined as $D_{i,j} = ml/al$, where ml denotes the number mismatches, including gaps (but ignoring end-gaps), in the optimal global alignment of R_i and R_j with respect to a given scoring system and al is the alignment length. The optimal global alignment of R_i and R_j can be computed with the dynamic programming (DP) based NW algorithm [13].

The values of ml and al can be found during the traceback procedure. If all genetic distances are computed using the NW algorithm, the overall amount of DP cells to be calculated is around $3l^2n^2/2$. Assuming input data sizes of $n = 250,000$ and $l = 400$, as well as a computing power of 10 GCUPS (Giga Cell Updates per Second), this procedure would take more than seventeen days. Furthermore, storing the genetic distance matrix would require 116.4 GBytes (using 4 bytes per entry) of memory.

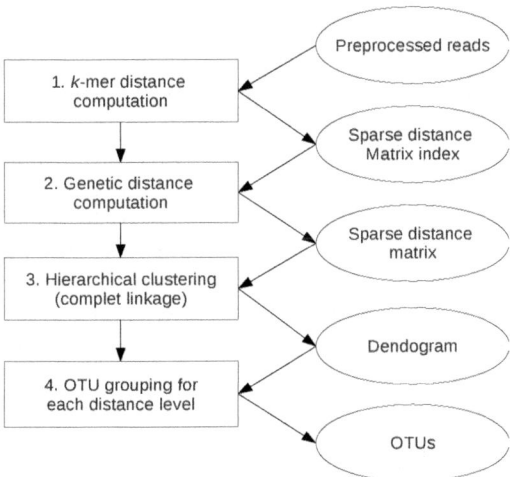

Fig. 1. The flowchart of the ESPRIT algorithm

ESPRIT [7] uses two techniques to reduce runtime and memory requirements:
1.Filtration. ESPRIT only computes genetic distances for read pairs which have corresponding k-mer distances below a given threshold θ_k. Given two reads R_i and R_j of length l_i and l_j and a positive integer k, their k-mer distance is defined as:

$$d_k(R_i, R_j) = 1 - \frac{\sum_{p=1}^{|\Omega|} \min(n_i[p], n_j[p])}{\min(l_i, l_j) - k + 1}$$

where Ω is the set of all substrings over Σ of length k enumerated in lexicographically sorted order and $n_i(p)$ and $n_j(p)$ are the numbers of occurrences of substring number p in R_i and R_j respectively. This approach is efficient since computation of all pairwise k-mer distances can be done in time $O(ln^2)$. It also relies on the assumptions that k-mer distance and genetic distance are correlated [5] and that a read pair with a large k-mer distance is usually not grouped into the same OTU. Lower values for θ_k would increase filtration efficiency but decrease sensitivity.

2.Sparse matrix representation. Filtration typically eliminates the majority of read pairs from further consideration. Thus, the k-mer distance matrix and the genetic distance matrix can both be efficiently stored in a sparse matrix format, which reduces memory requirements. Figure 1 shows the processing pipeline of the ESPRIT algorithm.

3   CRiSPy

3.1   Parallel k-mer Distance Computation

Although the computation of k-mer distances is around two orders of magnitude faster than the computation of genetic distances, it can still require a significant amount of time. Therefore, we have designed a sorting-based k-mer distance calculation method which can be efficiently parallelized with CUDA.

Initially, a so-called value array V_i is pre-computed for each input read R_i. It consists of all substrings of R_i of length k sorted in lexicographical order. For a pairwise k-mer distance, the two corresponding value arrays are scanned in ascending order. In each step, two elements are compared. If the two compared elements are equal, the indices to both value arrays are incremented and the corresponding counter $min(n_1(i), n_2(i))$ is increased by one. Otherwise, only the index pointing to the value array of the smaller element is incremented. The pairwise comparison stops when the end of one value array is reached. The sorting-based algorithm is illustrated in Figure 2. Obviously, it requires time and space $O(l)$ for two reads of length l each.

```
count:=0; i:=0; j:=0;
while (i < l₁ – k + 1) and (j < l₂ – k + 1) do
        if V₁(i) < V₂(j) then
                i++;
        elseif V₁(i) > V₂(j) then
                j++;
        else
                count++;  i++; j++;
        endif
endwhile
distance:=1 – count / (min(l₁,l₂) – k + 1);
```

Fig. 2. Sorting-based k-mer distance calculation for two reads R_1, R_2 of length l_1, l_2

The pair indices with k-mer distance smaller than a given threshold value are kept in a sparse index matrix for the subsequent processing stage. In this paper, we use the threshold $\theta_k = 0.3$ for large datasets and $\theta_k = 0.5$ for medium datasets.

3.2 Parallel Genetic Distance Computation

CRiSPy uses a simplified formula for global alignment with affine gap penalty based on the Needleman-Wunsch (NW) algorithm [13]. The scoring matrix of the alignment is computed using the following formula:

$$M(p,q) = \max \begin{cases} M(p-1, q-1) + sbt(R_i[p], R_j[q]) \\ M(p, q-1) + \alpha D(p, q-1) + \beta U(p, q-1) \\ M(p-1, q) + \alpha D(p-1, q) + \beta L(p-1, q) \end{cases}$$

where D, L and U are binary DP matrices to indicate which neighbor (diagonal, left or up) the maximum in cell $M(p,q)$ is derived from. Matrices D, L and U are defined as follows:

$U(p,q) = 0, L(p,q) = 0, D(p,q) = 1 \; if M(p,q) = M(p-1, q-1) + sbt(R_i[p], R_j[q])$
$U(p,q) = 0, L(p,q) = 1, D(p,q) = 0 \; if M(p,q) = M(p, q-1) + \alpha D(p, q-1) + \beta U(p, q-1)$
$U(p,q) = 1, L(p,q) = 0, D(p,q) = 0 \; if M(p,q) = M(p-1, q) + \alpha D(p-1, q) + \beta L(p-1, q)$

Note that $D(p,q) + L(p,q) + U(p,q) = 1$ for $p = 0, \ldots, l_i, q = 0, \ldots, l_j$.

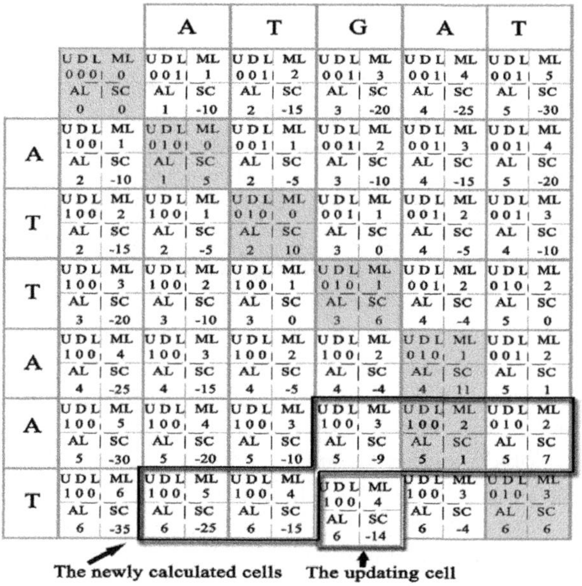

Fig. 3. DP matrices for two input reads ATGAT and ATTAAT with the scoring scheme: $sbt(x = y) = 5$, $sbt(x \neq y) = -4$, $\alpha = -10$, $\beta = -5$

To make the genetic distance calculation more suitable for parallelization, we have designed a trace-back-free linear space implementation by merging the ml and al calculation into the DP computation of the optimal global alignment score. To obtain the values ml and al, we introduce two more matrices ML and AL with the recurrent relations as follows:

$$ML(p,q) = U(p,q)ML(p,q-1) + L(p,q)ML(p-1,q) + D(p,q)ML(p-1,q-1)$$
$$- m(R_i[p], R_j[q]) + 1$$
$$AL(p,q) = U(p,q)AL(p,q-1) + L(p,q)AL(p-1,q) + D(p,q)AL(p-1,q-1) + 1$$

where $m(R_i[p], R_j[q]) = 1$ if $R_i[p] = R_j[q]$ and $m(R_i[p], R_j[q]) = 0$ otherwise. Initial conditions are given by $ML(0,0) = AL(0.0) = 0, ML(0,q) = AL(0,q) = q, ML(p,0) = AL(p,0) = p$ for $p = 1, \ldots, l_i, q = 1, \ldots, l_j$. Figure 3 illustrates an example for the computation of the DP matrices M, U, D, L, ML and AL. The dark shaded cells and arrows show the optimal global alignment path. Note that this is a score-only computation and therefore requires only linear space.

Furthermore, we have employed the banded alignment concept to reduce the number of computed DP matrix cells. In this approach, only cells within a narrow band along the main diagonal are calculated. Using eight test datasets in Section 5 and a band of width 1/5 (1/10) of the length of the shorter read, pairwise genetic distances up to the threshold $\theta_g = 0.2$ (0.1) can still be computed accurately while the computation runtime is reduced by 2.29 (3.58) compared to the full alignment. However, read pairs with a genetic distance larger than 0.2 (0.1), might get a higher (but never lower) distance value assigned.

Even though some of the larger distance values might change, the pairwise distances can still result in an identical OTU structure after clustering. This is due to the fact that for most datasets used for microbial community profiling, we are only concerned about the species or genus level OTU assignment which assumes a distance threshold of 0.03 or 0.05 between reads within the same cluster. Though these distinctions are still controversial amongst microbiologists [14], the threshold of 0.2 can still ensure the correctness of the estimation.

An overview of the CUDA implementation on a single GPU of the genetic distance computation is shown in Figure 4. The pair indices and input reads are transferred to CUDA global memory, whereby reads are represented as binary strings using two bits per base: A=00, T=01, G=11, C=10.

Multiple CUDA threads can calculate the pairwise distances in parallel. During the computation, one row of DP matrix values per pairwise alignment is stored in CUDA global memory which is accessed using coalesced data transfer to reduce transfer time. Moreover, each thread within a thread block computes a DP matrix block of size 16×16 using fast access shared memory, which reduces the costly accesses to global memory by a factor of 16 and makes the kernel compute-bound rather than memory-bound. At the end of the computation, each thread returns a distance value to a buffer located in CUDA global memory. The result buffer is then transferred to host memory and the CPU creates the final sparse genetic distance matrix.

For the GPU cluster version, the sparse matrix of pair indices from the k-mer distance module is divided equally amongst the GPUs in the cluster through the host nodes. On each GPU, multiple CUDA thread blocks can perform the genetic distance computation in parallel since they are independent.

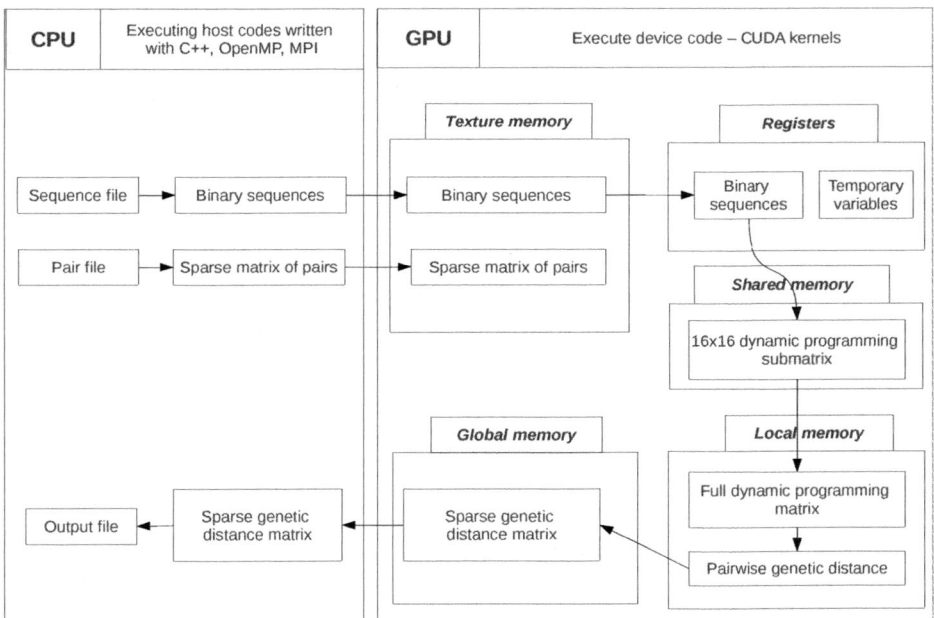

Fig. 4. CUDA implementation on GPU of the pairwise genetic distance computation

3.3 Space-Efficient Hierarchical Clustering

Hierarchical full linkage clustering is used for binning reads into OTUs from the linkage information provided by the sparse genetic distance matrix. Although the matrix is sparse, it is still of considerable size and often exceeds the amount of available RAM for large-scale datasets. Therefore, we have designed a memory-efficient hierarchical full linkage clustering implementation which can deal with sparse matrices of several hundred gigabytes in size. To reduce memory, ESPRIT [7] proposed the Hcluster algorithm using an "on-the-fly" strategy. Hcluster sorts the distances first and then shapes clusters by sequentially adding linkages. However, when the sparse matrix is very large the sorting procedure becomes a bottleneck. We have modified the Hcluster approach to make it more scalable.

Our approach first splits the sparse matrix into a number of smaller sub-matrices and then sorts each submatrix separately. We use a value queue to compensate for the fact that the entire sparse matrix is not fully sorted. Initially each submatrix file contributes one value (the smallest value) to the value queue. If the smallest value is removed from the queue by the clustering procedure, it is replaced by the next value from the corresponding file.

CRiSPy clustering is a procedure of shaping a binary tree, where each leave node presents a unique read. Pairwise distance values are scanned in ascending order. After a pairwise distance value is read from the value queue, the parent nodes of the two corresponding reads are located and the linkage information is added. Linkage information is only stored in parent nodes. This approach is memory efficient and achieves a constant search time for the parent nodes regardless of the sparse matrix size. At the end of the computation, the OTUs with respect to the given genetic distance cutoff as well as the node linkage information are outputted. Unlinked reads and reads with first linkage at high distance level are outputted as outliers.

4   Results

To evaluate the performance of CRiSPy, we have acquired eight 16S rRNA pyrosequencing datasets from the NCBI Sequence Read Archive (SRA) including three medium datasets and five large datasets. These raw datasets usually contain reads of low quality which can introduce a considerable number of false diversity into the species richness estimation. Hence, we have preprocessed these raw datasets to remove reads which contains ambiguous nucleotides (N) and reads with lengths that are not within 1 standard deviation from the average length. The numbers of reads before and after preprocessing are recorded in Table 1.

We benchmark the performance of CRiSPy against ESPRIT, a software package for estimating microbial diversities using 16S rRNA pyrosequencing data written in C++. ESPRIT package includes two different versions: a personal computer version (ESPRIT PC) and a computer cluster version (ESPRIT CC). ESPRIT PC implements sequential codes running on a single-CPU. ESPRIT CC takes a data parallel approach to parallelize both k-mer and genetic distance computation. The input data is split into several parts and distributed to compute nodes in a cluster through job submission. The partial distance matrices are then combined to form the full sparse distance matrix which is clustered by the same clustering module as ESPRIT PC.

Table 1. Runtime (in secs) and speedup comparison of k-mer distance computation between ESPRIT and CRiSPy

Dataset	Number of raw reads	Number of cleaned reads	ESPRIT PC T	ESPRIT CC T	S	CRiSPy MT T	S	Single-GPU T	S	Quad-GPU T	S
SRR029122	40864	15179	322	98	3.29	78	4.13	6.7	48	1.7	189
SRR013437	57902	22498	838	257	3.26	203	4.13	15.8	53	4.0	211
SRR064911	23078	16855	837	251	3.33	203	4.12	19.1	44	4.8	173
SRR027085	249953	131482	27600	8648	3.19	6014	4.59	473	58	119	232
SRR033879_81	1494071	300667	153000	48316	3.17	29130	5.25	3150	49	791	193
SRR026596_97	333630	178860	91300	29386	3.11	22432	4.07	1746	52	440	208
SRR058099	339344	162223	78000	25550	3.05	17920	4.35	1459	53	368	212
SRR043579	857248	256760	195000	66115	2.95	48221	4.04	4046	48	1015	192

Since the performance of ESPRIT CC depends on cluster setup and is subjected to communication overheads, we mainly use ESPRIT PC to benchmark the performance of CRiSPy. However, we also report the runtime and speedup of ESPRIT CC on a cluster of 4 CPUs to give readers a rough idea about the performance of ESPRIT CC. In this paper, we use the latest ESPRIT version released on February 2011. ESPRIT source code is available upon request to Dr. Sun Yijun (http://plaza.ufl.edu/sunyijun/ESPRIT.html).

We have implemented three different versions of CRiSPy:

1. a multithreaded C++ program with OpenMP (CRiSPy MT)
2. a CUDA program running on a single-GPU (CRiSPy single-GPU)
3. a CUDA program running on a multi-GPU cluster (CRiSPy multi-GPU)

We have conducted performance comparisons under the Linux OS with the following setup. ESPRIT PC and CRiSPy MT runtime are estimated on a Dell T3500 Workstation with a quad-core Intel Xeon 2.93 GHz processor, 4GB RAM. ESPRIT CC runtime is estimated on a cluster for four Dell T3500 Workstations connected via Ethernet. Condor High-Throughput Computing System is used for job submission. CRiSPy single-GPU runtime is measured on a Fermi-based Tesla S2050 GPU card with 3GB RAM. CRiSPy quad-GPU runtime is measured on a quad-GPU cluster which consists of two host nodes, each of which connected to two S2050 GPU cards. These nodes are connected by a high-speed Infiniband network. We use the following parameters for our experiments: $k = 6$, $\theta_k = 0.3$ for large datasets, $\theta_k = 0.5$ for medium datasets, $\theta_g = 0.2$, $sbt(x = y) = 5$, $sbt(x \neq y) = -4$, $\alpha = -10$ and $\beta = -5$.

4.1   Performance Comparison of k-mer Distance Computation

Table 1 records the runtime (in seconds) of the pairwise k-mer distance computation. For a dataset which contains n reads, the total number of pairs to be computed is $n(n-1)/2$. For both k-mer and genetic distance modules, T stands for runtime and S stands for speedup. T includes IO transfer time between host CPU and CPU, computation time of the algorithm and file IO time to output results to file.

CRiSPy MT is a multithreaded program written in C++ with OpenMP for a multi-core PC. It exploits the data parallelism of the pairwise distance matrix computation and achieves an average speedup 4.34 compared to ESPRIT PC. ESPRIT CC requires more time that CRiSPy MT mainly due to scheduling, communication overheads and data splitting stage. CRiSPy on a single-GPU (quad-GPU) runs 50.7(201.3) times faster than ESPRIT PC and 11.7 (46.4) times faster than CRiSPy MT.

4.2   Performance Comparison of Genetic Distance Computation

Table 2 shows the runtime (in minutes) of the genetic distance computation on the aforementioned datasets. The percentage of pairs p reported in Table 2 is defined as the number of read pairs with k-mer distance less than the threshold $\theta_k = 0.3$ divided by the total number of pairs which is $n(n-1)/2$ for a dataset of n reads. Hence, $1 - p$ is the percentage of pairs that k-mer distance module has effectively filtered out.

When running on multiple GPUs, the number of pair indices acquired from k-mer filtering step are divided equally amongst all the GPUs. Each GPU will then process its set of pair indices independently. As the runtime of ESPRIT PC, ESPRIT CC and CRiSPy MT are tremendous for large datasets, we sample representative parts of each dataset to get an estimated runtime.

Table 2. Runtime (in mins) and speedup comparison of genetic distance computation between ESPRIT and CRiSPy

Dataset	Average length	Percentage of pairs left p	ESPRIT PC	ESPRIT CC		CRiSPy MT		Single-GPU		Quad-GPU	
			T	T	S	T	S	T	S	T	S
SRR029122	239	11.04%	161	64	2.50	33	4.88	1.7	94	0.4	374
SRR013437	267	11.82%	462	184	2.51	94	4.93	4.6	101	1.1	402
SRR064911	524	56.73%	4700	1964	2.39	958	4.90	44	108	11	429
SRR027085	260	0.84%	1074	423	2.54	212	5.07	10	107	2.6	416
SRR033879_81	268	9.03%	64806	27029	2.40	12683	5.11	611	106	153	424
SRR026596_97	541	28.45%	2815	1128	2.50	559	5.04	26	110	6.5	434
SRR058099	531	6.76%	52430	22568	2.32	10551	4.97	479	109	120	437
SRR043579	556	8.37%	166160	73289	2.27	32627	5.09	1496	111	374	444

In comparison with ESPRIT PC, the runtime taken by CRiSPy MT reduces by the factor of 5.00 on average. Furthermore, average speedup gain by CRiSPy single-GPU (quad-GPU) is 105.6 (419.8) compared to ESPRIT PC and 21.1 (84.0) compared to CRiSPy MT. Similar to the filtration stage, ESPRIT CC encounters even more signification communication overheads since the amount of input and output data are much larger.

The speedup gain of the genetic distance module is much more significant compared to the k-mer distance module since genetic distance computation is more compute-intensive and hence it can utilize more processing powers of GPUs. Furthermore, the performance of this module increases in correspondence to the average length and the size of the input dataset.

4.3   Execution Time of CRiSPy Full Run

CRiSPy's processing pipeline includes three modules: parallel k-mer distance computation, parallel genetic distance matrix computation and sequential clustering of the resulting sparse distance matrix. Table 3 and Table 4 show the runtime of CRiSPy on a single-GPU and a quad-GPU cluster respectively for five large datasets.

For the genetic distance computation, we use three different global alignment schemes including full alignment, 1/5 banded alignment and 1/10 banded alignment. Please note that 1/5 (1/10) banded alignment reduces the runtime by a factor of 2.29 (3.58) compared to full alignment. We have observed from these experiments that 1/5 banded alignment produces identical OTUs as the full alignment and the 1/10 banded alignment results in only some minor difference in terms of outliers.

Table 5 shows the comparison between ESPRIT PC, CRiSPy MT and CRiSPy single-GPU full run for three medium datasets. We notice that by using banded alignment, CRiSPy single-GPU often requires two orders of magnitude less time than ESPRIT PC to execute the whole pipeline. Besides, the larger the dataset, the more significant speedup CRiSPy achieves.

Table 3. Runtime (in mins) of CRiSPy full run on a single-GPU

Dataset	k-mer distance computation	Genetic dist computation			Clustering		Total runtime		
		full	1/5	1/10	Sorting	Hcluster	full	1/5	1/10
SRR027085	7.9	10	4.8	3.3	2.6	3.0	24	18	17
SRR033879_81	53	611	286	195	162	285	1111	785	694
SRR026596_97	29	26	10	6.2	1.7	1.8	58	43	39
SRR058099	24	479	195	116	36	49	587	303	224
SRR043579	67	1496	607	360	102	152	1849	960	714

Table 4. Runtime (in mins) of CRiSPy full run on a quad-GPU cluster

Dataset	k-mer distance computation	Genetic dist computation			Clustering		Total runtime		
		full	1/5	1/10	Sorting	Hcluster	full	1/5	1/10
SRR027085	2.0	2.6	1.2	0.8	2.6	3.0	10	8.8	8.4
SRR033879_81	13	153	72	49	162	285	613	532	509
SRR026596_97	7.3	6.5	2.6	1.6	1.7	1.8	17	13	12
SRR058099	6.1	120	49	29	36	49	210	139	119
SRR043579	17	374	152	90	102	152	678	455	393

Table 5. Runtime (in mins) and speedup comparison between ESPRIT and CRiSPy

Dataset	ESPRIT PC	CRiSPy MT		Full band		1/5 band		1/10 band	
	T	T	S	T	S	T	S	T	S
SRR029122	167	35	4.79	2.4	68	1.6	106	1.3	125
SRR013437	477	98	4.85	6.1	78	3.7	130	3.0	157
SRR064911	4720	968	4.88	50	94	24	196	17	276

4.4 Assessment of Microbial Richness Estimation Accuracy by CRiSPy

To assess richness estimation accuracy and robustness of CRiSPy against sequencing errors, we have used several simulated datasets based on a dataset used by Huse et al. [15]. The simulated datasets have been generated by random sampling of reads from 43 known species and randomly introducing errors (insertion, deletion, mutation, ambiguous calls). Each of the simulated datasets contains ten thousand reads. A combination of error rates of 0.18% insertion, 0.13% deletion, 0.08% mutation and 0.1% ambiguous calls are used as standard error background. Besides 1/4, 1/2, 1 and 2 fold of the standard error rate are also introduced into the test dataset. Each test has been repeated ten times with newly simulated datasets and the average result is recorded.

Table 6 shows the estimated species richness for CRiSPy, ESPRIT, and MUS-CLE+MOTHUR [5,16] for a varying amount of sequencing errors at a 0.03 and 0.05 genetic distance cutoff. The results show that MUSCLE+MOTHUR estimates a significantly higher richness at higher error rates than CRiSPy and ESPRIT. Furthermore, at the 0.03 genetic distance cutoff level, CRiSPy shows better stability than ESPRIT.

Table 6. Comparison of richness estimation of CRiSPy, ESPRIT and MUS-CLE+MOTHUR for simulated datasets derived from 43 species

Tool	Genetic distance cutoff	Species estimated 1/4	1/2	1	2
CRiSPy	0.03	43	44	46	65
	0.05	42	42	43	43
ESPRIT	0.03	43	44	47	73
	0.05	42	42	43	43
MUSCLE+	0.03	45	49	84	168
MOTHUR	0.05	44	44	54	80

5   Conclusion

In this paper, we present CRiSPy - a scalable tool for taxonomy-independent analysis of large-scale 16S rRNA pyrosequencing datasets running on low-cost hardware. Using a PC with a single CUDA-enabled GPU, CRiSPy can perform species richness estimation of input datasets containing over three hundred thousand reads in less than half a day. Based on algorithms which are designed for massively parallel CUDA-enabled GPUs, CRiSPy achieves speedup of up to two orders of magnitude over the state-of-the-art ESPRIT software for the time-consuming genetic distance computation step. Since large-scale microbial community profiling becomes more accessible to scientists, scalable yet accurate tools like CRiSPy are crucial for research in this area.

Although CRiSPy is designed for microbial studies targeting DNA sequence analysis, the individual k-mer distance and genetic distance modules on GPUs can easily be extended to support protein sequence analysis and be used in general sequence analysis studies such as the usage of k-mer distance for fast, approximate phylogenetic tree construction by Edgar [17] or the utilization of pairwise genetic distance matrix in multiple sequence alignment programs such as ClustalW [18].

Availability: CRiSPy is available from the authors upon request.

References

1. Sogin, M.L., Morrison, H.G., Huber, J.A., et al.: Microbial diversity in the deep sea and the underexplored rare biosphere. PNAS 103(32), 12115–12120 (2006)
2. Turnbaugh, P., Hamady, M., Yatsunenko, T., et al.: A core gut microbiome in obese and lean twins. Nature 457(7228), 480–484 (2009)
3. Fabrice, A., Didier, R.: Exploring microbial diversity using 16S rRNA high-throughput methods. Applied and Environmental Microbiology 2, 074–092 (2009)
4. Hamady, M., Knight, R.: Microbial community profiling for human microbiome projects: Tools, techniques, and challenges. Genome Research 19(7), 1141–1152 (2009)
5. Edgar, R.C.: MUSCLE: multiple sequence alignment with high accuracy and high throughput. Nucleic Acids Research 32(5), 1792–1797 (2004)

6. Nawrocki, E.P., Kolbe, D.L., Eddy, S.R.: Infernal 1.0: inference of RNA alignments. Bioinformatics 25(10), 1335–1337 (2009)
7. Sun, Y., Cai, Y., Liu, L., et al.: ESPRIT: estimating species richness using large collections of 16S rRNA pyrosequences. Nucleic Acids Research 37(10), e76 (2009)
8. Edgar, R.C.: Search and clustering orders of magnitude faster than BLAST. Bioinformatics 26(19), 2460–2461 (2010)
9. Huse, S.M., Welch, D.M., Morrison, H.G., et al.: Ironing out the wrinkles in the rare biosphere through improved OTU clustering. Environmental Microbiology 12(7), 1889–1998 (2010)
10. Sun, Y., Cai, Y., Huse, S., et al.: A Large-scale Benchmark Study of Existing Algorithms for Taxonomy-Independent Microbial Community Analysis. Briefings in Bioinformatics (2011)
11. Liu, Y., Schmidt, B., Maskell, D.L.: CUDASW++2.0: enhanced Smith-Waterman protein database search on CUDA-enabled GPUs based on SIMT and virtualized SIMD abstractions. BMC Research Notes 3, 93 (2010)
12. Shi, H., Schmidt, B., Liu, W., et al.: A parallel algorithm for error correction in high-throughput short-read data on CUDA-enabled graphics hardware. Journal of Computational Biology 17(4), 603–615 (2010)
13. Needleman, S.B., Wunsch, C.D.: A general method applicable to the search for similarities in the amino acid sequence of two proteins. Journal of Molecular Biology 48(3), 443–453 (1970)
14. Schloss, P.D., Handelsman, J.: Introducing DOTUR a Computer Program for Defining Operational Taxonomic Units and Estimating Species Richness. Applied and Environmental Microbiology 71(3), 1501–1506 (2005)
15. Huse, S.M., Huber, J.A., Morrison, H.G., et al.: Accuracy and quality of massively parallel DNA pyrosequencing. Genome Biology 8(7), R143 (2007)
16. Schloss, P.D., Westcott, S.L., Ryabin, T., et al.: Introducing MOTHUR Open-Source Platform-Independent Community-Supported Software for Describing and Comparing Microbial Communities. Applied and Environmental Microbiology 75(23), 7537–7541 (2009), doi:10.1128/AEM.01541-09
17. Edgar, R.C.: Local homology recognition and distance measures in linear time using compressed amino acid alphabets. Nucleic Acids Research 32(1), 380–385 (2004)
18. Thompson, J.D., Higgins, D.G., Gibson, T.J.: CLUSTAL W: improving the sensitivity of progressive multiple sequence alignment through sequence weighting, position-specific gap penalties and weight matrix choice. Nucleic Acids Research 22(22), 4673–4680 (1994)

New Gene Subset Selection Approaches Based on Linear Separating Genes and Gene-Pairs

Amirali Jafarian, Alioune Ngom, and Luis Rueda

School of Computer Science, University of Windsor, Windsor, Ontario, Canada
{jafaria,angom,lrueda}@uwindsor.ca

Abstract. The concept of linear separability of gene expression data sets with respect to two classes has been recently studied in the literature. The problem is to efficiently find all pairs of genes which induce a linear separation of the data. It has been suggested that an underlying molecular mechanism relates together the two genes of a separating pair to the phenotype under study, such as a specific cancer. In this paper we study the *Containment Angle* (CA) defined on the unit circle for a linearly separating gene-pair (LS-pair) as an alternative to the paired *t*-test ranking function for gene selection. Using the CA we also show empirically that a given classifier's error is related to the degree of linear separability of a given data set. Finally we propose gene subset selection methods based on the CA ranking function for LS-pairs and a ranking function for linearly separation genes (LS-genes), and which select only among LS-genes and LS-pairs. Our methods give better results in terms of subset sizes and classification accuracy when compared to a well-performing method, on many data sets.

Keywords: Linearly Separating Features, Gene Expression, Microarray, Gene Selection, Feature Ranking, Filtering, Subset Selection.

1 Introduction

DNA microarrays give the expression levels for thousands of genes in parallel either for a single tissue sample, condition, or time point. Microarray data sets are usually noisy with a low sample size given the large number of measured genes. Such data sets present many difficult challenges for sample classification algorithms: too many genes are noisy, irrelevant or redundant for the learning problem at hand. Our present work introduces gene subset selection methods based on the concept of *linear separability* of gene expression data sets as introduced recently in [1]. We use their geometric notion of *linear separation* by pairs of genes (where samples belong to one of two distinct classes termed *red* and *blue* samples in [1]) to define a simple criterion for selecting (best subsets of) genes for the purpose of sample classification. Gene subset selection methods have received considerable attention in recent years as better dimensionality reduction methods than feature extraction methods which yield features that are difficult to interpret. The gene subset selection problem is to find a smallest subset of genes, whose expression values allow sample classification with the highest possible accuracy. Many approaches have been proposed in the literature to solve this problem. A simple and common method is the *filter approach* which first

M. Loog et al. (Eds.): PRIB 2011, LNBI 7036, pp. 50–62, 2011.

ranks single genes according to how well they each separate the classes (we assume two classes in this paper), and then selects the top r ranked genes as the gene subset to be used; where r is the smallest integer, which yields the best classification accuracy when using the subset. Many gene ranking criteria are proposed based on different (or a combination of) principles, including *redundancy* and *relevancy* [2], [5]. Filter methods are simple and fast, but they do not necessarily produce the best gene subsets; since there are gene subsets allowing better separation than the best subsets of top ranked genes. Other methods introduced in literature are the *wrapper approaches*, which evaluate subsets of genes irrespective of any possible ranking over the genes. Such methods are based on heuristics which directly search the space of gene subsets and guided by a classifier's performance on the selected gene subsets [8]. The best methods combine both gene ranking and wrapper approaches but they are computationally intensive.

Recently, some authors have considered pairs of genes as features to be used in filtering methods rather using than single genes. The motivation for using gene-pairs instead of single genes is that two single genes considered together may distinguish the classes much better than when they are considered individually; this is true even if one or both of the genes have low ranks from a ranking function defined for single genes. In other words, when we select only top-ranked single genes using such ranking function, some subsets of genes which have greater class distinguishing capability (than the subset of top-ranked genes) will not be selected due to the presence of low-ranked single genes. The authors of [2] devised the first gene selection method based on using pairs of genes as features. Given a gene-pair, they used *diagonal linear discriminant* (DLD) and compute the projected coordinate of each sample data on the DLD axis using only the two genes, and then take the two-sample t-statistic on these projected samples as the pair's score. The authors then devised two filter methods for gene subset selection based on the pair t-scores. Our approach in [10] was to use and evaluate linearly separating pairs of genes (LS-pairs) for the purpose of finding the best gene subsets. We proposed a simple ranking criterion for only LS-pairs and in order to evaluate how well each pair separates the classes. Additionally in order to find the best gene subsets, we devised a filter method, based on selecting only LS-pairs.

Our approach in this paper is to use both linearly separating singles genes (LS-genes) and linearly separating gene-pairs (LS-pairs) as features for the purpose of finding the best gene subsets. We propose ranking criteria for both LS-genes and LS-pairs in order to evaluate how well such features separate the classes then devise methods that select among top-ranked LS-genes and LS-pairs.

2   Linear Separability of Gene Expression Datasets

Recently, [1] proposed a geometric notion of *linear separation* by gene pairs, in the context of gene expression data sets, in which samples belong to one of two distinct classes, termed *red* and *blue* classes. The authors then introduced a novel highly efficient algorithm for finding all gene-pairs that induce a linear separation of the two-class samples. Let $m = m_1 + m_2$ be the number of samples, out of which m_1 are red and m_2 are blue. A gene-pair $g_{ij} = (g_i, g_j)$ is a *linearly separating* pair (LS-pair) if there exists a separating line L in the two-dimensional (2D) plane produced by the projection of the m samples according to the pair g_{ij}; that is, such that all the m_1 red samples are in one side of L and the remaining m_2 blue samples are in the other side of L, and no sample lies on L itself. Figure 1 and 2 show examples of LS and non-LS gene pairs, respectively.

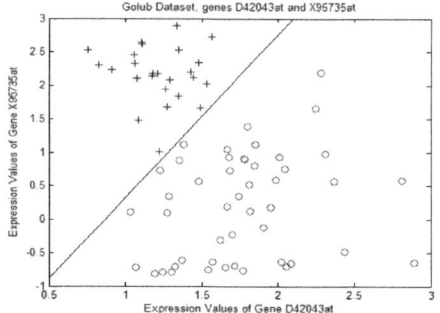
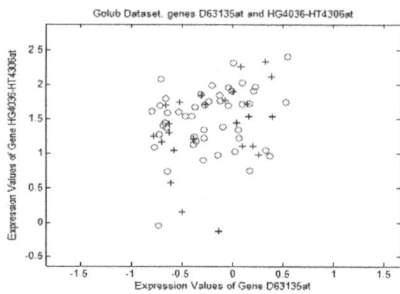

Fig. 1. An LS-pair taken from Golub (Leukemia) dataset

Fig. 2. A non LS-pair taken from Golub (Leukemia) dataset

In order to formulate a condition for linear separability, [1] first views the 2D points in a geometric manner. That is, each point of an arbitrarily chosen class, say red class, is connected by an arrow (directed vector) to every blue point. See Figures 3a and 4a, for example. Then the resulting $m_1 m_2$ vectors are projected onto the unit circle, as in Figures 3b and 4b, retaining their directions but not their lengths. The authors then proceed with a theorem proving that: *a gene pair $g_{ij} = (g_i, g_j)$ is an LS pair if and only if its associated unit circle has a sector of angle $\beta < 180°$ which contains all the $m_1 m_2$ vectors*. Figures 3 and 4 illustrate this theorem for pairs (x, y). Thus, to test for linear separability of pair g_{ij} one only needs to find the vector with the smallest angle and the vector with the largest angle and check whether the two vectors form a sector of angle $\beta < 180°$ containing all $m_1 m_2$ vectors.

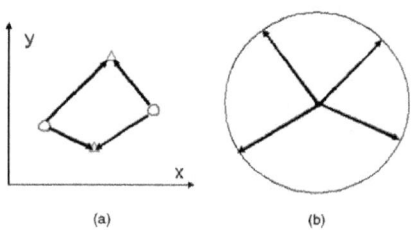
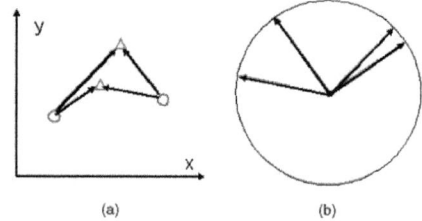

Fig. 3. A set of four non-separable points. (a) The construction of the vectors. (b) Their projection onto the unit circle [1].

Fig. 4. A set of four separable points producing vectors on the unit circle that are contained in a sector of angle $\beta < 180°$ [1]

Using the theorem above, [1] proposed a very efficient algorithm for finding all LS-pairs of a data set. Next, they derived a theoretical upper bound on the *expected number* of LS-pairs in a *randomly labeled* data set. They also derived, for a given data set, an empirical upper bound resulting from shuffling the labels of the data at random. The degree to which an actual gene expression is linearly separable, (in term of the actual number of LS-pairs in the data) is then derived by comparing with the theoretical and empirical upper bounds. Seven out of the ten data sets they have examined were highly separable and very few were not (see Table 4).

Let G be the set of genes, we generalize the definition of linear separation to apply to any t-tuple $g_{1...t} = (g_{i1}, g_{i2}, ..., g_{it})$ of genes where $1 \leq t \leq |G|$, $1 \leq j \leq t$, and $i_j \in \{1, ..., |G|\}$, and say that: $g_{1...t}$ is a linearly separating t-tuple (LS-tuple) if there exists a separating $(t-1)$-dimensional hyperplane H in the t-dimensional sub-space defined by the genes in $g_{1...t}$. It remains open to generalize the theorem of [1] to t-tuples of genes, $t \geq 1$, by considering projecting the $m_1 m_2$ vectors obtained from the t-dimensional points onto a unit $(t-1)$-sphere, and then determine a test for linearly separability of a t-tuple from the $(t-1)$-sphere. Clearly, the theorem is true for $t=1$: since a 0-sphere is a pair of points delimiting a line segment of length 2, and that the $m_1 m_2$ vectors point in the same direction (i.e., they form a sector of angle 0) if and only the single gene is linearly separable.

3 Feature Ranking Criteria

As said before, we will use LS-genes and LS-pairs as features to select from, and for the purpose of finding a minimal number of such features such that their combined expression levels allow a given classifier to separate the two classes as much as possible. Our approach in this paper is to first obtain all the LS-genes and LS-pairs of a given data set, rank these features according to some ranking criteria, and then apply a filtering algorithm in order to determine the best subsets of genes.

3.1 LS-Pair Ranking Criterion

The LS-pairs from given data sets were also used as classifiers in [1], using a standard training-and-test process with cross-validation. The authors compared the performance of these new classifiers with that of an SVM classifier applied to the original data sets without gene selection step. They found that highly separable data sets exhibit low SVM classification errors, while low to non-separable data sets exhibit high SVM classification errors. However, no theoretical proof exists showing the relation between SVM performance and the degree of separability of a data set; although this seems quite intuitive.

In [10], we investigated the relationship between the performance of a classifier applied to an LS-pair of a given data set and the β-sector of the LS-pair (discussed in Section 2, see Fig. 4b). We call β, the *Containment Angle*. Intuitively, the smaller is β for an LS-pair then the higher should be the accuracy of a classifier using the LS pair as input. This is because: the smaller is the angle β, the farther the samples are from the separating line L. Also for LS-pairs, the generalization ability of a classifier should decreases when β is close to $180°$ since some samples are very close to the separating line. To test this, we used the algorithm of [1] in [10] to generate all the LS pairs of a given data set and sorted them in increasing order of their angles β. We then proceeded as follows. For each LS pair, $g_{ij} = (g_i, g_j)$ of D, we applied a classifier with 10 runs of 10-fold cross-validation on D but using g_{ij} as the feature subset. We experimented on each separable data set examined in [1] and tried with many classifiers. From these experiments, we observed that the accuracy of the classifiers increased in general as the containment angle decreased from the bottom LS-pair (having largest angle) to the top LS-pair (having smallest angle). There were very few examples (see last row of Table 1, for instance), where the accuracy does not increase monotonously as the angle decreases within a consecutive sequence of LS-pairs. However, the average of the accuracies of the bottom ten LS-pairs were lower than that of the top ten LS-pairs. These experiments also show that using LS pairs is a better alternative than using the full set of genes for sample classification purpose, since classifying using pairs is much faster than using the gene set while still giving satisfactory performances. This enforces our intuition above while suggesting that one can use the Containment Angle as a measure of the quality of an LS-pair.

Table 1 shows the performance of SVM used on each of the top three LS-pairs for each data set, and compares with SVM used on all genes of the data sets (last column) with ten-fold cross-validation. In Table 1, we can see that applying SVM on top LS-pairs yields performance comparable to applying SVM on the full gene set; indeed better accuracies are obtained from the LS-pairs than from the full data. (in bold fonts).

3.2 LS-Gene Ranking Criterion

As mentioned earlier, a single gene is an LS-gene if and only if all the $m_1 m_2$ vectors in the corresponding zero-sphere point in the same direction (See Fig. 5 and 6 for a non LS-gene, an LS-gene and their projection in the Zero-sphere). We use a simple ranking criterion illustrated in Fig. 7: for each LS-gene, we compute the quantities A and B and use the ratio A/B as the score of the LS-gene.

4 Gene Subset Selection

Gene subset selection approaches based on gene pairs have been proposed in [2]. For a given gene pair, the authors used a two-sample t-statistic on projected data samples as the score of pairs (pair t-score), and then pairs are ranked according to their t-scores for the purpose of subset selection. They devised two subset selection algorithms which differ in the way gene pairs are selected for inclusion in a current subset. In their fastest method, they iteratively select the top-ranked gene g_i from the current list of genes, then find a gene g_j such that the t-score of the pair $g_{ij} = (g_i, g_j)$ is the maximum given all pairs $g_{ik} = (g_i, g_k)$, and then remove any other gene-pairs containing either g_i or g_j; this continues until r gene are selected. In their best but very slow method, they generate and rank all the possible gene pairs, and then select the top r ranked gene-pairs. The gene-pairs in [2] are not necessarily LS-pairs. In [10], we iteratively selected the top-ranked LS-pair until r genes are selected.

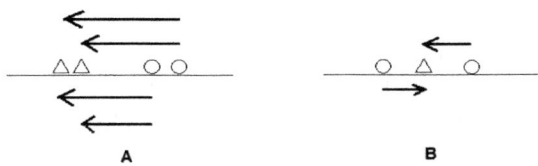

Fig. 5. A set of points causing Linear Separability (Left Panel) Vs. Non Linear Separability (Right Panel)

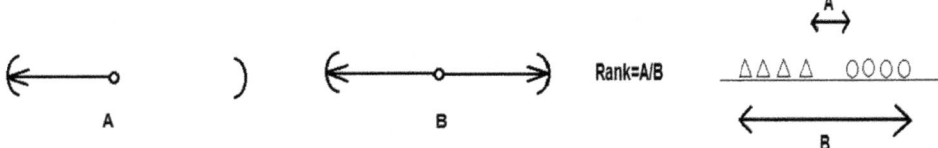

Fig. 6. The projection of vectors of LS Points in the Zero-Sphere (Left Panel) Vs. Non Linear Separability (Right Panel)

Fig. 7. Ranking Criterion for LS Genes

In this section, we propose gene subset selection approaches based on selecting only LS-genes and LS-pairs. The problem with this is that, initially, a data set may have a low degree of linear separability, and hence, not enough LS-features to select from. To overcome this problem, we first apply SVM with soft margin on the initial given data set before performing any gene selection method, and then sort the support vector (SV) samples in decreasing order of their Lagrange coefficients. When there are no more LS-features to select from during the process of gene selection, we then iteratively remove the current SV sample having the largest Lagrange coefficient, until the resulting data set contains LS-features; such SV samples are farthest from the separating maximum margin hyperplane and are probably misclassified by SVM. We devised two filtering methods to be discussed below.

Our first gene subset selection method (LSGP) proceeds by iteratively selecting in this order, from the LS-genes and then from the LS-pairs until a subset S of r genes is obtained. The LS-features are ranked according to the ranking criteria discussed above. In the LSGP algorithm, S is the subset to be found and r is the desired size of S, and G and P are respectively the sets of LS-Genes and LS-pairs. In lines 6.b and 7.c.ii, we apply a classifier to the currently selected subset S to keep track of the best subset $Best$-S of size $\leq r$. We use ten runs of ten-fold cross-validation on S, and the algorithm returns subsets S and $Best$-S and their performances. SV samples with largest Lagrange coefficients are iteratively removed from data set D, in line 8.a.i, whenever there are not enough LS-pairs in the current D. When an LS-gene g_i (resp., LS-pair g_{ab}) is selected, we also remove all LS-pairs containing g_i, (resp., g_a or g_b); see lines 7.b and 7.c.iii. This deletion is in order to minimize redundancy. That is, when LS-gene g_i is selected then any LS-pair containing g_i will be redundant. In [2] they select the gene top-ranked gene g_i and then find a gene g_j such that the pair g_{ij} has maximal pair t-score. Also in their slow approach (which yields better performance than their fast method) they iteratively select the top-ranked pairs in such a way that the selected pairs are mutually disjoint from each other. That is, they delete all of those pairs which intersect the currently selected subset of genes. This deletion is not true for selected LS-pairs, however. Assume an LS-pair $g_{ab} = (g_a, g_b)$ is selected and assume LS-pair $g_{bc} = (g_b, g_c) \in P$ not yet selected. If we remove g_{bc}, then the *possible* LS-triplet $g_{abc} = (g_a, g_b, g_c)$, which may yield a better subset S or a shorter subset *Best-S*, will be lost. Hence, in our second method (DF-LSGP) we consider the intersection graph $N = (P, E)$ where, the vertex set is the set of LS-pairs, P, in D and edges $(v_i, v_j) \in E$ if v_i and v_j have a gene in common. We then perform a graph traversal algorithm on N, which selects LS-pairs as the graph is being traversed.

Table 1. Accuracy on the top three LS-pairs versus accuracy on the full gene set, using SVM with hard margin

	TP1	TP2	TP3	Full Data
Small Beer	98.96%	98.96%	98.96%	**100%**
Beer	98.96%	98.96%	98.96%	**99.06%**
Squamous	**100%**	**100%**	**100%**	**100%**
Bhttacharjee	99.23%	**100%**	99.74%	98.08%
Gordon	99.83%	99.56%	**99.94%**	99.28%
Golub1	95.42%	**100%**	**100%**	98.61%

The differences between our two methods are in lines 7 up to but not including lines 8. In DF-LSGP, the LS-genes are selected first as in the first method. Then we iteratively select the best LS-pair vertex and its un-selected neighbors in a depth-first manner; see line 7.6 and thereafter. This continues until the desired number of genes, r, is obtained. We have also implemented a breadth-first traversal of the graph, BF-LSGP, where the neighbors of a selected LS-pair are sent to a queue starting from the top-ranked ones. In practice, we do not create an intersection graph N (line 7.4) given that P may be very large for some data sets; we simply push or enqueue the top-ranked LS-pair from the initial P onto the stack or queue (line 7.6.3) then simulate the graph-traversal algorithm.

LSGP - LS-Gene and LS-Pair Selection on *D*:

1. $S \leftarrow \{\}$
2. $r \leftarrow$ desired number of genes to select
3. $d \leftarrow 0$
4. $G \leftarrow$ set of LS-genes of *D*
5. $G \leftarrow G - \{g_i \text{ s.t. } g_i \in S\}$; '$-$' = *set-difference*
6. Repeat
    a. $S \leftarrow S + \{g_i \leftarrow \text{top-ranked LS-gene in } G\}$
        ; '$+$' = *union*
    b. Apply a classifier on *S* and update *Best-S*
    c. $G \leftarrow G - \{g_i\}$
    d. $d \leftarrow d + 1$

    Until $d = r$ or $G = \{\}$
7. If $d < r$ Then
    a. $P \leftarrow$ set of LS-pairs of *D*
    b. $P \leftarrow P - \{g_{ij} \text{ s.t. } g_i \in S \text{ or } g_j \in S\}$
    c. Repeat
        i. $S \leftarrow S + \{g_{ij} \leftarrow \text{top-ranked LS-pair in } P\}$
        ii. Apply a classifier on *S* and update *Best-S*
        iii. $P \leftarrow P - \{g_{ij} \text{ s.t. } g_i \in S \text{ or } g_j \in S\}$
        iv. $d \leftarrow d + 2$

        Until $d \geq r$ or $P = \{\}$
8. If $d < r$ Then
    a. Repeat
        i. $D \leftarrow D - \{\text{SV sample with largest Lagrange coefficient}\}$
        Until *D* contains LS-features
    b. Repeat from 4 with the resulting *D*
9. Return *S*, *Best-S*, and their performances

DF-LSGP - Graph-Based LSGP Selection on *D*:

1- $S \leftarrow \{\}$
2- $r \leftarrow$ desired number of genes to select
3- $d \leftarrow 0$
4- $G \leftarrow$ set of LS-genes of *D*
5- $G \leftarrow G - \{g_i \text{ s.t. } g_i \in S\}$; '$-$' = *set-difference*
    ; *remove already selected LS-genes*
6- Repeat
    a. $S \leftarrow S + \{g_i \leftarrow \text{top-ranked LS-gene in } G\}$
        ; '$+$' = *union*
    b. Apply a classifier on *S* and update *Best-S*
    c. $G \leftarrow G - \{g_i\}$
    d. $d \leftarrow d + 1$
    Until $d = r$ or $G = \{\}$
7- If $d < r$ Then
    7.1. $P \leftarrow$ set of LS-pairs of *D*
    7.2. $P \leftarrow P - \{g_{ij} \text{ s.t. } g_i \in G \text{ or } g_j \in G\}$
          ; *remove LS-pairs containing LS-genes*
    7.3. $P \leftarrow P - \{g_{ij} \text{ s.t. } g_i \in S \text{ and } g_j \in S\}$
          ; *remove already selected LS-pairs*
    7.4. Construct intersection graph $N = (P, E)$
    7.5. For each vertex g_{ij}: set *visited* $[g_{ij}] \leftarrow$ false
    7.6. While there are un-visited vertices and $d < r$ Do:
          7.6.1. $Stack \leftarrow \{\}$
          7.6.2. $g_{ij} \leftarrow$ top-ranked vertex in *N*
          7.6.3. Push g_{ij} onto *Stack*
          7.6.4. While $Stack \neq \{\}$ and $d < r$ Do:
                  7.6.4.1. Pop g_{ij} from *Stack*
                  7.6.4.2. If g_{ij} is un-visited Then
                            7.6.4.2.1. *visited*$[g_{ij}] \leftarrow$ true
                            7.6.4.2.2. $d \leftarrow |S + \{g_{ij}\}|$
                            7.6.4.2.3. $S \leftarrow S + \{g_{ij}\}$
                            7.6.4.2.4. If *S* has changed Then
                                        7.6.4.2.4.1. Apply a classifier on *S* and update *Best-S*
                            7.6.4.2.5. $P \leftarrow P - \{g_{ab} \text{ s.t. } g_a \in S \text{ and } g_b \in S\}$
                                        ; *delete already selected vertices from N*
                            7.6.4.2.6. Push all un-visited neighbors of g_{ij} onto *Stack* starting from the least-ranked ones.
8- If $d < r$ Then
    a. Repeat
        i. $D \leftarrow D - \{\text{SV sample with largest Lagrange coefficient}\}$
        Until the resulting *D* contains LS-features
    b. Repeat from 4 with the resulting *D*
9- Return *S*, Best-*S*, and their performances

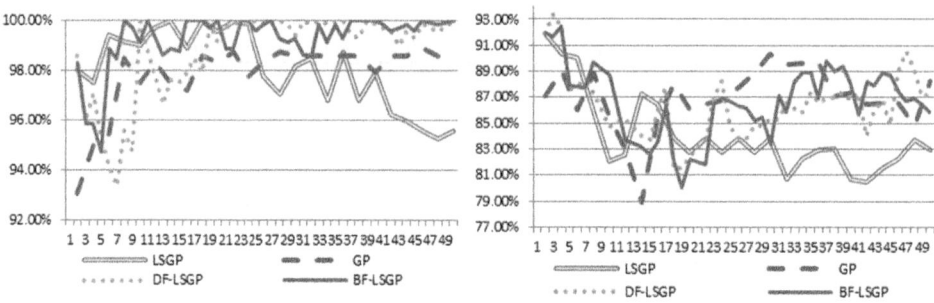

Fig. 8. Performance of SVM-Hard on Golub2

Fig. 12. Performance of SVM-Hard on Alon2

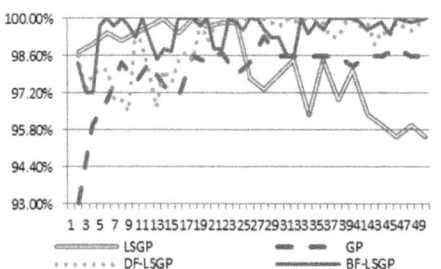

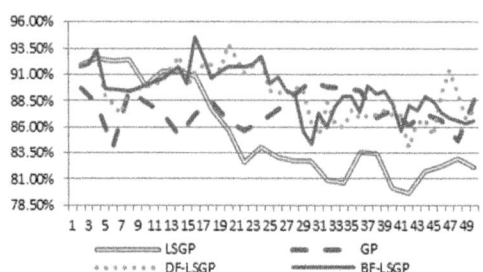

Fig. 9. Performance of SVM-Soft on Golub2

Fig. 13. Performance of SVM-Soft on Alon2

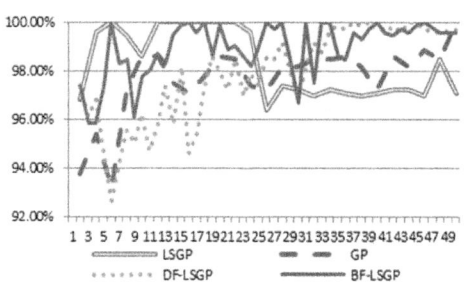

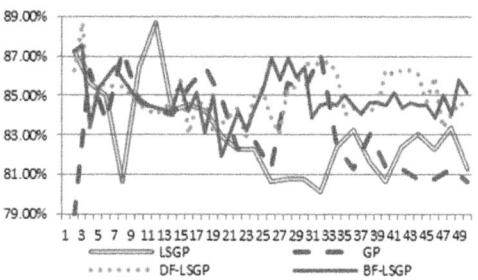

Fig. 10. Performance of KNN on Golub2

Fig. 14. Performance of KNN on Alon2

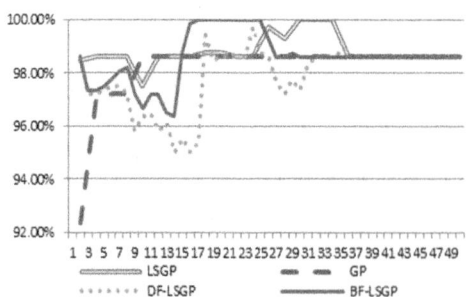

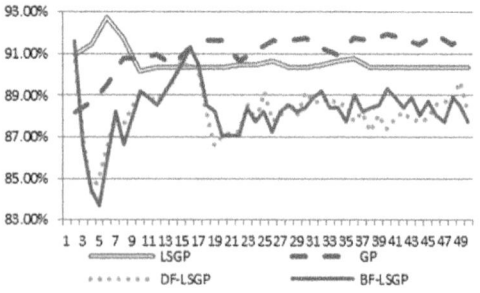

Fig. 11. Performance of DLD on Golub2

Fig. 15. Performance of DLD on Alon2

5 Computational Experiments

In the first set of experiments, we compared our three filtering approaches (LSGP, DF-LSGP, and BF-LSGP) with the *greedy-pair* (GP) method of [2]. We compared on the two publicly available data sets (Golub [3] and Alon [4]) used in [2]; which we have pre-processed in the same manner as in [2], and renamed as Alon2 and Golub2 to differentiate them with the Golub and Alon data sets used in [1] but pre-processed differently. In these experiments, we set the number of desired genes to $r = |S| \approx 50$ and also keep track of the best subset, *Best-S*, of size $\leq r$. Figures 8 to 15 show the results of our three filtering methods compared with the *greedy-pair* method of [2]. In this set of experiment four classifiers were applied using ten runs of ten-fold cross-validation, and we returned the average accuracy over the hundred folds for both the r-subset S and the best subset *Best-S*. The horizontal axis corresponds to the size of a selected gene subset and the vertical axis is the performance (classifier's accuracy) of the subset. Naturally, the four filtering methods performed best on the *highly separable* Golub2 data set (Figures 8 to 11) and performed worst on the *borderline separable* Alon2 data set (Figures 12 to 15). Our graph-based method, DF-LSGP and BF-LSGP performed better than LSGP and GP, in general; their curves are higher on average. LSGP performed the worst on average. The best subsets *Best-S* returned by our three methods are also smaller than those return by GP. Our graph-based methods make use and take advantage of the information or knowledge already present in the currently selected S subset in order to decide which LS-pairs to select next. Top-ranked LS-pairs which intersect S are always selected first, the advantage of which being the selection of t-tuples which are possibly linearly separating or which give better performances than arbitrarily selected LS-pairs. The selection of LS-pairs in GP and LSGP is somewhat arbitrary since it is based solely on their ranks.

Also we performed the second set of experiments in which the ranking and subset selection are performed on the training dataset within the framework of ten-fold cross-validation process. That is, we partition a data set D into ten distinct parts, and in each iteration of ten-fold cross validation process: 1) we perform feature ranking and selection on the nine-part training set; 2) train a classifier on this training set but using only the selected genes; and 3) estimate the performance of classification on the remaining one-part validation set. We did this set of experiments with our LSGP methods on the eight data sets of [1] (given in Table 4) and which are pre-processed as in [1] also.

The results for these experiments are shown in Tables 2 and 3 for subsets *Best-S* and S respectively. We show the performances of LSGP, DF-LSGP and BF-LSGP in terms of the average accuracy for both subsets S and *Best-S*. We must note that since feature ranking and selection is performed in each fold of the ten-fold cross-validation, then ten different subsets S and *Best-S* are obtained after the ten iterations of the cross-validation process. These subsets are not fixed as in our three sets of experiment above. Thus for subsets *Best-S*, in Table 2, we list in parenthesis the minimum, the average, and the maximum size of the hundred subsets *Best-S* obtained after the ten runs of ten-fold cross-validation, beside showing the average of the accuracies of the hundred subsets. For subsets S, an entry is the average of the accuracies of the hundred subsets of size $r = 50$ each. The averages in both Tables 2 and 3 are quite high, even for the least separable data sets Alon and Adeno Beer. In addition, for all data sets, we obtained a subset *Best-S* with the maximal accuracy of 100%.

6 Benchmark Datasets

To evaluate the performance of our proposed method, we have done extensive experiments on eight publicly available microarray gene expression datasets, namely, Golub [3], Alon [4], Gordon[9], Beer [6], Small Beer [6], AdenoBeer [6], Bhattacharjee [7] and Squamous [7] datasets shown in table 4.

For this research we used eight Datasets which are publicly available. For datasets we did the following preprocessing steps; similar to those dataset used in [1]):

> ➢ Trimming: All values lower than 100 were set to 100, and all values higher than 16,000 were set to 16,000, creating a range of 100-16,000.
> ➢ Logarithmic transformation: The natural logarithm $\ln(x)$ was taken for each value.
> ➢ Standardizing: Each sample was standardized to have a mean of 0 and a standard deviation of 1.

For two other datasets called Golub2 and Alon2 we did the same preprocessing steps, done in [2], in order to have a sound comparison between our Gene Subset returned by our approach and theirs. The preprocessing for these two datasets is as follows:

> ➢ Logarithmic transformation: Base 10 logarithmic transformation
> ➢ Standardizing: For each gene, subtract the mean and divide by standard deviation.

For Golub2 the following additional preprocessing step is done (Similar to [2]): thresholding with a floor of 1 and filtering by excluding genes with $max\ min \leq 500$. This leaves us with a dataset of 3,934 genes.

Due to limited space for the details of all of the datasets used in this research see [1].

Table 2. Accuracy of *Best-S* from [XX]-LSGP, with Ranking and Selection on Training Sets

	KNN			SVM-Soft			SVM-Hard		
	LSGP	**DF-LSGP**	**BF-LSGP**	**LSGP**	**DF-LSGP**	**BF-LSGP**	**LSGP**	**DF-LSGP**	**BF-LSGP**
Beer	100% (1,2.31,36)	99.80% (1,2.46,34)	99.78% (1,2.06,21)	100% (1,2.36,13)	99.60% (1,1.81,11)	99.90% (1,2.16,18)	100% (1,2.16,18)	99.90% (1,2.66,34)	100% (1,2.42,21)
Small Beer	99.18% (1,1.21,3)	98.96% (1,1.18,3)	98.96% (1,1.21,3)	99.18% (1,1.81,3)	98.96% (1,1.13,2)	98.96% (1,1.15,3)	99.69% (1,3.77,32)	99.27% (1,1.90,47)	99.07% (1,1.66,46)
Squamous	100% (1,1,1)	100% (1,1,1)	100% (1,1,1)	100% (1,1,1)	100% (1,1,1)	100% (1,1,1)	100% (1,1,1)	100% (1,1,1)	100% (1,1,1)
Gordon	99.61% (2,3.76,28)	99.70% (2,4.58,43)	99.77% (2,4.60,47)	99.56% (2,4.12,30)	99.32% (2,4.15,44)	99.52% (2,4.55,40)	99.50% (2,3.88,44)	99.32% (2,3.95,37)	99.84% (2,4.60,40)
Bhttacharjee	98.81% (1,3.41,14)	98.36% (1,2.97,31)	98.19% (1,3.15,44)	98.68% (1,2.68,16)	98.29% (1,2.43,32)	98.11% (1,2.48,44)	98.61% (1,3.06,18)	98.29% (1,2.79,26)	98.18% (1,3.10,46)
Golub	98.01% (2,5.41,42)	98.46% (2,7.79,43)	97.65% (2,6.14,45)	97.70% (2,4.65,48)	97.40% (2,4.96,40)	97.61% (2,5.83,45)	98.67% (2,5.35,30)	99.11% (2,6.1,49)	98.86% (2,6.71,48)
Alon	93.43% (2,7.1,50)	93.93% (2,7.67,45)	94.43% (2,8.26,47)	91.62% (2,6.82,48)	93.57% (2,5.37,36)	92.83% (2,6.24,44)	92.57% (2,6.98,48)	95.19% (2,5.49,35)	94.86% (2,6.99,47)
Adeno Beer	88.33% (2,12.12,50)	88.39% (2,13.53,50)	86.96% (1,10.62,50)	87.64% (1,10.64,48)	87.83% (2,12.52,49)	87.29% (2,12.85,48)	88.38% (2,13.64,48)	88.62% (2,16.07,47)	88.52% (2,14.72,48)

Table 3. Accuracy of S from [XX]-LSGP, with Ranking and Selection on Training Sets

	KNN			SVM-Soft			SVM-Hard		
	LSGP	DF-LSGP	BF-LSGP	LSGP	DF-LSGP	BF-LSGP	LSGP	DF-LSGP	BF-LSGP
Beer	99.26%	99.07%	98.94%	99.47%	99.18%	98.94%	99.69%	99.28%	100%
Small Beer	98.98%	98.96%	98.96%	98.98%	98.96%	98.96%	99.08%	99.27%	98.96%
Squamous	100%	100%	100%	100%	100%	100%	100%	100%	100%
Gordon	99.06%	99.09%	99.14%	99.23%	99.02%	98.89%	98.78%	98.72%	98.90%
Bhttacharjee	98.29%	97.33%	96.18%	98.29%	97.65%	96.96%	97.82%	97.63%	97.09%
Golub	95.89%	95.32%	95.31%	96.11%	93.92%	95.23%	96.84%	96.26%	95.99%
Alon	84.57%	85.95%	81.95%	80.57%	80.17%	79.95%	78.45%	80.55%	82.86%
Adeno Beer	75.04%	76.80%	74.83%	74.23%	75.47%	76.29%	73.84%	76.91%	76.17%

7 Conclusion

In this research we investigated the idea of using the concept of linear separability of gene expression dataset for the purpose of gene subset selection. We showed that the Containment Angle (CA) can be used to rank linearly separating pair of genes. We also introduced a new ranking criterion for ranking LS-genes. We proposed three different gene subset selection methods, LSGP, DF-LSGP and BF-LSGP, which select linearly separating features using our ranking criteria. Extensive experiments are carried out showing that our approaches are at least comparable to current filtering methods which are based selecting gene-pairs rather than only single genes.

Table 4. Gene Expression Datasets used

Dataset Name	Cancer Type	Nb of Genes	Nb of Samples	Nb of samples of Class1	Nb of samples of Class2	Degree of Separability
Beer	Lung	7129	96	86	10	High
Small Beer	Lung	4966	96	86	10	Very High
Squamous	Lung	4295	41	21	20	Very High
Gordon	Lug	12533	181	150	31	Very High
Bhttacharjee	Lung	4392	156	139	17	Very High
Golub	Leukemia	7129	72	47	25	High
Alon	Colon	2000	62	40	22	Border Line
Adeno Beer	Lung	4966	86	67	19	No

Our approaches are only *proof of concept* and we are currently studying wrapper methods based on selecting (not necessarily linearly separating) gene-pairs. In this regards, our graph-based methods, DF-LSGP and BF-LSGP, will be modified to back-track or continue the search depending on the classifier's error on the current subset. In this paper we devised ranking criteria applied only to LS-features, which is quite restrictive. Hence, we are devising general ranking criteria which will apply to *all* features, and in such a way that LS-features are ranked very high. As a future research, we plan to generalize the theorem of [1] for generating all linearly separating t-tuples $g_{1...t} = (g_{i1}, g_{i2}, ..., g_{it})$ from a given data set and for a given size $t \geq 3$. Finally, we have not cited reported gene subsets obtained by our approaches due to space constraint. In particular for our last set of experiments (Tables 2 and 3) for reporting/returning a *single* gene subset (S or *Best-S*) out of the hundred such subsets we obtain after the ten runs of ten-fold cross-validation, we can either 1) take the genes that appear most often in all hundred cross-validation folds, or 2) take the subset that is closest to all the other subsets (centroid) using an appropriate distance measure between subsets.

References

1. Unger, G., Chor, B.: Linear Separability of Gene Expression Datasets. IEEE/ACM Transactions on Computational Biology and Bioinformatics 7(2) (April-June 2010)
2. Bø, T.H., Jonassen, I.: New Feature Subset Selection Procedures for Classification of Expression Profiles. Genome Biology 3(4), 0017.1–0017.11 (2002)
3. Golub, T.R., Slonim, D.K., Tamayo, P., Huard, C., Gaasenbeeck, M., Mesirov, J.P., Coller, H., Loh, M.L., Downing, J.R., Caligiuri, M.A., et al.: Molecular classification of cancer: class discovery and class prediction by gene expression monitoring. Science 286, 531–537 (1999)
4. Alon, U., Barkai, N., Notterman, D.A., Gish, K., Ybarra, S., Mack, D., Levine, A.J.: Broad patterns of gene expression revealed by clustering analysis of tumor and normal colon tissues probed by oligonucleotide arrays. Proc. Natl. Acad. Sci. USA 96, 6745–6750 (1999)
5. Ding, C., Peng, H.: Minimum redundancy feature selection from microarray gene expression data. Journal of Bioinformatics and Computational Biology 3(2), 185–205 (2005)
6. Beer, D.G., et al.: Gene-Expression Profiles Predict Survival of Patients with Lung Adenocarcinoma. Nature Medicine 8(8), 816–824 (2002)
7. Bhattacharjee, A., et al.: Classification of Human Lung Carcinomas by mRNA Expression Profiling Reveals Distinct Adenocarcinoma Subclasses. Proc. Nat'l Academy of Sciences of the USA 98(24), 13790–13795 (2001)
8. Kohavi, R., John, G.: Wrapper for feature subset selection. Artificial Intelligence 97(1-2), 273–324 (1997)
9. Gordon, G.J., et al.: Translation of Microarray Data into Clinically Relevant Cancer Diagnostic Tests Using Gene Expression Ratios in Lung Cancer and Mesothelioma. Cancer Research 62(17), 4963–4967 (2002)
10. Jafarian, A., Ngom, A.: A New Gene Subset Selection Approach Based on Linear Separating Gene Pairs. In: IEEE International Conference on Computational Advances in Bio and medical Sciences (ICCABS 2011), Orlando FL, February 3-5, pp. 105–110 (2011)

Identification of Biomarkers for Prostate Cancer Prognosis Using a Novel Two-Step Cluster Analysis

Xin Chen[1,*], Shizhong Xu[2,*], Yipeng Wang[3], Michael McClelland[1,4],
Zhenyu Jia[1,**], and Dan Mercola[1,**]

[1] Department of Pathology and Laboratory Medicine, University of California, Irvine
[2] Department of Botany and Plant Sciences, University of California, Riverside
[3] AltheaDx Inc. San Diego
[4] Vaccine Research Institute of San Diego

xinc6@uci.edu,
shxu@ucr.edu,
ywang@altheadx.com,
mmcclelland@sdibr.org,
zjia@uci.edu,
dmercola@uci.edu

Abstract. Prognosis of Prostate cancer is challenging due to incomplete assessment by clinical variables such as Gleason score, metastasis stage, surgical margin status, seminal vesicle invasion status and preoperative prostate-specific antigen level. The whole-genome gene expression assay provides us with opportunities to identify molecular indicators for predicting disease outcomes. However, cell composition heterogeneity of the tissue samples usually generates inconsistent results for cancer profile studies. We developed a two-step strategy to identify prognostic biomarkers for prostate cancer by taking into account the variation due to mixed tissue samples. In the first step, an unsupervised EM clustering analysis was applied to each gene to cluster patient samples into subgroups based on the expression values of the gene. In the second step, genes were selected based on χ^2 correlation analysis between the cluster indicators obtained in the first step and the observed clinical outcomes. Two simulation studies showed that the proposed method identified 30% more prognostic genes than the traditional differential expression analysis methods such as SAM and LIMMA. We also analyzed a real prostate cancer expression data set using the new method and the traditional methods. The pathway assay showed that the genes identified with the new method are significantly enriched by prostate cancer relevant pathways such as the wnt signaling pathway and TGF-β signaling pathway. Nevertheless, these genes were not detected by the traditional methods.

---

[*] Xin Chen and Shizhong Xu are joint first authors.
[**] Corresponding author.

M. Loog et al. (Eds.): PRIB 2011, LNBI 7036, pp. 63–74, 2011.
© Springer-Verlag Berlin Heidelberg 2011

1   Introduction

Prostate cancer is the most frequently diagnosed male cancer and the second leading cause of cancer death in men in the United States [1]. Majority of the diagnosed cases are "indolent" that may not threaten lives. Radical prostatectomy provides excellent outcomes for patients with localized disease. It is the subsequent disease recurrence and metastatic spread of the cancer that accounts for most of the mortality. In order to improve disease management and benefit for the patients, reliable indicators are needed to distinguish the indolent cancer from the cancer that will progress. This information would guide treatment choices, avoid inappropriate radical prostatectomy and provide guidance to those who may profit from adjuvant therapy in the post-prostatectomy setting, a period that is seldom utilized as a treatment window. Much effort has been made to identify gene expression changes between aggressive cases and indolent cases [2,3,4]. Standard analytical approaches, such as t-test, significance analysis of microarray (SAM) [5] and linear models for microarray data (LIMMA) [6], have been applied to these studies. However, few accepted and clinically employed biomarkers have been developed owing to the lack of consistency among these studies, *i.e.*, the significant gene set identified in one study has little overlap with the significant gene set identified in other studies. Similar phenomenon has been observed in breast cancer research as well [10]. We noted that a major reason accounting for such inconsistency across studies is the heterogeneity in terms of cell composition, *i.e.*, the tissue samples used for assays were usually mixture of various cell-types with varying percentages [11,12,13]. For example, prostate samples are usually composed of tumor, stroma and BPH (benign prostatic hyperplasia). Therefore, the observed gene expression changes among samples could be merely due to the difference in cell composition of these samples. Nevertheless, such composition heterogeneity is rarely taken into account in biomarker studies since there is no straightforward way to deal with such variation through regular gene expression analyses, leading to false discoveries and inaccurate conclusions in individual studies.

In this paper, we first use a linear combination model to integrate cell composition data (obtained by pathological evaluation or in silico method [11]) as shown in our previous studies [11,13]. We then propose a two-step strategy to identify genes that are associated with disease outcomes based on the assumption that the potential prognostic biomarker is able to partition patients into groups of various levels of risk. Step 1: For each gene, unsupervised cluster analysis involving an EM algorithm [15] was used to categorize the subjects (patients) into several groups (for example, 2, 3, or 4 groups) solely based on the expression values for the gene across subjects. Note that the number of groups (C) needs to be specified before implementing the EM algorithm. Models with different C will be fitted with the data. The optimal number C (or the optimal model) is can be determined through the Bayesian information criterion (BIC). Step 2: Chi-square test is utilized to select genes with strong associations between the groups obtained at Step 1 and the frequency of classification of a subject as relapse or non-relapse based on the recorded outcomes. The pool of genes with significant Chi-square results defines a classifier with potential for predicting the risk of relapse for new subjects.

Our analyses of two simulated data sets using the traditional gene differential expression methods (SAM and LIMMA) consistently indicated that when gene expression data is substantially variable due to mixed cell-type composition, an improved method for identifying cases with high and low risk of relapse ought to be developed. By analyzing a real prostate cancer data set using the new method, we identified 648 genes which categorize prostate cancer cases into two groups of high and low risk for relapse. These genes are significantly enriched in a number of interested pathways disclosed by computer-assisted programs, such as DAVID [14]. Finally the method has potential for accounted for other sources of heterogeneity that bedevil the analysis of Prostate Cancer such as the polyclonal nature of this cancer.

We start the paper with describing the novel two-step approach biomarker identification. The new method was then compared to two commonly used methods by analyzing the simulated data sets as well as a real prostate cancer data set. This is followed by a detailed discussion of the desirable features of the new method and its application domains.

2   Theory and Method

2.1   Step 1: Unsupervised Cluster Analysis

Mixture Model of Gene Expression. Let N be the number of subjects (patients) for a microarray experiment. Let X be a $P \times N$ matrix for the tissue components percentage matrix determined by the pathologists, where P is the number of tissue types included in the model. Define $y = [y_1, \ldots, y_N]^T$ as an $N \times 1$ vector for the observed expression levels (the values transformed from the raw microarray data by logarithms) of a particular gene across N individuals. Note that the Gaussian error model is assumed as below. It is commonly accepted that the raw microarray data are log-normally distributed; therefore, logarithms can adequately transform microarray data to resemble a normal distribution [5,6,7,8,9]. We assume that each subject is sampled from one of C clusters of different levels of risk. The expression of the gene for individual i in the kth cluster $(k = 1, \ldots, C)$ can be described by following model:

$$y_i|_{Z_i=k} = X_i^T \beta_k + \epsilon_i, \tag{1}$$

where Z_i is the cluster indicator for the ith individual, β_k is the cell-type coefficient vector $(P \times 1)$ for the kth cluster, and $\epsilon_i \sim N(0, \sigma^2)$. Such MLR setting has been used in our previous studies [11,13]. Therefore, the likelihood for the Gaussian mixture can be constructed as follows,

$$L(\Theta; y, Z) = \prod_{i=1}^{N} \prod_{k=1}^{C} [f(y_i|Z_i = k; \beta_k, \sigma^2) P(Z_i = k)]^{\delta(Z_i, k)}, \tag{2}$$

where $\pi_k = P(Z_i = k)$ denotes the prior probability of $Z_i = k$, $\Theta = (\pi_1, \ldots, \pi_C, \beta_1, \ldots, \beta_C, \sigma^2)$ is a vector of parameters, $\delta(Z_i, k)$ is an indicator variable defined as

$$\delta(Z_i, k) = \begin{cases} 1, \text{ if } Z_i = k \\ 0, \text{ otherwise} \end{cases}, \tag{3}$$

and

$$f(y_i | Z_i = k; \beta_k, \sigma^2) = \frac{1}{\sqrt{2\pi}\sigma} \exp\left[-\frac{(y_i - X_i^T \beta_k)^2}{2\sigma^2} \right] \tag{4}$$

is the normal density.

EM Algorithm for Cluster Analysis. The EM algorithm [15] is a numerical algorithm commonly used for clustering analyses. Generally, the EM algorithm consists of two major routines: an expectation (E) routine in which the expectation of the log-likelihood is computed, and a maximization (M) routine in which the MLE is calculated for each parameter. The E-routine and M-routine alternate until some criterion of convergence is reached. In the current study, the EM algorithm is implemented as follows,

Procedure 0: Initializing all parameters, $\Theta = \Theta^{(0)}$.

Procedure 1 (E): Update the posterior membership probability,

$$\pi_{ik} = E(\delta(Z_i, k)) = \frac{\pi_k f(y_i | Z_i = k; \beta_k, \sigma^2)}{\sum_{k'=1}^{C} \pi_{k'} f(y_i | Z_i = k'; \beta_{k'}, \sigma^2)} \tag{5}$$

Procedure 2 (M1): Update the mixing probability,

$$\pi_k = \sum_{i=1}^{N} \pi_{ik} \tag{6}$$

Procedure 3 (M2): Update the expression coefficient,

$$\beta_k = \left(\sum_{i=1}^{N} \pi_{ik} X_i X_i^T \right)^{-1} \left(\sum_{i=1}^{N} \pi_{ik} y_i X_i \right) \tag{7}$$

Procedure 4 (M3): Update the residual variance,

$$\sigma^2 = \frac{1}{N} \sum_{i=1}^{N} \sum_{k=1}^{C} \pi_{ik} \left(y_i - X_i^T \beta_k \right)^2 \tag{8}$$

Procedure 5: Repeat procedure 1 through procedure 4 until a certain criterion of convergence is reached.

2.2 Step 2: Gene Identification Based on Correlation Analysis

The posterior probabilities $(\pi_{i1}, \ldots, \pi_{iC})$ are calculated for subject i where $i = 1, \ldots, N$. The $\tau_i = \max(\pi_{i1}, \ldots, \pi_{iC})$ determined the membership for subject i given $\tau_i \geq 0.6$. If $\tau_i < 0.6$, the subject cannot be assigned to any of the C clusters. In this study, we only consider genes that definitively determine the membership for $\geq 60\%$ of the subjects. Therefore, genes that do not meet this criterion will be considered irrelevant to the disease progression. A contingency table is constructed for each gene based on the membership indicator obtained by cluster analysis and the observed clinical outcomes, *i.e.*, relapse or non-relapse. A chi-square test is then performed for each contingency table to evaluate the association between these two variables. Genes that are strongly associated with the observed outcome are selected if their p-values are less than a pre-selected cutoff point, *e.g.*, 0.001, where p-values are calculated from the central χ^2 distribution.

3 Application

3.1 Simulation Study

To demonstrate the efficiency of the new method, we carried out two simulation studies each with 1000 genes and 200 subjects. Among the 200 subjects, 1 to 100 were designated as non-relapse patients and 101 to 200 were designated as relapse patients. We randomly generated three tissue types (tumor, stroma and BPH) with various component percentages for each subject. In the first simulation, the first 150 genes were set to be associated with the disease outcomes. The expression coefficients $\beta_{intercept}$, β_{tumor}, β_{stroma} and β_{BPH} for each gene represent the strength of gene-outcome association. They were generated as follows: (1) $\beta_{intercept}$ and β_{BPH} were not altered between relapse and non-relapse patients; (2) for genes 1 - 50, the relapse and non-relapse patients had different β_{tumor} but the same β_{stroma}, *i.e.*, these genes are differentially expressed in tumor between relapse and non-relapse groups (tumor related signatures); for genes 51 - 100, the relapse and non-relapse patients had different β_{stroma} but the same β_{tumor}, *i.e.*, these genes are differentially expressed in stroma between relapse and non-relapse groups (stroma related signatures); (3) for genes 101 - 150, both β_{tumor} and β_{stroma} were different between the relapse and non-relapse patients, *i.e.*, these genes are differentially expressed in both tumor and stroma between the relapse and non-relapse groups (tumor and stroma related signatures). Table 1 (column 1) shows the expression coefficients for the first 150 genes. We deliberately generated random dichotomous variables for the 200 patients. These random dichotomous variables are similar to but unrelated to the outcome variable. We let genes 151 - 300 to be associated with these simulated dichotomous variables, yielding a special group of control genes. The remaining 700 genes were set to have no association with any indicator variables. Residual errors were sampled from a normal distribution with mean 0 and variance 0.01. The two-cluster EM algorithm was performed for each gene as described below.

First, the 200 subjects were randomly (0.5 : 0.5) assigned into the relapse and non-relapse groups. A simple regression analysis was used to generate the two initial expression coefficients sets $\beta_{10}, \beta_{11}, \beta_{12}, \beta_{13}$ and $\beta_{20}, \beta_{21}, \beta_{22}, \beta_{23}$ for these two groups. The average of the estimated residual errors for the 200 subjects was used as the initial value of σ^2. The EM algorithm was then applied until the difference between successive log-likelihoods was smaller than 0.0001. We constructed a 2×2 contingency table and calculated the p-value based on the central χ^2 distribution. Using the cutoff point (p < 0.001) mentioned in Theory and Method, we detected 148 differentially expressed genes with no false detection. The two missing genes were gene 5 and 29, owing to the fact that less than 60% of the 200 subjects had been definitively clustered. However, the p-values of the χ^2 test for the two genes are still significant for the successfully clustered subjects.

Table 1. The number of the genes (out of the first 150 genes) identified by SAM, LIMMA, and the new method in the first simulation experiment. The first column represents the simulation scenarios: (+) and (-) represents positive and negative expression coefficient, respectively.

Gene		SAM	LIMMA	New method
1-30	Tumor(+)	23	21	28
31-50	Tumor(-)	14	16	20
51-80	Stroma(+)	30	30	30
81-100	Stroma(-)	17	18	20
101-110	Tumor(+) Stroma(+)	10	10	10
111-130	Tumor(+) Stroma(-)	3	3	20
131-140	Tumor(-) Stroma(-)	10	10	10
141-150	Tumor(-) Stroma(+)	4	6	10
1-150		111	114	148

We also analyzed the simulated data using two commonly used differential expression analysis software packages, SAM [5] and LIMMA [6]. False discovery rate (FDR) ≤ 0.05 was chosen as the criterion for gene detection. SAM identified 111 differentially expressed genes with no false detection. LIMMA identified 114 differential expressed genes with no false detection. Table 1 (columns 3, 4 and 5) shows the numbers of genes identified by SAM, LIMMA, and the proposed new method, suggesting the new method has improved the accuracy for gene identification over the two existing methods. Another advantage of the new method is that it can identify the cell origin of the observed expression changes, i.e., the new method can determine if such expression changes occurred in tumor cells or stroma cells. The contingency test in Table 2 shows that our method can successfully identify 98% tumor-cell-related signatures, 100% stroma-related signatures, and 80% tumor-stroma related signatures. Ten tumor-stroma related genes were misclassified into stroma related genes. There are two possible explanations: (1) For these ten genes, the stroma expression coefficients (average 1.28) are much greater than the tumor expression coefficients (average 0.63), and (2)

The stroma percentage data are more variable than the tumor percentage data (about 50% of the simulated subjects had 0% tumor cells). Thus this simulation study reveals several advantages of the new method and highlights the improved efficiency.

Table 2. The performance of the new method in the first simulation experiment

# of estimate \ # of true	Tumor	Stroma	Tumor+stroma	Null
Tumor	48	0	0	0
Stroma	0	50	10	0
Tumor+stroma	0	0	40	0
Null	2	0	0	700

The second simulation experiment was based on the assumption that subtypes of tumors which are possibly polyclonal variants may exist for relapse patients, *i.e.*, different types of tumors may be subject to different levels of recurrence risk. In this simulation experiment, we divided relapse patients into low risk (subjects 101 - 150) and high risk relapse patients (subjects 151 - 200). We used the same X matrix as in the first simulation experiment. We let the first 150 genes (1-150) be associated with non-relapse/relapse status (2 clusters) but the next 150 genes (151-300) be associated with non-relapse/low risk/high risk status (3 clusters). The clusters were assigned different expression coefficients. Genes 301 - 350 and 351 - 450 were controlled by other randomly generated dichotomous and trichotomous variables respectively. The last 550 genes were not association with any indicator variables. The residual errors were sampled from normal distribution with mean 0 and variance 0.01. We used the two-cluster EM algorithm (EM2) and the three-cluster EM algorithm (EM3) to analyze the simulation data. A similar strategy was used for initializing parameters for each model. Comparison of these two models is shown in Figure 1. For genes 1 - 150, both EM2 and EM3 identified 145 of them, but each method identified two additional genes, making a shared proportion of 97.3% (see Figure 1(a)). Among the 149 identified genes, 132 genes were selected by the EM2 algorithm (2-cluster model) and the remaining 17 genes were selected by the EM3 algorithm (3-cluster model) based on the BIC scores (Model with lower BIC score was preferred.). For genes 151-300, 138 genes were identified by both the EM2 and EM3 algorithms (shared proportion of 92.6%, see Figure 1(b)). The two algorithms together identified 149 genes, among which 133 genes were selected by EM3 (3-cluster model) and 16 genes were selected by EM2 (2-cluster model) based on the BIC scores. These results are consistent with the setting of the simulation experiment, which shows that subtypes within a risk group may be identified by increasing the number of clusters in EM algorithm.

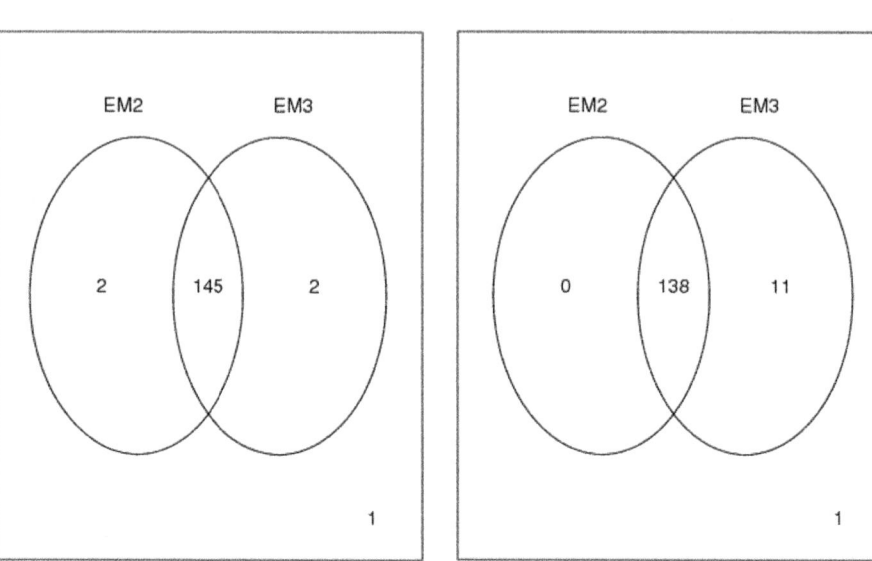

Fig. 1. The number of genes detected by EM2 and EM3 algorithms in the second simulation experiment. (a)Venn diagram of the detected genes for gene 1 - 150. For this analysis, EM2 identified 147 (2 + 145) genes and EM3 identified 147 (2 + 145) genes, with 145 genes identified by both models. Only 1 gene (bottom-right corner) has been missed by two models. (b)Venn diagram of the detected genes for gene 151 - 300.

3.2   Real Data Analysis

A publicly available prostate cancer data set [12] was analyzed here. A total 136 postprostatectomy frozen tissue samples were obtained from 82 subjects by informed consent using Institutional Review Board (IRB)-approved and HIPPA-compliant protocols. All tissues were collected at surgery and escorted to pathology for expedited review, dissection and snap freezing in liquid nitrogen. The tissue components (tumor epithelial cells, stroma cells, epithelial cells of BPH) were estimated by four pathologists. RNA samples prepared from the frozen tissue samples were hybridized to Affymetrix U133A GeneChip array. The data have been deposited in the Gene Expression Omnibus (GEO) database with accession number GSE08218. Out of the 136 samples, 80 samples were from relapsed patients, 50 samples from non-relapsed patients, and 6 samples from normal subjects.

The EM2 and EM3 algorithms were applied to the prostate cancer data following the same strategy as in the simulation studies. We only used 130 patient samples for the analysis. The cutoff point we used for gene selection was p-value ≤ 0.005 (χ^2 test), which resulted in 648 detected genes (215 genes by EM2, 324 genes by EM3, and 109 genes by both algorithms). A computer-assisted analysis of these 648 genes using DAVID Bioinformatics [14] indicated that these genes are significantly enriched in several prostate cancer pathways, such as ECM-receptor interaction, wnt signaling pathway, focal adhesion, TGF-β signaling pathway and gap junction (see Table 3). For comparison, we also analyzed the same data using SAM and LIMMA. The 650 highly ranked genes (similar in size to the detected genes by the new method) in each analysis were selected and then analyzed using DAVID Bioinformatics. The pathway analysis of the genes from SAM and LIMMA did not pull out relevant connections between these genes from literatures.

4 Discussion

We have developed a new gene differential expression method for identifying reliable prognostic signatures for prostate cancer. The new method consists of two steps - cluster analysis and correlation analysis. Two major contributions of the new method include (1) the use of cell-type distribution information, and (2) avoiding direct use of relapse/non-relapse states information, which is often not definitive due to data censoring. It is very common in the tumor marker literature that different labs produce inconsistent results [16,17]. One possible explanation among others is that tumor samples used for gene expression assays are highly heterogeneous in terms of cell composition. We have solved the problem by incorporating tissue component percentage into the analysis. Another advantage of incorporating cell-type distribution into the data analysis is that we were able to identify the cell origin of the observed gene expression changes. This provides a clue for identification of desirable therapeutic targets or better understanding the mechanism of cancer biology.

By using two sets of simulated data, we demonstrated that the new method is more desirable than the commonly used differential expression methods when the samples are highly heterogeneous in cell composition. Moreover, the new method can also identify the potential subtypes of tumors based on the gene expression profiles (represented by the genes detected in EM3 model). Given larger sample size, more complex models, for example EM4 or EM5, could be applied to identify finer tumor subtypes represented by special expression signatures. It has been hypothesized that these subtypes of tumor likely arise from the polyclonal nature of prostate cancer and they may have distinct potentials to progress [13,18,19,20]. Therefore, the gene signatures that are distinctive of these subtypes of prostate tumors may be useful for predicting patients' clinical outcomes. The pathway analysis based on the real prostate cancer data demonstrated that the new method identified genes significantly enriched or associated with prostate cancer related pathways. Genes selected by the traditional methods, however, did not show relevant connectivity in biological functions. The failure of the traditional methods might be explained by (1) the substantial variation caused by mixed tissue or (2) the incomplete recorded relapse/non-relapse data due to censoring.

Table 3. Pathway analysis of detected genes by three methods

EM clustering	SAM	LIMMA
ECM-receptor interaction	Huntington's disease	Huntington's disease
Wnt signaling pathway	Oxidative phosphorylation	Oxidative phosphorylation
Focal adhesion	Pyruvate metabolism	Pyruvate metabolism
TGF-β signaling pathway	Alzheimer's disease	Alzheimer's disease
Lysosome	Parkinson's disease	Parkinson's disease
RIG-I-like receptor	Valine,leucine, and	Valine, leucine, and
signaling pathway	isoleucine degradation	isoleucine degradation
Biosynthesis of	Glycosylphosphatidylinositol	Glycosylphosphatidylinositol
unsaturated fatty acids	(GPI)-anchor biosynthesis	(GPI)-anchor biosynthesis
Ribosome	Fatty acid metabolism	Fatty acid metabolism
Regulation of autophagy	Cardiac muscle contraction	Cardiac muscle contraction
Gap junction	Propanoate metabolism	Tryptophan metabolism
Tryptophan metabolism		Terpenoid backbone biosynthesis
Oocyte meiosis		Endocytosis
Dilated cardiomyopathy		

In this study, we aim to identify expression changes that are associated with bad outcomes of prostate cancer and hope to translate these gene signatures to clinical use. Therefore, genes that do not contribute to accurate classification may not be of clinical use even though they may be biologically related to the disease. One reviewer suggested further Gene Ontology study or Enrichment analysis of the genes that have not been selected in gene identification process owing to the fact that they do not meet the membership criterion in step (2). We agree that such analysis would help gain valuable information conveyed by those "ignored" genes. We will definitely look into the biological functions of those genes in the subsequent biological validation. However, these efforts do not improve the performance of the classifier that we are trying to develop in the study.

To deal with uncertain issue related to data censoring, we first developed an unsupervised EM clustering method to cluster samples and then constructed a contingency table between our cluster variables and the recorded outcome variables (relapse/non-relapse) to select genes. Our new method avoided the direct use of recorded relapse/non-relapse information to identify reliable signatures for disease prognosis. To validate that the unsupervised clustering method is superior to the supervised clustering, we did another experiment. The classification obtained for the 113 significant genes (p-value ≤ 0.001 in the χ^2 test) were used for deriving the outcome variables for the 130 patient samples *via* majority voting, yielding 16 non-relapse samples and 114 relapse samples. We then simply reanalyzed the prostate cancer data using SAM and LIMMA with the outcome variable estimated from the 113 genes. The pathway analysis based on the most significant 650 genes from the SAM and LIMMA analyses identified relevant pathways, *e.g.*, the wnt signaling pathway from both methods. Other important cancer related pathways, such as ECM-receptor interaction, focal adhesion and

TGF-β signaling pathway were identified by SAM using the "corrected" outcome information. We conclude that the new method has potential for classification of prostate cancer into risk groups for relapse and non-relapse outcome and is worthy of further development. In particular it will be important to test the method by comparison to other methods like the forest of trees approach [21] which does not require cell-type express and by testing the method on additional real data sets with known or calculated cell type distributions. These studies are in progress.

Acknowledgments. This work was supported by NIH grants U01 CA114810-01 and UO1 CA152738-01.

References

1. A.C.S: American Cancer Society: Cancer Facts and Figures 2011 [online] (2011)
2. Barwick, B.G., Abramovitz, M., Kodani, M., Moreno, C.S., Nam, R., Tang, W., Bouzyk, M., Seth, A., Leyland-Jones, B.: Prostate cancer genes associated with TMPRSS2-ERG gene fusion and prognostic of biochemical recurrence in multiple cohorts. Br. J. Cancer 102, 570–576 (2010)
3. Bibikova, M., Chudin, E., Arsanjani, A., Zhou, L., Garcia, E.W., Modder, J., Kostelec, M., Barker, D., Downs, T., Fan, J.B.: Expression signatures that correlated with Gleason score and relapse in prostate cancer. Genomics 89, 666–672 (2007)
4. Bickers, B., Aukim-Hastie, C.: New molecular biomarkers for the prognosis and management of prostate cancer-the post PSA era. Anticancer Res. 29, 3289–3298 (2009)
5. Tusher, V.G., Tibshirani, R., Chu, G.: Significance analysis of microarrays applied to the ionizing radiation response. Proc. Natl. Acad. Sci. U. S. A. 98, 5116–5121 (2001)
6. Smyth, G.K.: Linear models and empirical Bayes methods for assessing differential expression in microarray experiments. Stat. Appl. Genet. Mol. Biol. 3, Article3 (2004)
7. Ibrahim, J., Chen, M.-H., Gray, R.: Bayesian models for gene expression with dna microarray data. J. Am. Stat. Assoc. 97, 88–99 (2002)
8. Ishwaran, H., Rao, J.: Detecting differentially expressed gene in microarrays using bayesian model selection. J. Am. Stat. Assoc. 98, 438–455 (2003)
9. Lewin, A., Bochkina, N., Richardson, S.: Fully Bayesian mixture model for differential gene expression: Simulations and model checks. Stat. Appl. Genet. Mol. Biol. 6, 1–36 (2007)
10. Fan, C., Oh, D.S., Wessels, L., Weigelt, B., Nuyten, D.S., Nobel, A.B., Van't Veer, L.J., Perou, C.M.: Concordance among gene-expression-based predictors for breast cancer. N. Engl. J. Med. 355, 560–569 (2006)
11. Wang, Y., Xia, X.Q., Jia, Z., Sawyers, A., Yao, H., Wang-Rodriquez, J., Mercola, D., McClelland, M.: In silico Estimates of Tissue Components in Surgical Samples Based on Expression Profiling Data. Cancer Res. 70, 6448–6455 (2010)
12. Stuart, R.O., Wachsman, W., Berry, C.C., Wang-Rodriguez, J., Wasserman, L., Klacansky, I., Masys, D., Arden, K., Goodison, S., McClelland, M.: In silico dissection of cell-type-associated patterns of gene expression in prostate cancer. Proc. Natl. Acad. Sci. U. S. A. 101, 615–620 (2004)

13. Jia, Z., Wang, Y., Sawyers, A., Yao, H., Rahmatpanah, F., Xia, X.Q., Xu, Q., Pio, R., Turan, T., Koziol, J.A.: Diagnosis of prostate cancer using differentially expressed genes in stroma. Cancer Res. 71, 2476–2487 (2011)
14. Dennis Jr, G., Sherman, B.T., Hosack, D.A., Yang, J., Gao, W., Lane, H.C., Lempicki, R.A.: DAVID: database for annotation, visualization, and integrated discovery. Genome Biol. 4, R60 (2003)
15. Dempster, A.P., Laird, N.M., Rubin, D.B.: Maximum likelihood from incomplete data via the EM algorithm. J. R. Stat. Soc. B. 39, 1–38 (1977)
16. Woodson, K., Tangrea, J.A., Pollak, M., Copeland, T.D., Taylor, P.R., Virtamo, J., Albanes, D.: Serum insulin-like growth factor I: tumor marker or etiologic factor? A prospective study of prostate cancer among Finnish men. Cancer Res. 63, 3991–3994 (2003)
17. Xu, L., Tan, A.C., Naiman, D.Q., Geman, D., Winslow, R.L.: Robust prostate cancer marker genes emerge from direct integration of inter-study microarray data. Bioinformatics 21, 3905–3911 (2005)
18. Sutcliffe, P., Hummel, S., Simpson, E., et al.: Use of classical and novel biomarkers as prognostic risk factors for localised prostate cancer: a systematic review. Health Technol. Assess 13, 5 (2009)
19. Mucci, L.A., Pawitan, Y., Demichelis, F., et al.: Testing a multigene signature of prostate cancer death in the Swedish Watchful Waiting Cohort. Cancer Epidemiol. Biomarkers Prev. 17, 1682–1688 (2008)
20. Tomlins, S.A., Bjartell, A., Chinnaiyan, A.M., et al.: ETS gene fusions in prostate cancer: from discovery to daily clinical practice. Eur. Urol. 56, 275–286 (2009)
21. Díaz-Uriarte, R., de Andrés, A.: Gene selection and classification of microarray data using random forest. BMC Bioinformatics 7, 3 (2006)

Renal Cancer Cell Classification Using Generative Embeddings and Information Theoretic Kernels

Manuele Bicego[1], Aydın Ulaş[1], Peter Schüffler[2], Umberto Castellani[1]
Vittorio Murino[1,3], André Martins[4,6], Pedro Aguiar[5,6], and Mario Figueiredo[4,6]

[1] University of Verona, Department of Computer Science, Verona, Italy
[2] ETH Zürich, Department of Computer Science, Zürich, Switzerland
[3] Istituto Italiano di Tecnologia, Genova, Italy
[4] Instituto de Telecomunicações, Lisboa, Portugal
[5] Instituto de Sistemas e Robótica, Lisboa, Portugal
[6] Instituto Superior Técnico, Technical University of Lisbon, Portugal

Abstract. In this paper, we propose a hybrid generative/discriminative classification scheme and apply it to the detection of renal cell carcinoma (RCC) on tissue microarray (TMA) images. In particular we use *probabilistic latent semantic analysis* (pLSA) as a generative model to perform generative embedding onto the *free energy score space* (FESS). Subsequently, we use information theoretic kernels on these embeddings to build a kernel based classifier on the FESS. We compare our results with support vector machines based on standard linear kernels and RBF kernels; and with the nearest neighbor (NN) classifier based on the Mahalanobis distance using a diagonal covariance matrix. We conclude that the proposed hybrid approach achieves higher accuracy, revealing itself as a promising approach for this class of problems.

1 Introduction

The computer-based detection and analysis of cancer tissues represents a challenging, yet unsolved, task for researchers in Medicine, Computer Science and Bioinformatics. The complexity of the data, as well as the intensive labor needed to obtain them, makes the development of such automatic tools very problematic. In this paper, we consider the problem of classifying cancer tissues from tissue microarray (TMA) data, a technology which enables studies associating molecular changes with clinical endpoints [13]. In particular we focus on the specific case of renal cell carcinoma (RCC). One keypoint in the automatic TMA analysis for renal cell carcinoma is the nucleus classification. In this context, the main goal is to automatically classify cell nuclei into cancerous or benign, which is typically done by trained pathologists by visual inspection. Clearly, prior to classification, the nucleus needs to be detected and segmented in the image, as illustrated in Fig. 1.

In this paper, the classification problem described in the previous paragraph is addressed by using hybrid generative-discriminative schemes [12,14]. The underlying idea is to combine the best of the generative and discriminative paradigms

M. Loog et al. (Eds.): PRIB 2011, LNBI 7036, pp. 75–86, 2011.

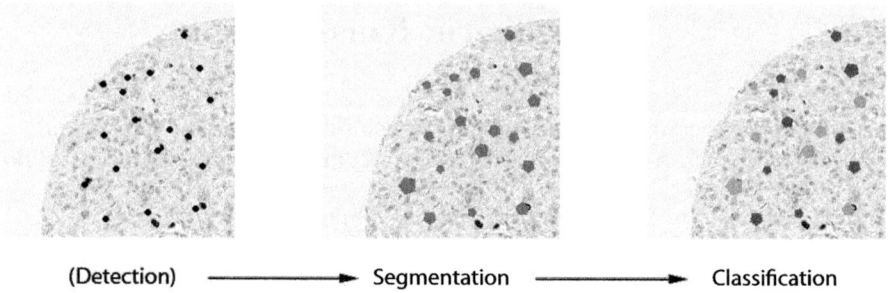

Fig. 1. The nuclei classification pipeline: detection, segmentation and classification into benign or cancerous

– the former being based on probabilistic class models and *a priori* class probabilities, learnt from training data and combined via Bayes law to yield posterior probabilities, the latter aiming at learning class boundaries or posterior class probabilities directly from data, without relying on probabilistic class models [17,20]. In the hybrid generative-discriminative scheme, the typical pipeline is to learn a generative model – able to properly model the data in hand – from the data, and then use it to project every object onto a feature space (the so-called generative embedding space), where a discriminative classifier may be trained. This class of approaches has been successfully applied in many different scenarios, especially with non-vectorial data (strings, trees, images) [24,5,19].

In particular, concerning the generative model, we adopt the *probabilistic latent semantic analysis* (pLSA) [11], a powerful methodology introduced in the text understanding community for unsupervised topic discovery in a corpus of documents, and subsequently largely applied by the computer vision community [9,5] as well as in medical informatics [1,8,2]. Given the trained generative model, two generative embedding spaces have been considered: the posterior distribution over topics (as in [5,8]); the very recently proposed *free energy score space* (FESS) [19,18]. The latter has been shown to outperform other generative embeddings (including those in [12] and [24]) in several applications [19,18].

Typically, the feature vectors resulting from the generative embedding are used to feed some kernel-based classifier, namely, a *support vector machine* (SVM) with linear or radial basis function (RBF) kernels. In this paper, we follow an alternative route. Instead on relying on standard kernels, we investigate the use of the recently introduced information theoretic (IT) kernels [15]. The rationale behind this choice is that these kernels can exploit the probabilistic nature of the generative embeddings, possibly improving the classification results of hybrid approaches. In particular, we investigate a particular class of IT kernels, based on a non-extensive generalization of the classical Shannon information theory, and defined on normalized (probability) or unnormalized measures. In [15], these IT kernels were successfully used for text categorization, based on multinomial (bag-of-words type) text representations. Here, the idea is to consider the points of the generative embedding as multinomial distributions, thus valid arguments for the information theoretic kernels.

The proposed approach has been tested on a dataset composed by 474 cell nuclei images, employing different features as well as different IT kernels, in comparison with standard kernels and nearest neighbor classifiers on the original feature space and the score spaces created by the generative embeddings. The results are encouraging, showing that this is a promising research direction.

The remainder of the paper is organized as follows: in Section 2, we explain the tissue micro array pipeline and how the features are extracted; Section 3 introduces our methods, while the experimental results are reported in Section 4. Section 5 concludes the paper.

2   The Tissue Microarray (TMA) Pipeline

In this section, the TMA pipeline is briefly summarized; for more details, please refer to [21]. In particular, we first describe how TMA are obtained, followed by the image normalization and patching (how to segment the nuclei). Finally, the image features that we employed are described.

2.1   Tissue Micro Arrays

A TMA is a microscope slide containing a set of small round tissue spots of (possibly cancerous) tissue, adequate for microscopic histological analysis. The diameter of the spots is of the order of 1mm and the thickness corresponds to one cell layer. Eosin staining is used to make the morphological structure of the cells visible, so that the cell nuclei appear. Immunohistochemical staining for the proliferation protein MIB-1 (Ki-67 antigen) makes the nuclei of cells in division status appear brown.

For subsequent computer analysis, the TMA slides are scanned into three-channel color images at a resolution of $0.23~\mu m$/pixel. The spots of single patients are collected into images of size 3000×3000 pixels.

The data set used in this paper consists of the top left quarter of eight tissue spots from eight patients, therefore, each image shows a quarter of a spot, with $100 \sim 200$ cells per image (see Figure 2). In order to have a ground truth, these TMA images were independently labeled by two pathologists [10], retaining only those nuclei on which the two pathologists agree on the label.

2.2   Image Normalization and Patching

The images were adjusted to minimize illumination variations among the scans. To classify the nuclei individually, patches of dimension 80x80 pixels were extracted from the whole image, such that each patch contains one nucleus approximately in the center (see Figure 3). The locations of the nuclei were known from the labels of the pathologists. Both procedures drastically improved the following segmentation of cell nuclei.

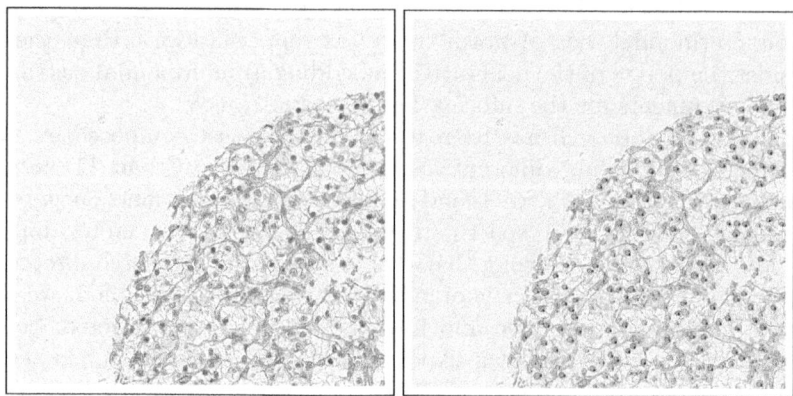

Fig. 2. Left: One 1500×1500 pixel quadrant of a TMA spot from an RCC patient. **Right:** A pathologist exhaustively labeled all cell nuclei and classified them into malignant (black) and benign (red).

2.3   Segmentation

Segmentation of cell nuclei was performed using the *graph cuts* approach [7], with the gray levels used in the unary potentials. The binary potentials were linearly weighted based on their distance to the center, to encourage roundish objects lying in the center of the patch (see Figure 3). The contour of the segmented object was used to calculate several shape features as described in the following section.

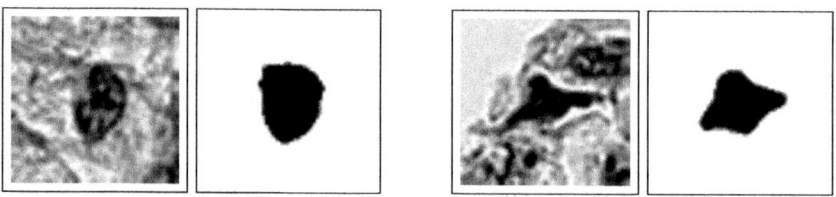

Fig. 3. Two examples of nucleus segmentation using the graph cuts algorithm with the potentials described in the text (the size of the patches in 80×80 pixels)

2.4   Feature Extraction

Given the patch image, several features are extracted, inspired by several intuitive guidelines used by pathologists to visually classify the nuclei [21]. In this work, we employed *pyramid histograms of oriented gradients* (PHOG, see [6] for details) – calculated over a 2-level pyramid on gray-scaled patches – which have been previously proven to be the most informative [22].

3   The Proposed Nuclei Classification Scheme

In this section, the proposed hybrid generative-discriminative approach to classify the nuclei is presented. After a brief overview, each step is thoroughly described.

3.1   Overview

Given the characterization of each nucleus by the features described in the previous section, the general scheme may be summarized as follows:

1. **Generative model training:** given the training set, a generative model is trained. In particular we employ the pLSA model, for the reasons explained below.
2. **Generative embedding:** in this step, all the objects involved in the problem (namely training and testing patterns) are embedded, using the learned model, in a vector space. Here we use two types of embedding: the posterior distribution over topics (of pLSA) and the FESS embedding.
3. **Discriminative classification:** in this step, the objects in the generative embedding space are classified. In particular, we consider information theoretic kernels, to be used in SVM and nearest neighbor techniques.

The following subsections describe each of these step in detail.

3.2   Generative Model Training

The generative model adopted is based on pLSA [11], which was introduced in the text understanding community for unsupervised topic discovery in a corpus of documents, and subsequently largely applied by the computer vision community [9,5], as well as in bioinformatics [1,8,2].

The basic idea underlying pLSA – and in general the class of the so-called topic models (of which another well-known example is the *latent Dirichlet allocation* model [4]) – is that each document is characterized by the presence of one or more topics (e.g. sport, finance, politics), which may induce the presence of some particular words. From a generative probabilistic point of view, pLSA generates a set of co-occurrences of the form (d, w), where each of these pairs specifies the presence of a given word w in a document d (as in *bag-of-words* descriptions of documents). The generative model underlying these co-occurrence pairs is as follows: (i) obtain a sample z from the distribution over the topics $P(z)$; (ii) given that topic sample, obtain a word sample from the conditional distribution of words given topics $P(w|z)$; (iii) given that topic sample, obtain a document sample (independently from the word sample) from the conditional distribution of documents given topics $P(d|z)$. The resulting distribution is

$$P(d, z) = \sum_{z} P(z)P(d|z)P(w|z),$$

where the sum ranges over the set of topics in the model. The parameters of this generative model may be obtained from a dataset using an expectation-maximization (EM) algorithm; for more details, the reader is referred to [11].

In our approach, we simply assume that the visual features previously described are the words in the pLSA model, while the nuclei are the documents. The pLSA model learned from this data can be seen as defining *visual topics*. The representation of *documents* and *words* with topic models has one clear advantage: each topic is individually interpretable, providing a probability distribution over words that picks out a coherent cluster of correlated terms. This may be advantageous in the cancer detection context, since the final goal is to provide knowledge about complex systems, and provide possible hidden correlations.

3.3   Generative Embedding

In this step, all the objects involved in the problem (namely training and testing patterns) are projected, through the learned model, onto a vector space. Different approaches have been proposed in the past, each one with different characteristics, in terms of interpretability, efficacy, efficiency, and others. Here we employ two schemes: the posterior distribution $P(z|d)$ – which was the first generative embedding based on pLSA models that was considered – and the *free energy score space* (FESS) – a novel method whose efficacy has been shown in different contexts [19,18].

In the posterior distribution embedding, a given nucleus (or document) d is represented by the vector of posterior topic probabilities, obtained via the function ψ defined as

$$\psi(d) = [P(z = 1|d), \ldots, P(z = T|d)] \in \mathbb{R}^T, \tag{1}$$

where we are assuming that the set of topics is indexed from 1 to T (the total number of topics). The intuition is that the co-occurrence of visual features is different between healthy and cancerous cells and that these co-occurrences are captured by the topic distribution $P(z|d)$, which should thus contain meaningful information for discrimination. This representation with the topic posteriors has been already successfully used in computer vision tasks [9,5] as well as in medical informatics [8,2].

The FESS embedding [19,18] has been shown to outperform other generative embeddings (including those in [12] and [24]) in several applications. This embedding expresses how well each data point fits different parts of the generative model, using the variational free energy as a lower bound on the negative log-likelihood. It has been shown that the FESS embedding yields highly informative for discriminative representations that lead to state-of-the-art results in several computational biology and computer vision problems (namely, scene/object recognition) [19,18]. Due to lack of space, the details of the FESS embedding are not reported here – please refer to [19,18] for a detailed presentation. The only important fact that needs to be pointed out here is that (as the posterior distribution embedding), the components of the FESS embedding of any object are all non-negative.

3.4   Discriminative Classification

In a typical hybrid generative-discriminative classification scenario, the feature vectors resulting from the generative embedding are used to feed some kernel-based classifier, namely, a *support vector machine* (SVM) with simple linear or radial basis function (RBF) kernels. Here, we take a different approach. Instead of relying on standard kernels, we investigate the use of the recently introduced information theoretic (IT) kernels [15] as a similarity measure between objects in the generative embedding space. The main idea is that, with such kernels, we can exploit the probabilistic nature of the generative embeddings, improving even more the classification results of the hybrid approaches – this has been already shown in other classification contexts [3,16].

More in details, given two probability measures p_1 and p_2, representing two objects, several information theoretic kernels (ITKs) can be defined [15]. The Jensen-Shannon kernel (will be referred to as JS) is defined as

$$k^{\mathrm{JS}}(p_1, p_2) = \ln(2) - JS(p_1, p_2), \qquad (2)$$

with $JS(p_1, p_2)$ being the Jensen-Shannon divergence

$$JS(p_1, p_2) = H\left(\frac{p_1 + p_2}{2}\right) - \frac{H(p_1) + H(p_2)}{2}, \qquad (3)$$

where $H(p)$ is the usual Shannon entropy.

The Jensen-Tsallis (JT) kernel (will be referred to as JT) is given by

$$k_q^{\mathrm{JT}}(p_1, p_2) = \ln_q(2) - T_q(p_1, p_2), \qquad (4)$$

where $\ln_q(x) = (x^{1-q} - 1)/(1 - q)$ is a function called the q-logarithm,

$$T_q(p_1, p_2) = S_q\left(\frac{p_1 + p_2}{2}\right) - \frac{S_q(p_1) + S_q(p_2)}{2^q} \qquad (5)$$

is the Jensen-Tsallis q-difference, and $S_q(r)$ is the Jensen-Tsallis entropy, defined, for a multinomial $r = (r_1, ..., r_L)$, with $r_i \geq 0$ and $\sum_i r_i = 1$, as

$$S_q(r_1, ..., r_L) = \frac{1}{q-1}\left(1 - \sum_{i=1}^{L} r_i^q\right).$$

In [15], versions of these kernels applicable to unnormalized measures were also defined as follows. Let $\mu_1 = \omega_1 p_1$ and $\mu_2 = \omega_2 p_2$ be two unnormalized measures, where p_1 and p_2 are the normalized counterparts (probability measures), and ω_1 and ω_2 arbitrary positive real numbers (weights). The weighted versions of the JT kernels are defined as follows:

– The weighted JT kernel (version A, will be referred to as JT-W1) is given by

$$k_q^A(\mu_1, \mu_2) = S_q(\pi) - T_q^\pi(p_1, p_2), \qquad (6)$$

where $\pi = (\pi_1, \pi_2) = \left(\frac{\omega_1}{\omega_1 + \omega_2}, \frac{\omega_2}{\omega_1 + \omega_2}\right)$ and

$$T_q^\pi(p_1, p_2) = S_q(\pi_1 p_1 + \pi_2 p_2) - (\pi_1^q S_q(p_1) + \pi_2^q S_q(p_2)).$$

– The weighted JT kernel (version B, will be referred to as JT-W2) is defined as

$$k_q^B(\mu_1, \mu_2) = \left(S_q(\pi) - T_q^\pi(p_1, p_2)\right)(\omega_1 + \omega_2)^q. \tag{7}$$

The approach herein proposed consists in defining a kernel between two observed objects x and x' as the composition of the generative embedding function ψ (the posterior embedding or the FESS embedding) with one of the JT kernels presented above. Formally,

$$k(x, x') = k_q^i\big(\psi(x), \psi(x')\big), \tag{8}$$

where $i \in \{JT, A, B\}$ indexes one of the Jensen-Tsallis kernels (4), (6), or (7), and $\psi(x)$ is the generative embedding of object x. Notice that this kernel is well defined because all the components of ψ are non-negative, as is clear from (1) for the posterior probability embedding and was mentioned above for the FESS embedding. Once the kernel is defined, SVM learning can been applied. Recall that positive definiteness is a key condition for the applicability of a kernel in SVM learning. It was shown in [15] that k_q^A is a positive definite kernel for $q \in [0, 1]$, while k_q^B is a positive definite kernel for $q \in [0, 2]$. Standard results from kernel theory [23, Proposition 3.22] guarantee that the kernel k defined in (8) inherits the positive definiteness of k_q^i, thus can be safely used in SVM learning algorithms. Moreover, we also employ nearest neighbor (NN) classifiers, in order to clearly assess the suitability of the derived kernels.

4   Experiments

In this section, we give details about the experimental setup and present the results obtained.

The classification experiments have been carried using a subset of the data presented in [21]. We selected a subset of three patients preserving the cancerous/benign cell ratio. From the labeled TMA images, we extracted 600 nuclei-patches of size 80×80 pixels. Each patch shows a cell nucleus in the center (see Figure 3). In 474 (79 %) of the 600 nuclei, the two pathologists agree on the label, with the following proportions: 321 (67 %) benign and 153 (33 %) malignant; all the experiments are performed on this set of 474 nuclei images, which is divided into ten folds (with stratification). For each fold, we learn a pLSA model from the training set and apply it to the test set. The number of topics has been chosen using leave-another-fold-out (of the nine training folds, we used 9-fold cross validation to estimate the best number of topics) cross validation procedure on the training set. We applied the same partitioning scheme also to choose the q parameter in IT kernels. All reported accuracies are percentual accuracies and are the averages over 10 folds. In all experiments the standard errors around the mean were inferior to 0.02.

4.1   One Model for Both Classes

In this setup, pLSA is trained in an unsupervised way, *i.e.*, we learn the pLSA model ignoring the class labels. Table 1 presents the results using the posterior distribution (referred to as PLSA) and the FESS embedding with SVM classification; these results show that in the proposed hybrid generative-discriminative approach, the IT kernels outperform linear and RBF kernels. The first and second columns show the classification results of ψ and FESS scores classified using linear and RBF kernels which allows us to show the contribution of the IT kernels.

The results of the NN classifier are shown in Table 2. Although NN is not a good choice for this experiment (baseline NN accuracy using Mahalanobis distance on the original data is 64.57%), we still see the advantage of the IT kernels on the generative approach. We can achieve 72.74% and 72.53% using pLSA and FESS embeddings, respectively, when we use the similarities computed by the IT kernels in the NN classifier.

Table 1. Average accuracies (in percentage) using pLSA and FESS embeddings with SVMs. ORIG shows the baseline accuracies on the original feature space.

	LIN	RBF	JS	JT	JT-W1	JT-W2
PLSA	76.78	76.99	79.31	80.17	74.22	80.17
FESS	77.41	76.17	73.21	78.87	72.31	79.96
ORIG	75.45	76.55		N/A		

4.2 One Model Per Class

In our second experimental setup, we apply pLSA in a supervised manner, *i.e.* the training set is split into the two classes and one pLSA model is trained for each class. The final feature space embedding is formed by the concatenation of the embeddings based on each of the two models. The accuracies obtained with SVM classification using this embeddings are shown in Table 3. Again the first column is the result of classifying the score spaces wihout the IT kernels. Although the accuracies obtained in this case with the linear kernel are better than those obtained with a single pLSA model, the IT kernels yield a smaller improvement of this linear kernel. We believe that this may be due to curse of dimensionality; when we use pLSA in a supervised way, we concatenate the outputs of each pLSA doubling the number of features. We can still achieve 78.48% accuracy with posterior distribution. With NN classification, comparing Table 4 with Table 2, we see that the accuracies increase except for FESS with JT-W2.

Table 2. Average accuracies (in percentage) using pLSA and FESS embeddings with NN classifiers. ORIG shows the baseline accuracies on the original feature space with Mahalanobis distances.

	MB	JS	JT	JT-W1	JT-W2
PLSA	66.41	68.97	72.53	72.74	68.75
FESS	67.11	67.08	72.53	71.27	71.08
ORIG	64.57		N/A		

Table 3. Average accuracies (in percentage) using pLSA and FESS embeddings with SVMs in the supervised pLSA learning setup

	LIN	RBF	JS	JT	JT-W1	JT-W2
PLSA	78.20	74.47	78.48	78.48	74.47	78.26
FESS	78.38	73.20	79.73	79.73	73.41	79.73
ORIG	75.45	76.55		N/A		

Table 4. Average accuracies (in percentage) using pLSA and FESS embeddings with NN classification in the supervised pLSA learning setup

	MB	JS	JT	JT-W1	JT-W2
PLSA	67.29	68.97	73.61	73.38	68.97
FESS	68.26	68.33	72.76	71.69	68.33
ORIG	64.57		N/A		

5   Conclusions

In this paper, we have presented a hybrid generative-discriminative classification approach that combines generative embeddings based on probabilistic latent semantic analysis (pLSA) with kernel-based discriminative learning based on a class of recently proposed information theoretic kernels. We applied the proposed approach to the diagnosis of Renal Cell Carcinoma on tissue micro array (TMA) images. We have seen that coupling the generative capabilities of pLSA with the discriminative capabilities of the information theoretic kernels yields higher classification accuracies than previous approaches based on linear and RBF kernels.

Acknowledgements. We acknowledge financial support from the FET programme within the EU FP7, under the SIMBAD project (contract 213250).

References

1. Bicego, M., Lovato, P., Ferrarini, A., Delledonne, M.: Biclustering of expression microarray data with topic models. In: Proceedings of the International Conference on Pattern Recognition, pp. 2728–2731 (2010)
2. Bicego, M., Lovato, P., Oliboni, B., Perina, A.: Expression microarray classification using topic models. In: ACM Symposium on Applied Computing (Bioinformatics and Computational Biology track) (2010)
3. Bicego, M., Perina, A., Murino, V., Martins, A., Aguiar, P., Figueiredo, M.: Combining free energy score spaces with information theoretic kernels: Application to scene classification. In: Proceedings of the IEEE International Conference on Image Processing, pp. 2661–2664 (2010)
4. Blei, D., Ng, A., Jordan, M.: Latent Dirichlet allocation. Journal of Machine Learning Research 3, 993–1022 (2003)

5. Bosch, A., Zisserman, A., Muñoz, X.: Scene Classification Via pLSA. In: Leonardis, A., Bischof, H., Pinz, A. (eds.) ECCV 2006. LNCS, vol. 3954, pp. 517–530. Springer, Heidelberg (2006)
6. Bosch, A., Zisserman, A., Munoz, X.: Representing shape with a spatial pyramid kernel. In: Proceedings of the 6th ACM International Conference on Image and Video Retrieval, pp. 401–408 (2007)
7. Boykov, Y., Veksler, O., Zabih, R.: Efficient approximate energy minimization via graph cuts. IEEE Transactions on Pattern Analysis and Machine Intelligence 20(12), 1222–1239 (2001)
8. Castellani, U., Perina, A., Murino, V., Bellani, M., Brambilla, P.: Brain morphometry by probabilistic latent semantic analysis. In: International Conference on Medical Image Computing and Computer Assisted Intervention (2010)
9. Cristani, M., Perina, A., Castellani, U., Murino, V.: Geo-located image analysis using latent representations. In: IEEE Conference on Computer Vision and Pattern Recognition, pp. 1–8 (2008)
10. Fuchs, T., Wild, P., Moch, H., Buhmann, J.: Computational pathology analysis of tissue microarrays predicts survival of renal clear cell carcinoma patients. In: International Conference on Medical Image Computing and Computer Assisted Intervention (2008)
11. Hofmann, T.: Unsupervised learning by probabilistic latent semantic analysis. Machine Learning 42(1-2), 177–196 (2001)
12. Jaakkola, T., Haussler, D.: Exploiting generative models in discriminative classifiers. In: Advances in Neural Information Processing Systems, pp. 487–493 (1999)
13. Kononen, J., Bubendorf, L., Kallionimeni, A., Bärlund, M., Schraml, P., Leighton, S., Torhorst, J., Mihatsch, M., Sauter, G., Kallionimeni, O.: Tissue microarrays for high-throughput molecular profiling of tumor specimens. Nature Medicine 4, 844–847 (1998)
14. Lasserre, J., Bishop, C., Minka, T.: Principled hybrids of generative and discriminative models. In: Proceedings of the IEEE Conference on Computer Vision and Pattern Recognition, New York (2006)
15. Martins, A., Smith, N., Xing, E., Aguiar, P., Figueiredo, M.: Nonextensive information theoretic kernels on measures. Journal of Machine Learning Research 10, 935–975 (2009)
16. Martins, A.F.T., Bicego, M., Murino, V., Aguiar, P.M.Q., Figueiredo, M.A.T.: Information Theoretical Kernels for Generative Embeddings Based on Hidden Markov Models. In: Hancock, E.R., Wilson, R.C., Windeatt, T., Ulusoy, I., Escolano, F. (eds.) SSPR&SPR 2010. LNCS, vol. 6218, pp. 463–472. Springer, Heidelberg (2010)
17. Ng, A., Jordan, M.: On discriminative vs generative classifiers: A comparison of logistic regression and naive Bayes. In: Advances in Neural Information Processing Systems (2002)
18. Perina, A., Cristani, M., Castellani, U., Murino, V., Jojic, N.: Free energy score space. In: Advances in Neural Information Processing Systems (2009)
19. Perina, A., Cristani, M., Castellani, U., Murino, V., Jojic, N.: A hybrid generative/discriminative classification framework based on free-energy terms. In: Proceedings of the International Conference on Computer Vision (2009)

20. Rubinstein, Y.D., Hastie, T.: Discriminative vs informative learning. In: Proceedings of the Third International Conference on Knowledge Discovery and Data Mining, pp. 49–53. AAAI Press (1997)
21. Schüffler, P.J., Fuchs, T.J., Ong, C.S., Roth, V., Buhmann, J.M.: Computational TMA Analysis and Cell Nucleus Classification of Renal Cell Carcinoma. In: Goesele, M., Roth, S., Kuijper, A., Schiele, B., Schindler, K. (eds.) Pattern Recognition. LNCS, vol. 6376, pp. 202–211. Springer, Heidelberg (2010)
22. Schüffler, P., Ulaş, A., Castellani, U., Murino, V.: A multiple kernel learning algorithm for cell nucleus classification of renal cell carcinoma. In: Proceedings of the 16th International Conference on Image Analysis and Processing (2011)
23. Shawe-Taylor, J., Cristianini, N.: Kernel Methods for Pattern Analysis. Cambridge University Press (2004)
24. Tsuda, K., Kawanabe, M., Rätsch, G., Sonnenburg, S., Müller, K.R.: A new discriminative kernel from probabilistic models. Neural Computation 14, 2397–2414 (2002)

Integration of Epigenetic Data in Bayesian Network Modeling of Gene Regulatory Network

Jie Zheng[1], Iti Chaturvedi[1], and Jagath C. Rajapakse[1,2,3]

[1] Bioinformatics Research Centre, School of Computer Engineering, Nanyang Technological University, Singapore 639798
[2] Department of Biological Engineering, Massachusetts Institute of Technology, Cambridge, MA 02142, USA
[3] Singapore-MIT Alliance, Singapore
{zhengjie,iti,asjagath}@ntu.edu.sg

Abstract. The reverse engineering of gene regulatory network (GRN) is an important problem in systems biology. While gene expression data provide a main source of insights, other types of data are needed to elucidate the structure and dynamics of gene regulation. Epigenetic data (e.g., histone modification) show promise to provide more insights into gene regulation and on epigenetic implication in biological pathways. In this paper, we investigate how epigenetic data are incorporated into reconstruction of GRN. We encode the histone modification data as prior for Bayesian network inference of GRN. Bayesian framework provides a natural and mathematically tractable way of integrating various data and knowledge through its prior. Applying to the gene expression data of yeast cell cycle, we demonstrate that integration of epigenetic data improves the accuracy of GRN inference significantly. Furthermore, fusion of gene expression and epigenetic data shed light on the interactions between genetic and epigenetic regulations of gene expression.

Keywords: Bayesian networks, gene regulatory network, epigenetics, histone modification, priors, gene expression, yeast cell cycle.

1 Introduction

Reconstruction of gene regulatory networks (GRN) is one of the most important problems in systems biology. Despite intense research, there remain many open problems in this area, partly due to the limited data available and the inherent noise and complexity of biological processes. On the other hand, thanks to the advances on data collection technologies such as next generation sequencing, new types of biological data are emerging, providing new insights and opportunities for GRN reconstruction.

Among the new data, epigenetic data are receiving more attention recently. Epigenetics is the study of changes in phenotypes (especially gene expression) caused by mechanisms other than the changes in DNA sequences (due to mechanisms of central dogma of molecular biology). Such data come from various

M. Loog et al. (Eds.): PRIB 2011, LNBI 7036, pp. 87–96, 2011.
© Springer-Verlag Berlin Heidelberg 2011

epigenetic processes, such as histone modification, DNA methylation, interferences by micro RNA, etc. It is believed that epigenetic control of gene expression represents an important layer of regulation beyond genes in DNA sequences. Analogous to genetic code for gene translation, it is hypothesized that there is an "epigenetic code" for controlling gene expression. The decoding of "epigenetic code" and an increased understanding of the mechanisms of epigenetic regulation of gene expression will bring about new breakthroughs in systems biology and translational medicine. For instance, epigenetic regulation plays an important role in the development of embryonic stem cells, as well as in reprogramming of induced pluripotent stem cells.

Nevertheless, the mechanisms of epigenetic regulation of gene expression remain poorly understood. To elucidate the interaction between genetic and epigenetic regulation of transcription, there is a need for incorporation of epigenetic information in the gene regulatory networks. This paper is motivated by our belief that by considering epigenetic information, the accuracy of GRN reconstruction can be improved. Furthermore, it will shed light on the interactions between genetic and epigenetic regulations.

Recently, there have been some attempts to infer causal relations between epigenetic features (especially histone modifications) and gene expression, and to elucidate the "epigenetic code". Yu et al. built a Bayesian network to model the combinatorial relationships among histone modifications and their effects on gene expression [1]. Cheng et al. gave a machine learning framework to predict gene expression from chromatin features [2]. They observed that chromatin features contribute a significant proportion of gene expression variation. The two papers both looked at the "global" effects of epigenetic features on gene expression while it is desirable to study the effects on individual genes. In particular, it would be highly interesting to examine what patterns occur in epigenetic features between a regulatory gene and its regulated gene. Ha et al. [3] used gene ontology analysis on plant genes to show that genes tend to have similar distribution patterns of histone modifications in the same functional classes but have different epigenetic patterns across different classes. This observation implies that genes involved in the same regulatory pathways have similar patterns of epigenetic features. Thus, the correlation of epigenetic feature distribution among genes can be used as additional information for the reconstruction of GRN.

There are mainly four types of approaches to GRN modeling, namely, information theory models, Boolean networks, differential equations, and Bayesian networks (see the review of [4] and references therein). Among the approaches, Bayesian networks have been the most mature framework for integration of heterogeneous data although analogous integration methods are being developed for other approaches as well. The Bayesian strategy of integration is realized by presenting additional information in the form of prior probability of network. This strategy has been developed to increase the accuracy of GRN reconstruction [5],[6]. The prior knowledge includes protein-protein interactions, transcription factor-DNA binding, sequence motifs, pathways, literature mining, etc. However, to our knowledge, no epigenetic features have been integrated in this framework.

In this paper, we integrate epigenetic features as prior knowledge in Bayesian network learning, based on the framework outlined in [6]. This approach is applied to gene expression data in [7] and histone modification (ChIP-Chip) data in [8] from the yeast genome. We first show that the histone modification profiles between regulators and target genes are more strongly correlated than a random pair of genes. Second, by comparing with benchmark regulatory networks identified from experiments [9], we demonstrated that the use of epigenetic prior can improve sensitivity by more than 10%. Interestingly, it is also observed that histone modification data alone can be used to infer GRN with lower false positive than using gene expression data. To the best of our knowledge, this is the first use of epigenetic information for the reverse engineering of GRN.

2   Methods

Bayesian network has long been used for GRN reconstruction. One of the strengths of Bayesian network is its ability to incorporate additional knowledge through the priors. In the following we will describe the construction of prior matrix B from epigenetic data of histone profiles obtained from [8].

The histone profiles of each gene i consist of a matrix $H_i = \{h^i_{l,h}\}_{t \times m}$ of positive float numbers, where m is the number of histone types and t is the number of loci assessed for the gene. Each row of H_i corresponds to a genomic locus near or within the gene (e.g. promoter, middle of transcribed region, etc.); each column of H_i corresponds to a type of histone modification (e.g. histone H3 lysine 9 acetylation (H3K9ac), histone H3K4 trimethylation (H3K4Me3), etc.). That is, $h^i_{l,h}$ represents the enrichment of the hth histone modification at the lth measured locus of gene i. Thus the genome wide histone dataset in [8] is represented as a 3-dimensional matrix $H = [H_t]^n_{t=1}$, where n is the number of genes.

To simplify analysis, for each gene we calculate the average enrichment of a histone type across t different loci of a gene, and obtain a vector of m float numbers each measuring the level of a certain type of histone modification. These vectors of histone features represent the epigenetic information of genes. Formally, the vector f_i of histone features is calculated for gene i as

$$f_i = \frac{1}{t}\left[\sum_{l=1}^{\tau} h^i_{l,l} \sum_{l=1}^{\tau} h^i_{l,2} \cdots \sum_{l=1}^{\tau} h^i_{l,m}\right] \tag{1}$$

Following [6], we define the biological prior knowledge matrix $B = \{b_{i,j}\}_{n \times n}$ as follows. Let matrix element $b_{i,j} \in [0.0, 1.0]$ represent the correlation between the histone modification patterns of gene i and gene j. If $b_{i,j} = 0.5$, we do not have any prior knowledge about the presence or absence of the edge (i, j); if $b_{i,j} < 0.5$, we have prior evidence of the absence; if $b_{i,j} > 0.5$, we have prior evidence of the presence of the edge. For more details, please see [6] and [5].

To estimate the epigenetic association between gene i and gene j, $b_{i,j}$ is defined by using the Pearson correlation coefficient ρ between the epigenetic profiles of the two genes. While ρ is a number between -1 and 1, the Bayesian prior need to be between 0 and 1. Hence, we scale to between 0 and 1 linearly to define the Bayesian prior for epigenetic features as

$$b_{i,j} = \frac{1}{2}[\rho(f_i, f_j) + 1] \tag{2}$$

Then, we integrate the priors in B matrix into Bayesian network learning as follows. To construct a gene network G that fits the data the best, the posterior probability of the network is maximize

$$P(G|D) \propto P(D|G)P(G) \tag{3}$$

where $P(G)$ is the prior probability of network G. Let matrix $C = \{c_{i,j}\}_{n \times n}$ be the connectivity matrix of the GRN G, i.e. $c_{i,j} = 1$ if the edge (i, j) is present in G and $c_{i,j} = 0$ otherwise. Following [6], the energies associated with the presence and absence of edges are defined as

$$E(G) = \sum_{i,j=1}^{n} |b_{i,j} - c_{i,j}| \tag{4}$$

Then, the prior probability $P(G)$ is modeled by the Gibbs distribution

$$P(G) = \frac{1}{Z} e^{-\beta E(g)} \tag{5}$$

Here Z is a normalizing partition function defined as

$$Z = \sum_{G \in S} e^{-\beta E(g)} \tag{6}$$

where S is the set of all possible GRNs, and β is a positive number as hyper parameter. To find networks of high posterior probabilities, we search for edges that minimize the energy $E(G)$, thus taking into account the prior knowledge in matrix B.

There are many heuristic and stochastic algorithms for learning Bayesian networks: e.g., genetic algorithms, simulated annealing, Markov Chain Monte Carlo, etc. Since our goal is to integrate epigenetic prior knowledge to improve the accuracy of GRN reconstruction and the effect of a good prior approach should be independent of specific learning algorithms used, we choose a greedy learning algorithm for the present study. As a future work, other structure learning algorithms will be explored.

Starting from a random network (i.e. the entries in the connectivity matrix C are initialized to 0's and 1's at random), the algorithm searches for networks with good fitness with the given data, through iterations of the following two basic steps: (1) make local change to an existing network by adding, deleting or replacing edges to generate a new network; (2) evaluate the posterior probability as the score of the proposed network. If the score is improved by local change in step (1), then the new network is accepted; otherwise it is rejected. The above two steps are iterated until some stopping criteria are met, e.g., the maximum running time or the maximum number of iterations are reached. The prior probabilities serve as weights in the evaluation of posterior probability of a proposed network. The result consists of a set of network structures (i.e. directed edges) and their corresponding posterior probability scores. In the end, the network with the highest score is output.

3   Experiments and Results

The above method was implemented as a Python program based on the open source library of Pebl [10]. Compared with other implementations of Bayesian network inference, a unique feature of Pebl is to allow easy implementation of soft prior constraints in the form of an energy matrix. However, Pebl is limited to static Bayesian networks and can handle only networks of small number of nodes. Here, we used the greedy learning algorithm implemented by Pebl.

First we use real data to show that the epigenetic profiles are more correlated between a regulatory pair of genes than a random pair. To this end, we downloaded a list of 87 confirmed regulator-gene interactions [11] and the histone modification profiles [8], both in the yeast genome (*S.cerevisiae*). There are totally 88 genes in the confirmed list, out of which 85 genes have histone profiles available, as shown in Figure 1. We did not use the edges in Figure 1 for verification because these are confirmed regulatory interactions but they do not represent a complete regulatory network (e.g. there is no feedback loop) .

The energy values of the 87 edges in the confirmed regulator-gene interactions as in Figure 1 are compared with the energy values of 87 random pairs of genes. As shown in Figure 2, the confirmed regulatory edges have significantly higher energy values (box on the right) than random edges (p-value < 0.01).

The result of Figure 2 shows that epigenetic information correlates with regulatory relationships, which supports our approach of using histone features as prior knowledge in Bayesian network inference. To further verify this approach, we compare Bayesian network algorithms with and without the energy matrix from histone profiles on an experimentally identified regulatory network (Figure 3) consisting of 9 genes related with yeast cell cycle [9].

The greedy algorithm as described in Section 2 for static Bayesian network inference is applied on the expression data of yeast cell cycle [7], with and without the prior of histone data. Moreover, we apply our method on histone data only. For each method, we compare the predicted edges and the established edges as in Figure 3, taking into account the edge directions. True positive (FP) is counted as the number of predicted edges matching the established edges (i.e. with the same end nodes and direction) and similarly for false positive (FP) and

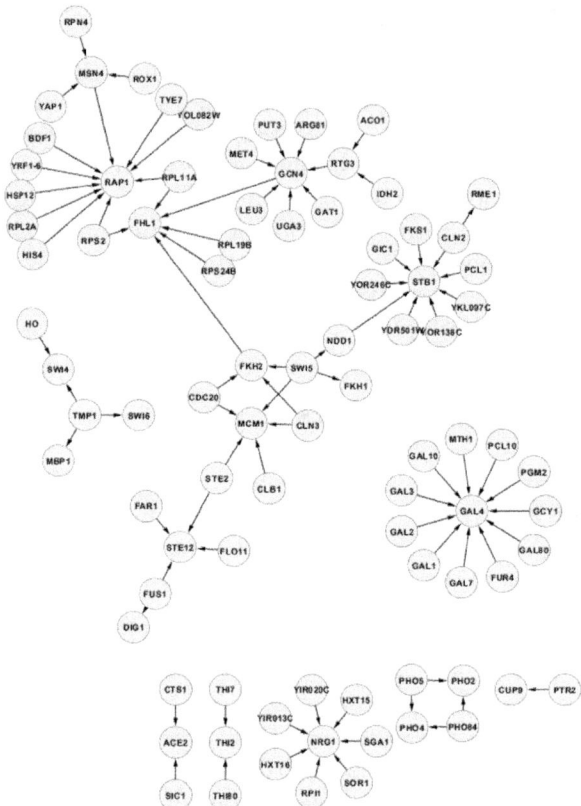

Fig. 1. Confirmed edges of regulator-gene interactions in yeast cell cycle

false negative (FN). Since there are 9 nodes and edge directions are taken into account, the true negative (TN) should be 72 - (FP+FN+TP).

As shown in Table 1, in Experiment A, only expression data and no prior is used; in Experiment B, only histone profiles are used as input to Bayesian network inference; in Experiment C, both expression data and histone prior are used. The result of C has the highest sensitivity, and compared with result of A the use of prior in C improved the sensitivity by more than 10%. The specificity of C is slightly lower than A due to one more false positive. Interestingly, when only histone data are used, the specificity is the highest among the three experiments. The histone data, when used alone, can reduce FP to 23; however, when used as prior with gene expression in experiment C, both FP and TP increase.

Now, let us look at the predicted GRNs from the three experiments more carefully. As shown in Figure 4, a big fraction of false negative edges in experiment A are due to wrong predicted directions (the T-shaped lines), while in experiment B there are more missing edges (dashed lines). Note that several edges missed in experiment A have been detected in experiment B (e.g. FKH2 to SWI5, MCM1

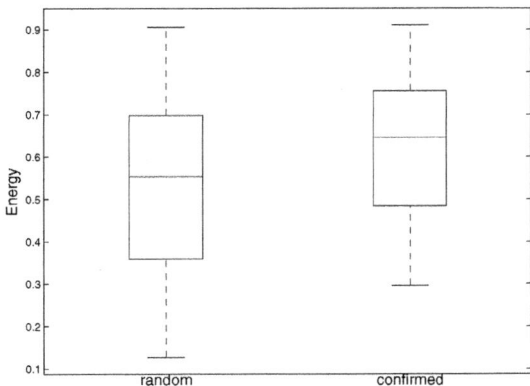

Fig. 2. Comparison between the energy values of random pairs of genes vs. confirmed pairs of genes in yeast (Wilcoxon test p-value < 0.01)

Table 1. Comparison of predicted networks with the benchmark network in three experiments. It shows the improvement of Bayesian network performance due to the use of epigenetic data as prior

Experiment	TP	TN	FP	FN	Sensitivity(%)	Specificity(%)
A. Expression	5	26	29	12	29.41	47.27
B. Histone	4	32	23	13	23.53	58.18
C. Expression & histone	7	25	30	10	41.18	45.45

to SWI5), and vice versa. This may explain why the result of experiment C, considering both gene expression and histone data, has a higher sensitivity than the other two experiments. To elucidate the effect of epigenetic data on the performance of GRN inference, however, larger datasets and networks will be needed.

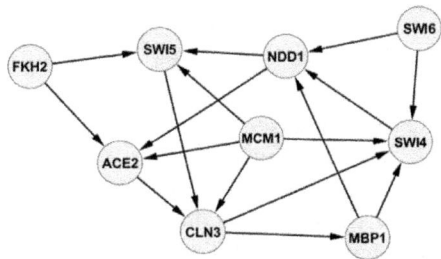

Fig. 3. The gene regulatory network of yeast cell cycle identified in [9]

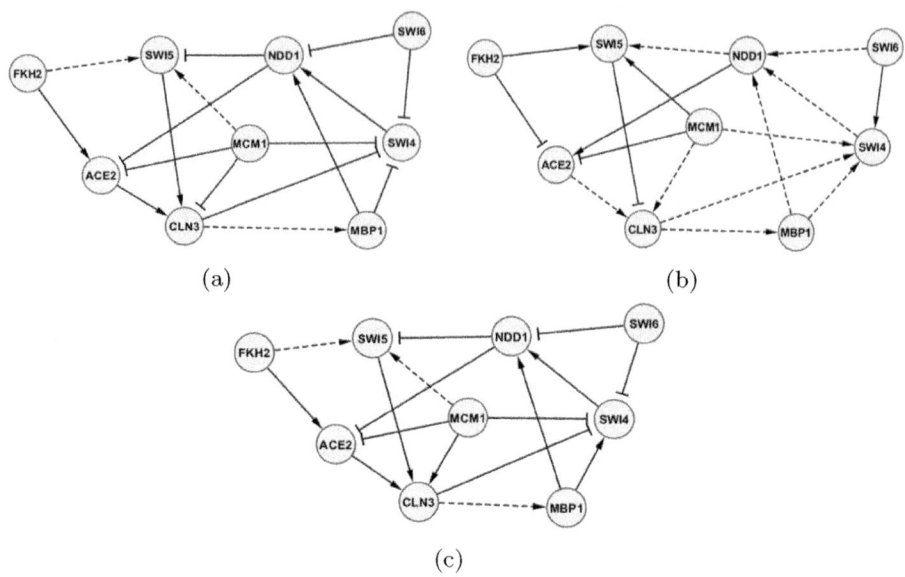

Fig. 4. BN modelling (a) Gene expression data only (b) Histone data only (c) Expression and histone data

4   Conclusion

In this paper, we proposed to integrate epigenetic data as prior knowledge of Bayesian network model for reconstruction of GRN. The approach has been applied to gene expressions of cell cycle and histone modification data of yeast genome. First, it was shown that the correlation of histone modification features between genes with experimentally confirmed regulatory-target gene pairs are stronger than the correlation between random pairs of genes. This suggests that histone features are associated with gene regulatory relation, and hence supports the rationale of our approach. Second, we demonstrated that histone data can also be used to reconstruct regulatory networks with performance comparable to gene expression data. Third, as demonstrated on an experimentally verified network from yeast cell cycle data, epigenetic prior improves the accuracy of Bayesian network inference GRN. As far as we know, this is the first paper to reconstruct GRN by incorporating histone modification data, which shows promise for pursuing further research.

However, as this is only a preliminary study in this direction, there remains much room for improvement. The major goal of this paper is to show that the fusion of gene expression with epigenetic data can improve the accuracy of GRN reconstruction. This goal has been achieved with the straightforward methods of GRN reconstruction and Bayesian integration [6]. It is desirable to develop data fusion approaches for more sophisticated GRN reconstruction methods under more realistic conditions in the future. For instance, one can implement a similar prior for dynamic Bayesian network (DBN) which is more compatible with the time-course gene expression data of yeast cell cycle than the static BN used here. Due to scalability of the greedy algorithm of Pebl, the accuracy is relatively low. More powerful learning algorithms with higher computational efficiency will be implemented. Then we can test the approach on larger expression and epigenetic datasets and benchmark networks. Despite the promising results in this paper, our model of epigenetic information is quite simplified, and we should model more realistic relations. For example, the two types of epigenetic data (i.e., histone acetylation and histone methylation) which we integrated here actually have different enrichment patterns along genes: acetylation tends to be enriched at the beginning of genes and methylation tends to be within transcribed regions. There are also combinatorial interactions among histone themselves, which has been modeled also using Bayesian network [1]. In this study we have performed the validation of our method only on a small network (Figure 3). This is mainly because of two reasons : (1) only a few experimentally verified GRNs are available so far due to the difficulty of GRN reconstruction and complexity of data; (2) we need to benchmark GRNs that have also high-throughput epigenetic information. It is still difficult to find a benchmark data that satisfy both criteria (i.e. experimentally verified and with epigenetic data). An important future work is to look for more such data, which can be used for either method evaluation or exploratory data analysis. Last but not least, as there are only a few gold-standard GRNs available, one can experiment with synthetic networks and data taking into account epigenetic regulation.

References

1. Yu, H., Zhu, S., Zhou, B., Xue, H., Han, J.-D.J.: Inferring causal relationships among different histone modifications and gene expression. Genome Research 18(8), 1314–1324 (2008)
2. Cheng, C., Yan, K.-K., Yip, K., Rozowsky, J., Alexander, R., Shou, C., Gerstein, M.: A statistical framework for modeling gene expression using chromatin features and application to modencode datasets. Genome Biology 12(2), R15 (2011)
3. Ha, M., Ng, D.W.-K., Li, W.-H., Chen, Z.J.: Coordinated histone modifications are associated with gene expression variation within and between species. Genome Research 21(4), 590–598 (2011)
4. Hecker, M., Lambeck, S., Toepfer, S., van Someren, E., Guthke, R.: Gene regulatory network inference: Data integration in dynamic models–a review. Biosystems 96(1), 86–103 (2009)
5. Imoto, S., Higuchi, T., Goto, T., Tashiro, K., Kuhara, S., Miyano, S.: Combining microarrays and biological knowledge for estimating gene networks via bayesian networks. Journal of Bioinformatics and Computational Biology 2(1), 77–98 (2004)
6. Husmeier, D., Werhli, A.V.: Bayesian integration of biological prior knowledge into the reconstruction of gene regulatory networks with bayesian networks. Computational systems bioinformatics 6, 85–95 (2007)
7. Spellman, P.T., Sherlock, G., Zhang, M.Q., Iyer, V.R., Anders, K., Eisen, M.B., Brown, P.O., Botstein, D., Futcher, B.: Comprehensive identification of cell cycle-regulated genes of the yeast saccharomyces cerevisiae by microarray hybridization. Mol. Biol. Cell 9(12), 3273–3297 (1998)

8. Pokholok, D.K., Harbison, C.T., Levine, S., Cole, M., Hannett, N.M., Lee, T.I., Bell, G.W., Walker, K., Rolfe, P.A., Herbolsheimer, E., Zeitlinger, J., Lewitter, F., Gifford, D.K., Young, R.A.: Genome-wide map of nucleosome acetylation and methylation in yeast. Cell 122(4), 517–527 (2005)
9. Simon, I., Barnett, J., Hannett, N., Harbison, C.T., Rinaldi, N.J., Volkert, T.L., Wyrick, J.J., Zeitlinger, J., Gifford, D.K., Jaakkola, T.S., Young, R.A.: Serial regulation of transcriptional regulators in the yeast cell cycle. Cell 106(6), 697–708 (2001)
10. Shah, A., Woolf, P.: Python environment for bayesian learning: Inferring the structure of bayesian networks from knowledge and data. Journal of Machine Learning Research 10(2), 159–162 (2009)
11. Lee, T.I., Rinaldi, N.J., Robert, F., Odom, D.T., Bar-Joseph, Z., Gerber, G.K., Hannett, N.M., Harbison, C.T., Thompson, C.M., Simon, I., Zeitlinger, J., Jennings, E.G., Murray, H.L., Gordon, D.B., Ren, B., Wyrick, J.J., Tagne, J.-B., Volkert, T.L., Fraenkel, E., Gifford, D.K., Young, R.A.: Transcriptional regulatory networks in saccharomyces cerevisiae. Science 298(5594), 799–804 (2002)

Metabolic Pathway Inference from Time Series Data: A Non Iterative Approach

Laura Astola[1,2], Marian Groenenboom[1,2], Victoria Gomez Roldan[3],
Fred van Eeuwijk[1,2], Robert D. Hall[3,4], Arnaud Bovy[3,4], and Jaap Molenaar[1,2]

[1] Biometris, Wageningen University and Research Centre,
Droevendaalsesteeg 1, Wageningen, The Netherlands
[2] Netherlands Consortium for Systems Biology, Amsterdam, The Netherlands
[3] Bioscience, Plant Research International, Wageningen University
and Research Centre, Wageningen, The Netherlands
[4] Centre for Biosystems Genomics, P.O. Box 98, Wageningen, The Netherlands

Abstract. In this article, we present a very fast and easy to implement method for reconstruction of metabolic pathways based on time series data. To model the metabolic reactions, we use the well-established setting of ordinary differential equations. In the present article we consider a network leading to the accumulation of quercetin-glycosides in tomato (*Solanum lycopersicum*). Quercetin belongs to a group of plant secondary metabolites, generally referred to as flavonoids, which are extensively being studied for their variety of important functions in plants as well as for their potentially health-promoting effects on human. We use time series measurements of metabolite concentrations of quercetin derivatives. In the present setting, the observed concentrations are the variables and the reaction rates are the unknown parameters. A standard method is to solve the parameters by reverse engineering, where the ordinary differential equations (ODE) are solved repeatedly, resulting in impractical computation times. We use an alternative method that estimates the parameters by least squares minimization, and which is, in the order of hundred times faster than the iterative method. Our reconstruction method can incorporate an arbitrary *a priori* known network structure as well as positivity constraints on the reaction rates. In this way we can avoid over-fitting, which is another often encountered problem in network reconstruction, and thus obtain better estimates for the parameters. We test the presented method by reconstructing artificial networks and compare it with the more conventional method in terms of residuals between the observed and fitted concentrations, computing times and the proportion of correctly identified edges in the network. Finally we exploit this fast method to statistically infer the kinetic constants in the flavonoid pathway. We remark that the method as such is not limited to metabolic network reconstructions, but can be used with any type of time-series data that is modeled in terms of linear ODE's.

Keywords: Metabolic network inference, flavonoid pathway reconstruction.

M. Loog et al. (Eds.): PRIB 2011, LNBI 7036, pp. 97–108, 2011.
© Springer-Verlag Berlin Heidelberg 2011

1 Introduction

Flavonoids are a class of secondary metabolites in plants, most commonly found in fruits and flowers. They are involved in various processes, for example in flower, fruit and seed pigmentation, plant growth, protection against UV radiation, and interaction with micro-organisms [1]. Daily dietary consumption of these compounds has been associated with human health promotion and disease prevention, in particular reducing cardiovascular diseases, certain cancers and other age related diseases [2,3]. In tomato, many genes involved in common flavonoid biosynthetic pathways have been identified. Nevertheless the molecular basis of the structural modifications in flavonoid glycosylation and methylation pathways is still relatively unknown. Glycosylation is an enzymatic process that modifies solubility, chemical stability and the biological properties of flavonoids. It is also crucial for flavonoid accumulation. Several glycosylated flavonoids have been reported in tomato fruits, most of them being derivatives of the flavonol quercetin [4]. In this work we consider the quercetin biosynthetic pathway in tomato seedlings.

Many popular models inferring metabolic reaction networks, rely on ordinary differential equations [5,6,7], although this approach has its limitations, especially when using the conventional approach [8]. In the conventional approach, one starts with an initial guess of the parameters (reaction rates), solves the ODE's and compares the resulting solution curves at discrete time points with the observed values at corresponding time points. If they are not sufficiently similar, one adjusts the parameters and repeats the comparisons until the solutions are close enough to the measurements. Although efficient optimization algorithms are available in most mathematical software packages, this approach is inherently time-consuming, due to the fact that one needs to solve ODE's repeatedly [9,6]. This poses a major problem, especially if one wants to perform a large number of simulations, e.g., to study the effect of perturbations or noise. In such case the computation time of a single reconstruction becomes critical. In this paper, we overcome this by presenting a method for fast reconstruction of metabolic networks from observed metabolite concentration data. In [10], Schmidt et al. introduced a method to infer interactions of a small genetic network via computing the Jacobian of the kinetic equations in the vicinity of a steady state. They build on an example given by Kholodenko et al. [11], proposing improvements by considering a series of constant perturbations. We apply a similar method to time series measurements of flavonoids in tomato seedlings. We adjust and extend their method to allow for constrains in the variables and so that a priori known non-existing interactions can be excluded from the network. This method can also be used to estimate unknown constant influxes from the ambient metabolic system.

This paper is organized as follows. In Sect. 2 we derive our reconstruction framework that modifies and extends the ideas given in [10], using only elementary calculus. In Sect. 3 we perform reconstructions using both, the conventional reverse engineering and the proposed method. As data for this comparison we take time series generated from artificial networks. We compare the differences in terms of residuals, computing times and the accuracy of the network topology. In Sect. 4, we statistically infer the quercetin glycosylation network by exploiting the fast reconstruction scheme. We finish with conclusions in Sect. 5.

2 Metabolic Network Reconstruction

In this section we specify the mathematical model for the dynamics of metabolic reactions, and derive a fast method for the reconstruction of metabolic networks.

2.1 Modeling Metabolic Interactions

Metabolic pathways are often visualized as graphs, where each node or vertex represents the molar concentration of the substrate participating in the reactions, and the edges represent the mass fluxes between the nodes. To reconstruct such a graph, i.e., to infer the metabolic pathway, we estimate the reaction rates from time-series measurements of concentrations of the compounds involved. A popular and powerful mathematical model for metabolic networks consists of a set of ordinary differential equations, depending on the initial concentrations and the reaction rates [5,6,10,12]. Our present task is to find estimates for the reaction rates such that:

- The model yields a good fit to the observations
- The model is not too sensitive to perturbations/noise
- The number of parameters is as small as possible

We note that in the case of flavonoid pathways, we cannot explicitly measure the concentrations of some boundary(input/output) nodes, due to the extremely fast conversion of one substrate into another. This hampers for example the use of graphical models for initial analysis, since we have missing data. Here, we show that we can still estimate these hidden substrates by including them as constants in the ODE system.

Let us first look at an example of a putative flavonoid network (see Fig. 1) and the corresponding mathematical model. Denoting the concentration of substrate i at time t as $X_i(t)$, $(i = 1, \ldots, 6)$, we can mathematically model this as

$$\dot{X}_1(t) = -k_{10}X_1(t) - k_{12}X_1(t) - k_{13}X_1(t) + k_{21}X_2(t) + k_{31}X_3(t) + k_{01}$$
$$\dot{X}_2(t) = -k_{21}X_2(t) - k_{24}X_2(t) - k_{25}X_2(t) + k_{12}X_1(t) + k_{42}X_4(t) + k_{52}X_5(t)$$
$$\dot{X}_3(t) = -k_{31}X_3(t) - k_{36}X_3(t) + k_{13}X_1(t) + k_{63}X_6(t)$$
$$\dot{X}_4(t) = -k_{42}X_4(t) + k_{24}X_2(t)$$
$$\dot{X}_5(t) = -k_{52}X_5(t) + k_{25}X_2(t)$$
$$\dot{X}_6(t) = -k_{63}X_6(t) + k_{36}X_3(t)$$

$$(1)$$

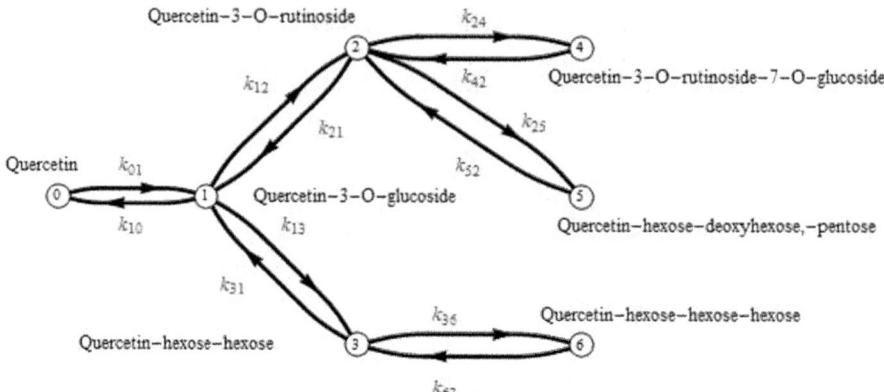

Fig. 1. A putative network model for quercetin glycosylation. This network has one vertex with number 0, which is connected to a larger metabolic network and from which a chemical precursor flows into the network.

In our example pathway, the substrate X_0 cannot be directly measured and has to be estimated. This substrate corresponds to the vertex number 0 connecting the quercetin pathway to the rest of the ambient metabolic network. In Fig. 1 we have drawn those edges that are considered to be relevant from a biological point of view. This putative network is based on the work of Iijima et al. [13] and the pathway model in KEGG [14].

Network inference is concerned with finding those edges that are most consistent with the given data. This may imply that one starts with the assumption that all possible edges are present and subsequently concludes that some rates k_{ij} are zero. A more general formulation of a linear ODE model is

$$\dot{X}_i(t) = -\sum_{j \neq i} k_{ij} X_i(t) + \sum_{j \neq i} k_{ji} X_j(t) + b_i , \qquad (2)$$

where $(i = 1, \ldots, n)$. To simplify the notation, we introduce a matrix A with components given by

$$\begin{cases} A_{ij} &= k_{ji}, \quad i \neq j \\ A_{ii} &= -\sum_{j \neq i} k_{ij}, \end{cases} \qquad (3)$$

Then, (2) becomes

$$\dot{X}_i(t) = \sum_{j=1}^{n} A_{ij} X_j(t) + b_i , \qquad (4)$$

with corresponding homogeneous system

$$\dot{X}_i(t) = \sum_{j=1}^{n} A_{ij} X_j(t) . \qquad (5)$$

For the present reconstruction algorithm we need the concentrations $X_i(t)$ at equidistant time points $t = t_0, t_1, \ldots, t_n$, with $n \geq N$, where N is the number of nodes in the network.

2.2   Derivation of the Objective Function

To reconstruct a metabolic network from time-series measurements, we have to estimate the reaction rates k_{ij}, which give the weights of the edges in the network. Due to (3), it is sufficient to estimate A. In what follows, we present a step by step derivation leading to the minimization problem (16), whose minimizer gives the required estimate for A. We denote the data, i.e., measured concentrations of substrate i at time point t_j, as $\mathbb{X}_{i,j}$.

We start from the well known property that for any solution of a homogeneous linear ODE with constant coefficients, such as the one in (5), it holds that

$$X(t + \Delta t) = \exp(A\Delta t)X(t) , \tag{6}$$

where $\exp(M)$ denotes the matrix exponential of M and Δt is some time step.

Now we construct matrices $\mathbb{X}_{\mathrm{new}}$ and $\mathbb{X}_{\mathrm{old}}$ as follows

$$\mathbb{X}_{\mathrm{new}} = \begin{pmatrix} \mathbb{X}_{1,n} & \mathbb{X}_{1,n-1} & \cdots & \mathbb{X}_{1,1} \\ \mathbb{X}_{2,n} & \mathbb{X}_{2,n-1} & \cdots & \mathbb{X}_{2,1} \\ \vdots & & \cdots & \vdots \\ \mathbb{X}_{n,n} & \mathbb{X}_{n,n-1} & \cdots & \mathbb{X}_{n,1} \end{pmatrix}, \qquad \mathbb{X}_{\mathrm{old}} = \begin{pmatrix} \mathbb{X}_{1,n-1} & \mathbb{X}_{1,n-2} & \cdots & \mathbb{X}_{1,0} \\ \mathbb{X}_{2,n-1} & \mathbb{X}_{2,n-2} & \cdots & \mathbb{X}_{2,0} \\ \vdots & & \cdots & \vdots \\ \mathbb{X}_{n,n-1} & \mathbb{X}_{n,n-2} & \cdots & \mathbb{X}_{n,0} \end{pmatrix} . \tag{7}$$

If the data would perfectly follow the model, we would have that

$$\mathbb{X}_{\mathrm{new}} = \exp(A\Delta t)\mathbb{X}_{\mathrm{old}} , \tag{8}$$

where $\Delta t = t_{i+1} - t_i$. We assume the measurement times to be equidistant. Taking the matrix logarithm we find an estimate for A

$$A = \frac{1}{\Delta t} \log \left(\mathbb{X}_{\mathrm{new}} \mathbb{X}_{\mathrm{old}}^{-1} \right) . \tag{9}$$

One may often encounter difficulties in inverting $\mathbb{X}_{\mathrm{old}}$. As a remedy one may regularize the matrix using Tikhonov regularization (or ridge regression) [15]. For this, one solves for some small $\alpha > 0$

$$A = \frac{1}{\Delta t} \log \left(\mathbb{X}_{\mathrm{new}} \left(\mathbb{X}_{\mathrm{old}} + \alpha\mathbf{I} \right)^{-1} \right) . \tag{10}$$

For an optimal choice of parameter α one may consult, e.g., [16].

We now turn to estimate A from the nonhomogeneous system (4). We append the scalar one to the vector X:

$$X^*(t) = \begin{pmatrix} X_1(t) \\ \vdots \\ X_n(t) \\ 1 \end{pmatrix} , \tag{11}$$

and to matrix A, we append the column of influx vectors $\mathbf{b}$.

$$A^* = \begin{pmatrix} A_{1,1} & A_{1,2} & \cdots & A_{1,n} & b_1 \\ \vdots & & \cdots & & \vdots \\ A_{n,1} & A_{n,2} & \cdots & A_{n,n} & b_n & \cdot \end{pmatrix} \tag{12}$$

Then we may concisely write (4) as

$$\dot{X}_i(t) = \sum_j A_{ij}^* X_j^*(t) \ . \tag{13}$$

The essence of the approach is that we are incorporating the data directly into our expression and that we have a homogeneous structure. Usually, the number of measurements is not equal to the number of unknowns. Thus having a square matrix is more the exception than the rule. As a consequence, typically we cannot solve linear ODE's using (9) or (10). Therefore, we approximate the derivatives with finite differences.

$$\dot{X}_{i,j} \approx \frac{X_{i,j+1} - X_{i,j}}{\Delta t} \approx \sum_k A_{i,k}^* X_{k,j}^* \ , \tag{14}$$

thus in terms of the data matrices introduced in (7) we get the estimate for A^* using pseudo-inverse:

$$A^* = \frac{1}{\Delta t} \left(\mathbb{X}_{\text{new}} - \mathbb{X}_{\text{old}} \right) \mathbb{X}_{\text{old}}^T \left(\mathbb{X}_{\text{old}} \mathbb{X}_{\text{old}}^T \right)^{-1} \ . \tag{15}$$

It goes without saying that this is very fast since it involves only matrix manipulations. On the other hand it can result in over-fitting, since all possible edges are included in the modeled network. Another serious shortcoming of this approach is the fact that we cannot control the positivity of the reaction rates. Although in [10], positive(negative) coefficients were interpreted as activation(inhibition) of the compounds, in many biological pathways, negative coefficients are not allowed. This also holds for the example we will give in Sect. 4. Thus we need a more general approach that does allow sparse networks, where one can exclude all irrelevant edges that are not contained in any biologically feasible model, and in which one can constrain the reaction rates to be positive, without substantially compromising computation time.

To this end, we note that the formula in (15) provides in fact an explicit solution of the following minimization problem

$$\underset{A^*}{\arg\min} \left(\left\| A^* \mathbb{X}_{\text{old}} - \frac{1}{\Delta t} \left(\mathbb{X}_{\text{new}} - \mathbb{X}_{\text{old}} \right) \right\|^2 \right) \ . \tag{16}$$

This alternative formulation allows inclusion of expert knowledge in a simple way. E.g., we can at will put $A_{ij}^* = 0$, when an edge from node i to node j can not exist. Nearly all mathematical software packages (Mathematica, Matlab, Maple etc.) can numerically find the minimizer A^* (and thus the reaction rates k_{ij}) with the constraint that $k_{ij} \geq 0$.

3 Experiments with Artificial Data

In the conventional reverse engineering method the parameters k_{ij} are estimated, using optimization algorithms to minimize the sum of squares between the ODE solutions and the concentration measurements. This involves repeated solving of the ODE's, which is the major time consuming part in the process [6]. We compare our direct inference method with this conventional reverse engineering method. As an example we present here the case with six nodes, where five nodes X_1, X_2, X_3, X_4, X_5 correspond to measurable concentrations and one node X_0 is a boundary node that connects the network to the surroundings. We generated artificial networks in which both, the positions and weights of the edges were randomly chosen. For such a random network the ODE's in (2) were solved. We sampled these solution curves and subsequently reconstructed the original network based on these samples.

We generated these artificial networks as follows. We chose a (uniformly distributed) random integer to determine the number of zeros in an adjacency matrix for nodes X_2, X_3, X_4, X_5, X_6. After determining the topology of the network in this way, we assigned a (uniformly distributed) random real number between zero and one as the weight for each edge independently.

We compared the reconstructions using both, the conventional method and our proposed method, first by using exact samples and then by adding $\pm 10\%$ (uniformly distributed) noise to the samples. Finally we did reconstructions assuming that the topology of the network is known *a priori*. This can be compared to the situation when reconstructing real metabolic networks, since one usually has some putative information on the possible connections between substrates. A typical result using 20 sample points is plotted in Fig. 2. We observe that although the conventional method is tuned to closely approximate the solution curves, the resulting networks are not necessarily closer to the original. While it is obvious that the fast reconstruction method based on first order approximation will generally give larger residual with respect to the original data, another question is, what does this mean in terms of reconstructed networks. To answer this question experimentally, we generated random networks as described before. Then, to simulate a typical reconstruction situation, where only a minimum amount of data is available, we did repeated reconstructions using only six sample points. From each reconstruction, we recorded the residuals (i.e., the distances between the reconstruction and the original function at sample points) and the computation times in seconds. We plotted them in logarithmic scale to be able to include large values in the picture. In addition to this we compared the topologies of the network adjacency matrices. That is we counted all those edges that were missing or redundant compared to the adjacency matrix of the original network. The results for 100 reconstructions are shown in Fig. 3. It seems that the iterative method, while demanding a lot of computing time and indeed resulting in better fit, does not necessarily deliver better results in terms of network reconstruction. For illustration of this matter see Fig. 4.

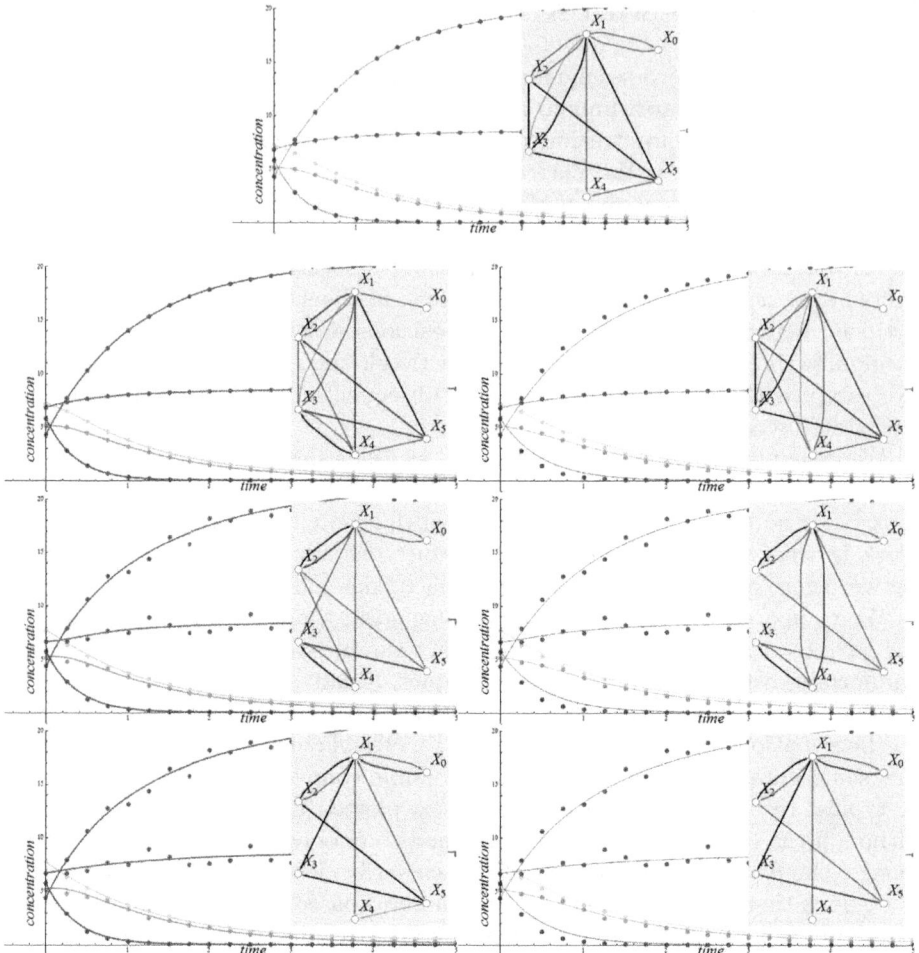

Fig. 2. On the top: the original artificial network and the corresponding ODE solution curves. Left column: reconstructions with the iterative method. Right column: reconstructions with the fast method described in this paper. Top row, reconstructions from exact samples. Middle row, reconstructions from samples with ±10% noise. Bottom row, reconstructions from the same noisy data, when the network topology is known *a priori.*

4   Experiments with Flavonoid Data

The high efficiency of the present method allows a statistical strategy to discriminate between relevant and redundant edges. The idea is to perform repeated

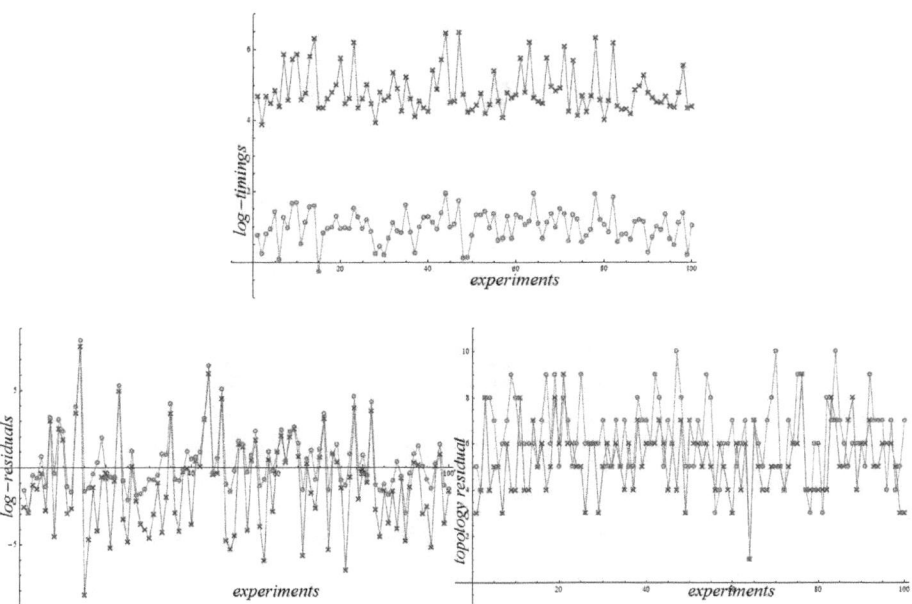

Fig. 3. Red crosses correspond to the iterative method and the blue circles to the fast method described in this paper. Top: logarithms of the computation times, in seconds. Bottom left: logarithms of the residuals (in substrate concentrations) with respect to the original concentration curves. Bottom right: the numbers of missing/redundant edges compared to the original network.

network reconstructions using the putative network in Fig. 1, meanwhile adding random noise to the measurements. If the reconstructions consistently assign a zero value to a parameter k_{ij}, we can suspect that the corresponding edge is not likely to exist in a network derived from an ODE model. In our experiments we took the substrate concentration data of the metabolites involved in the putative quercetin pathway. These concentrations were measured from tomato seedlings during days 5 to 9 after germination [17]. Subsequently, we performed 1000 reconstructions using formula (16), while adding ±10% random noise to the data. The resulting distributions for parameters k_{ij} can be seen in Fig. 5. The number of bars in the histograms is approximately the square root of the number of reconstructions. This kind of simulation can give a significant clue to whether the nodes i and j are connected or not and also provide insight on how sensitive the parameters are w.r.t. noise. From this result we could for example conclude that the edge from node 1 to node 0 and the edge from node 3 to node 6 are redundant. The exact criteria for discarding edges depend on the context of the network.

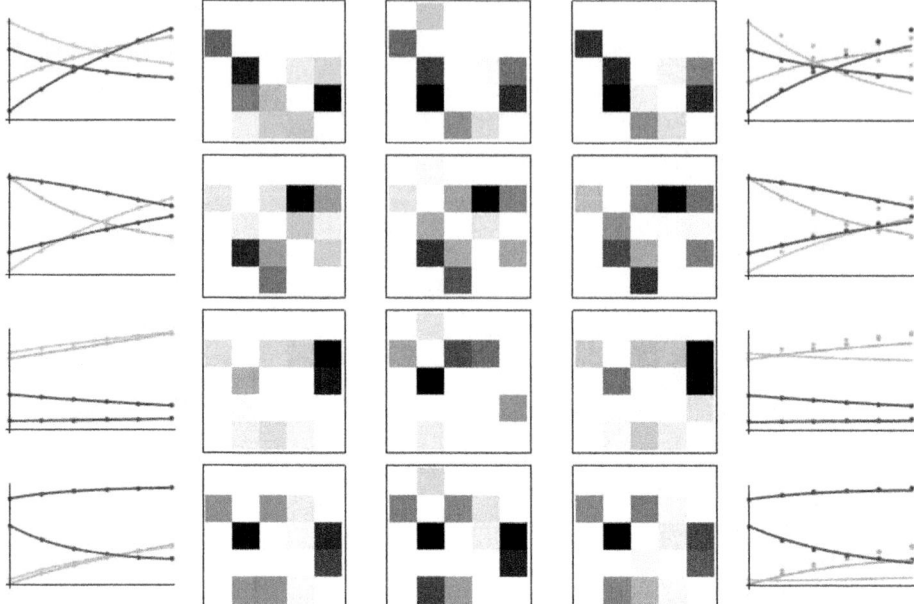

Fig. 4. Columns from left to right: Reconstructed solution curves from samples using iterative method, reconstructed network matrices using iterative method, original network matrices with which the data was created, reconstructed network matrices using the fast method described in this paper, and correspondingly the solution curves reconstructed with the fast method.

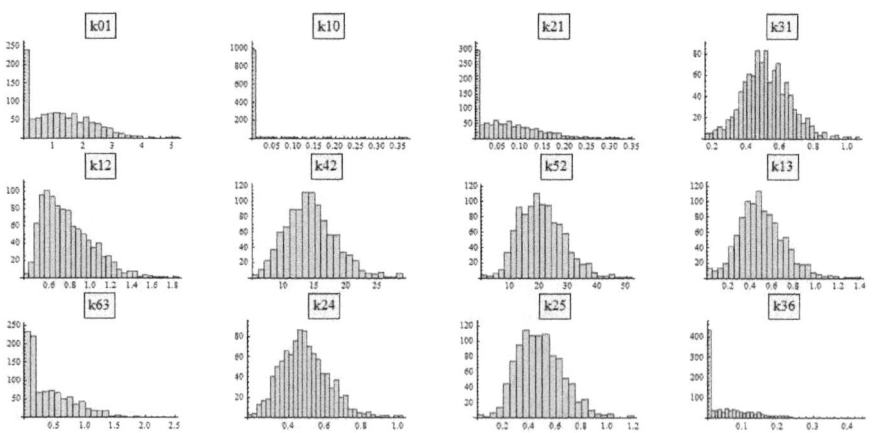

Fig. 5. Histograms showing the distributions of reaction rates k_{ij} estimated in a series of 1000 reconstructions. A pre-selection of relevant edges was done based on the putative model in Fig. 1. In each reconstruction the concentration data were randomly perturbed with $\pm 10\%$ noise. From the results, we can immediately distinguish those coefficients k_{ij} that are distributed around zero and those which in turn accumulate around a positive value.

5 Conclusions

We have experimented with a method for network reconstruction that is very fast compared to the conventional approach. The only requirements are that time-series data are available, the dynamics of the network can be modeled with ODE's, and that the number of measurements $n \geq N + 1$, where N is the number of nodes. Although our approach is inspired by [10], the application is different, since their work considers the approximation of the Jacobian of kinetic equations such as those in [11], in the vicinity of steady state and their time series consists of *in silico*, constant rate perturbations to a maximal enzyme rate. We, on the other hand model *in vivo* measurements, where the unknown influx rates correspond to the constant rate perturbation. In either case the formula (15) to estimate Jacobian is well known in numerical mathematics. We have modified this to a minimization problem (16) to adjust it to our model, where the kinetic constants have to be positive and where we have to be able to exclude nonsensical edges from the network.

The main advantage of this method is that, though it is slightly less accurate than the iterative method that minimizes the residual between the ODE-solutions and measurements, it is significantly faster allowing one to do statistical analysis that require large number of simulations. From the simulations in Sect. 3 we see that, it is around hundred times faster than the conventional iterative method and thus highly suitable for repeated reconstructions. We remark that the residual between the solution curves is not the best measure of successful network reconstruction. We have also experimentally observed (see Fig. 3) that in terms of the network structure, i.e., the adjacency matrix of the nodes, the proposed method performs similarly to the residual-based iterative method.

Acknowledgements. This work results from a collaboration between plant biologists, statisticians and mathematicians, initiated by the Netherlands Consortium for Systems Biology (NCSB) and Centre for Biosystems Genomics (CBSG) and financed by the Netherlands Genomics Initiative.

References

1. Koes, R., Quattrocchio, F., Mol, J.: The flavonoid biosynthetic pathway in plants: Function and evolution. The American Journal of Clinical Nutrition 16(2), 123–132 (1994)

2. Martin, C., Butelli, E., Petroni, K., Tonelli, C.: How can research on plants contribute to promoting human health? Plant Cell (May 2011), doi:10.1105/tpc.111.083279

3. Bovy, A., Schijlen, E., Hall, R.: Metabolic engineering of flavonoids in tomato Solanum lycopersicum: the potential for metabolomics. Metabolomics 3(3), 399–412 (2007)

4. Slimestad, R., Fossenn, T., Verheul, M.: The flavonoids of tomatoes. J. Agric. Food Chem. 56(7), 2436–2441 (2008)

 5. Hatzimanikatis, V., Floudas, C., Bailey, J.: Analysis and design of metabolic reaction networks via mixed-integer linear optimization. AIChE Journal 42(5), 1277–1292 (1996)
 6. Chou, I.-C., Voit, E.: Recent developments in parameter estimation and structure identification of biochemical and genomic systems. Math Biosci. 219(2), 57–83 (2009)
 7. Zhan, C., Yeung, L.: Parameter estimation in systems biology models using spline approximation. BMC Systems Biology 5(14) (2011)
 8. Hendrickx, D., Hendriks, M., Eilers, P., Smilde, A., Hoefsloot, H.: Reverse engineering of metabolic networks, a critical assessment. Molecular BioSystems 7, 511–520 (2011)
 9. Kimura, S., Nakayama, S., Hatakeyama, M.: Genetic network inference as a series of discrimination tasks. Bioinformatics 25(7), 918–925 (2009)
10. Schmidt, H., Cho, K.-H., Jacobsen, E.: Identification of small scale biochemical networks based on general type system perturbations. The FEBS Journal 272, 2141–2151 (2005)
11. Kholodenko, B., Kiyatkin, A., Bruggeman, F., Sontag, E., Westerhoff, H., Hoek, J.: Untangling the wires: A strategy to trace functional interactions in signaling and gene networks. Proc Natl. Acad. Sci. USA 99(20), 12841–12846 (2002)
12. Jha, S., van Schuppen, J.: Modelling and control of cell reaction networks. Pnar0116, CWI, Amsterdam (2001)
13. Iijima, Y., Nakamura, Y., Ogata, Y., Tanaka, K., Sakurai, N., Suda, K., Suzuki, T., Suzuki, H., Okazaki, K., Kitayama, M.: Metabolite annotations based on the integration of mass spectral information. The Plant Journal 54(5), 949–962 (2008)
14. Kyoto Encyclopedia of Genes and Genomes: Flavone and Flavonol Biosynthesis (2010), http://www.genome.jp/kegg/pathway/map/map00944.html
15. Golub, G., Hansen, P., O'Leary, D.: Tikhonov regularization and total least squares. SIAM J. Matrix Anal. & Appl. 21(1), 185–194 (1999)
16. Hansen, P., O'Leary, D.: The use of the l-curve in the regularization of discrete ill-posed problems. SIAM J. Sci. Comput. 14(6), 1487–1503 (1993)
17. Gomez-Roldan, M.V., Bovy, A., de Vos, R., Groenenboom, M., Astola, L.: LC-MS metabolite profiling on tomato seedlings in a systems biology approach. In: Metabomeeting, Helsinki, Finland (September 2011)

Highlighting Metabolic Strategies Using Network Analysis over Strain Optimization Results

José Pedro Pinto[1,2], Isabel Rocha[2], and Miguel Rocha[1]

[1] Department of Informatics / CCTC - University of Minho
mrocha@di.uminho.pt
[2] IBB - Institute for Biotechnology and Bioengineering,
Centre of Biological Engineering, University of Minho,
Campus de Gualtar, 4710-057 Braga, Portugal
irocha@deb.uminho.pt

Abstract. The field of Metabolic Engineering has been growing, supported by the increase in the number of annotated genomes and genome-scale metabolic models. *In silico* strain optimization methods allow to create mutant strains able to overproduce certain metabolites of interest in Biotechnology. Thus, it is possible to reach (near-) optimal solutions, i.e. strains that provide the desired phenotype in computational phenotype simulations. However, the validation of the results involves understanding the strategies followed by these mutant strains to achieve the desired phenotype, studying the different use of reactions/ pathways by the mutants. This is quite complex given the size of the networks and the interactions between (sometimes distant) components. The manual verification and comparison of phenotypes is typically impossible.

Here, automatic methods are proposed to analyse large sets of mutant strains, by taking the phenotypes of a large number of possible solutions and identifying shared patterns, using methods from network topology analysis. The topological comparison between the networks provided by the wild type and mutant strains highlights the major changes that lead to successful mutants. The methods are applied to a case study considering *E. coli* and aiming at the production of succinate, optimizing the set of gene knockouts to apply to the wild type. Solutions provided by the use of Simulated Annealing and Evolutionary Algorithms are analyzed. The results show that these methods can help in the identification of the strategies leading to the overproduction of succinate.

Keywords: Metabolic Engineering, Strain optimization, Metabolic networks, Network visualization.

1 Introduction

Recent efforts in Bioinformatics and Systems Biology allowed the development of genome-scale metabolic models for several microorganisms [1]. These models have been used to guide biological discovery promoting the comparison between

M. Loog et al. (Eds.): PRIB 2011, LNBI 7036, pp. 109–120, 2011.

predicted and experimental data, to foster Metabolic Engineering (ME) efforts in finding appropriate genetic modifications to synthesize desired compounds, to analyze global network properties and to study bacterial evolution [2].

The most popular approach to phenotype simulation considers the cell to be in steady-state and takes reaction stoichiometry/ reversibility in a constraint-based framework to restrict the set of possible values for the reaction fluxes. Cellular behaviour is thus predicted using for instance Flux Balance Analysis (FBA), based on the premise that microorganisms have maximized their growth along natural evolution [3]. Using FBA, it is possible to predict the behaviour of microbes under distinct environmental/ genetic conditions.

The combination of reliable models with efficient simulation methods has been the basis for different strain optimization algorithms. Their goal is to find the set of genetic modifications to apply to a given strain, to achieve an aim, typically related with the industrial production of a metabolite of interest.

In previous work, an approach based in the use of metaheuristics, such as Evolutionary Algorithms (EAs) and Simulated Annealing (SA), has been proposed to solve the optimization task of reaching an optimal (or near optimal) subset of reaction deletions to optimize an objective function related with the production of a given compound [9]. The idea is to force the microbe to synthesize a desired product, while keeping it viable.

The next logical step is to validate these results in the lab, a task that given its associated costs should be preceded by a thorough analysis of the solutions provided using computational methods. This screening process could identify more promising approaches and, thus, save resources in wet lab experiments.

In a first stage, the validation of the results involves the understanding of the strategies followed by these mutant strains to achieve the desired phenotype, by studying the different use of reactions/ pathways to achieve the desired metabolite and still keep the strain viable. This becomes quite complex given the size of the networks involved in genome-scale models and the interactions between (sometimes distant) components. The manual verification and comparison of the phenotypes of different mutants is typically impossible.

In this work, the major aim is the development of automatic methods to analyse large sets of mutant strains for specific ME problems. These methods take the phenotypes of a large number of possible solutions obtained by running strain optimization algorithms and attempt to identify shared patterns, taking advantage of methods from network topology analysis. The topological comparison between the networks provided by the wild type and mutant strains highlights the major changes, thus highly contributing to elucidate the strategies that lead to successful mutants.

The methods are applied to a case study considering *Escherichia coli* as the host and aiming at the production of succinate, by optimizing the set of gene knockouts to apply to the wild type. Large sets of solutions (mutants) provided by the use of SA and EAs are analysed. To provide for large sets of possible solutions, the strain optimization algorithms were modified to keep all interesting solutions found during their execution.

The paper is organized as follows: next, a description of the computational methods is provided; this is followed by a description of the case study, the results obtained and its discussion; finally, conclusions and further work are provided.

2   Methods

2.1   Overall Workflow

In this work, the workflow used in the experiments can be summarized in the following steps:

- **Inputs:** a genome scale metabolic model of a host organism; a set of currency metabolites; a metabolite of interest to be overproduced;
- **Step 1:** the strain optimization algorithms (EA and SA) are executed with the provided configuration (see section 2.3); each algorithm is executed a given number of runs and the result from each run is a set of solutions (mutant strains) of interest;
- **Step 2:** the solution sets from the previous set are merged in a single set and filtered (see details in section 2.4);
- **Step 3:** each solution in the set from step 2 is simulated using FBA (section 2.2) and the corresponding network is created according to the methods described in section 2.5;
- **Step 4:** each of the networks from step 3 is compared to the wild type network, as described in section 2.6;
- **Step 5:** the comparisons from step 4 are analysed for common patterns of variability analysis (see details in section 2.7;
- **Step 6:** the results from the previous step are compiled in a sub-network that can be also visualized and manually analysed.

2.2   Flux Balance Analysis

In this work, FBA was used as the phenotype simulation method in the strain optimization tasks and to provide for the network filtering. FBA is based on a steady state approximation to the concentrations of internal metabolites, which reduces the corresponding mass balances to a set of linear homogeneous equations [4]. For a network of M metabolites and N reactions, this is expressed as:

$$\sum_{j=1}^{N} S_{ij}v_j = 0 \qquad (1)$$

for every metabolite i, where S_{ij} is the stoichiometric coefficient for this metabolite in reaction j and v_j is the flux over the reaction j. The maximum/minimum values of the fluxes can be set by additional constraints in the form $\alpha_j \leq v_j \leq \beta_j$, also used to specify nutrient availability.

The set of linear equations obtained usually leads to an under-determined system, for which there exists an infinite number of feasible flux distributions that satisfy the constraints. However, if a given linear function over the fluxes is chosen to be maximized, it is possible to obtain a solution by applying standard algorithms (e.g. *simplex*) for Linear Programming. The most common flux chosen for maximization is the biomass given the premise of optimal evolution that underlies FBA.

2.3   Strain Optimization

The problem addressed in this work consists of selecting, from a set of reactions in a microbe's genome-scale model, a subset to be deleted to maximize a given objective function. The encoding of a solution is achieved by a variable size set-based representation, where each solution consists of a set of reactions from the model that will be deleted. For all reactions deleted, the flux will be constrained to 0, therefore disabling them from the metabolic model. The process proceeds with the simulation of the mutant using the chosen phenotype simulation method (in this work, FBA). The output of these methods is the set of fluxes for all reactions, that are then used to compute the fitness value, given by the objective function.

Here, the objective function used is the Biomass-Product Coupled Yield (BPCY) [6], given by: $BPCY = \frac{PG}{S}$, where P stands for the flux representing the excreted product; G for the organism's growth rate (biomass flux) and S for the substrate intake flux. Besides optimizing for the production of the desired product, this function also allows to select for mutants that exhibit high growth rates. To address this task, we will use Simulated Annealing (SA) and Evolutionary Algorithms (EAs) as proposed previously in [9], where full details can be found regarding operators and other configuration details.

The implementation of the original EA and SA methods was only modified in order to keep not only the best solutions obtained during the run, but rather the whole set of solutions deemed to be interesting for analysis. This does not change the optimization process, but stores more intermediate results. Therefore, each run of the algorithms generated a *solution set* containing all solutions where both the value of P and G where larger than 0. This set includes only simplified solutions, i.e. solutions where the removal of a given knockout would reduce the fitness function value. All solutions are simplified by removing unnecessary knockouts before entering this set.

2.4   Solution Set Pre-processing

The first step in the pre-processing is to merge all the solutions coming from each individual run of the optimization. In this task, all duplicate solutions are removed. Also, the final solution set is checked for the existence of solutions where the set of knockouts is a superset of other solutions and these are only kept if its fitness value is higher.

The next step is to filter this solution set, since using all the solutions can be an undesirable option because in many cases they do not provide acceptable results from the biological standpoint. Also, the comparison process can be computationally heavy if the solution sets are too large. So, the number of solutions used was reduced filtering solutions: (i) with a low growth, by setting a minimal threshold for the biomass production flux; (ii) with a low production of the desired flux, by setting a minimal threshold for the flux associated with the excretion of the metabolite; (iii) filtering solutions that lead to the same set of reactions in the network after simulation filtering (see next section). In the experiments, the thresholds for (i) and (ii) correspond to 40% of the maximum value (this value was empirically set to keep solutions near to the best values obtained).

2.5   Network Representation and Simulation Filtering

The metabolic networks used in this project are directed bipartite graphs, where the nodes represent either metabolites or reactions and the edges the consumption or production of metabolites, pointing from a metabolite to a reaction in the former case and from a reaction to a metabolite in the latter.

The first step in this project was to create a network with all reactions and metabolites contained in the metabolic model, thus creating a network which can function as a map of the organism's metabolism, or more precisely a network with all the possibilities which can occur in a simulation. This network is called the *base network*.

All other networks were derived from the base network using a method denoted as *simulation filtering*. This is a process which creates a sub-network starting with the base network, taking the results of a phenotype simulation, and executing the following steps:

1. nodes which correspond to reactions with a flux of zero in the simulation are removed;
2. nodes which correspond to currency metabolites are removed (the list of currency metabolites is provided by the user for each model);
3. nodes which are left isolated (with no neighbours) after step 1 and 2 are removed.

The result of the simulation filtering is a network which is a "snapshot" of how the metabolism behaves according to the results provided by a phenotype simulation method.

2.6   Network Comparison

The network comparison process is basically a series of operations in which each mutant network is compared with the wild type network used as a reference. These operations are typically followed by some global variation analysis to identify the most common patterns of network variation (see next section).

During the development of the methods and the tools used in this project it was noticed that the majority of the viable mutant networks are very similar to the wild type network, which limits the use of many global metrics for graph topology, such as centrality values, shortest path analysis or clustering coefficients as comparison metrics. This led us to define some novel network comparison metrics adapted to the purposes of the work and focused on the identification of local patterns of interest in metabolic networks. The set of analyses conducted and the metrics used are defined next.

Exclusivity. This comparison metric used is based on the set of exclusive nodes, i.e. those existing in one of the networks but not the other. In this case, for each mutant-wild type comparison two lists are created containing the set of nodes exclusive to each of the networks. After all mutants are compared to the wild type these lists are used to determine the frequency of each node in the exclusive lists.

Decision points. When analysing metabolic networks it is important to look not only at the topology but also to understand the flows over the network. For instance in the case of linear pathways with no splits, the existence of a flow in a reaction may be determined by another upstream reaction that can be distant. The decision point concept was thought as a way to determine the upstream network metabolite of pathways that exist on one of the networks and not the other. The first step in identifying decision points is to identify the decision metabolites, i.e. that are common to both networks, but that are consumed by different sets of reactions. The next step is to identify which reactions consuming these metabolites are present in one of the networks and not in the other.

Inversions. Many reactions in the metabolic model are classified as reversible, meaning that they can occur on both directions. The inversion metric is based in the fact that manipulations of the metabolism can result in a flux changing signs, meaning the reaction changes its direction. All the inversions that occur in the mutants compared to the wild type are identified and their frequency is calculated.

2.7   Variation Analysis

After all network comparisons between the mutants and the reference are conducted, the results obtained are used to identify common patterns. As a first step, the reactions that are exclusive in a significant part of the comparisons are identified. In this case, all reactions that are in the mutant exclusive lists with a frequency exceeding a threshold (in this work 80%) are identified. This group includes reactions typically used in mutant phenotype but not on the wild type strains. The same process is conducted for the wild type exclusive lists, thus identifying reactions used in the wild type but typically absent from the mutant strains resulting from the strain optimization algorithms.

Each of these sets of reactions is used to create a network also including the metabolites involved in those reactions. This typically creates several independent modules not connected by any metabolite. This network can be visualized using, for instance, the Cytoscape tool (http://www.cytoscape.org). This allows to color nodes differently highlighting important nodes that are decision points or typical knockouts identified by the strain optimization algorithms.

In some cases, the networks obtained are manually modified to include nodes and data which could not be identified by purely computationally process but which are nonetheless important for the analysis. The final result is a network that contains the parts of the metabolism which are more commonly altered when the organism is manipulated to produce the metabolite of interest. This network is named the *variation network*.

The last step of our methodology is the analysis of the variation network. This analysis was based in the observation of each module to determine how its existence relates with the knockouts conducted in the mutants, to the production of the target metabolite and to the production of biomass precursors.

During the analysis, the variation network was also compared with pathway maps obtained from KEGG and EcoCyc to determine the relationship of the variations with the organism metabolism as a whole and if they were related with any known important metabolic cycles.

3   Experiments and Results

3.1   Case Study and Experimental Setup

The implementation of the proposed algorithms was performed by the authors in *Java*, within the OptFlux open-source ME platform (http://www.optflux.org) [8]. Some of the methods in network analysis have been added as a plug-in for this platform and the workflow described here will be added in the future.

The case study considered uses the microorganism *Escherichia coli* and the aim is to produce succinate with glucose as the limiting substrate. Succinate is one of the key intermediates in cellular metabolism and therefore an important case study for ME [5]. It has been used to synthesize polymers, as additives and flavouring agents in foods, supplements for pharmaceuticals, or surfactants.

The genome-scale model used is given in [7] and includes a total of $N = 1075$ fluxes and $M = 761$ metabolites. A number of pre-processing steps were conducted to simplify the model and reduce the number of targets for the optimization (see [9] for details) leaving the simplified model with $N = 550$ and $M = 332$; 227 essential reactions are identified, leaving 323 variables to be considered when performing strain optimization.

In this study, we have used both EA and SA executing each algorithm for 30 runs. Each solution has 6 knockouts, since this was the minimum number of knockouts able to provide high quality solutions and an increase in this value does not provide significant gains. After the preprocessing was done we had a total of 4949 distinct networks from an initial batch of 8018 solutions.

3.2   Results and Discussion

The results of the first stage of the analysis, i.e. the results of the comparison between mutant networks and the wild type is provided in a spreadsheet provided as supplementary material in http://darwin.di.uminho.pt/prib2011.

After conducting the variation analysis, the final sub-network was composed by three apparently independent modules. Subsequent observations revealed that these were not in fact independent changes but in fact were all directly or indirectly related to a common set of alterations which occur in practically all mutant organisms. To show the results and address the discussion we will analyse these three modules in more detail. Figure 1 shows the meaning of the colour code used in the following figures to identify the different nodes coming from the analysis.

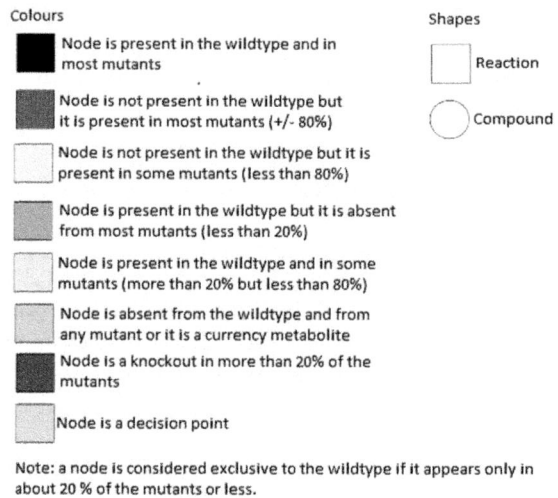

Fig. 1. Colour code used in the analysis of the variation network.

Main Knockouts and Succinate Production. This module, shown in Figure 2, is particularly interesting because it contains two of the most frequently knocked out reactions (SUCD4 and SUCD1i) and it is directly related with the production of succinate. Of all the modules in the variation network this is the one whose analysis is the most straightforward, since the reaction SUCD1i, central of this module, is the only consumer of succinate in the wild type. Thus, to achieve the production of succinate, the most direct method is to remove it. SUCD1i is one of the reactions with the higher value of wild type exclusivity and also of frequency of selection as a knockout in the solutions. SUCD1i is also the first fumarate production reaction mentioned in the dGDP Consumption

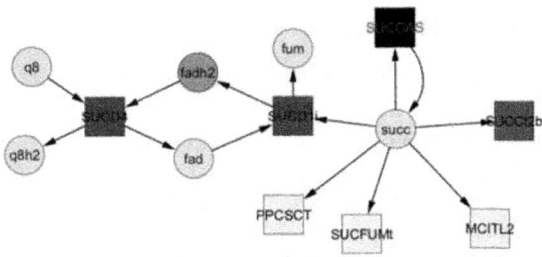

Fig. 2. Sub-network for the main knockouts in succinate production

Module (shown next). Its indirect effect on the former module illustrates how alterations in the metabolism can have unexpected effects.

Besides the direct removal of SUCD1i, another simple way of suppressing this reaction is the knockout of SUCD4. This reaction is the main way in which FADH2 is consumed and SUCD1i is the only producer of FADH2. The removal of SUCD4 will necessarily lead to the suppression of SUCD1i.

Another reaction of note is the transporter SUCCt2b which excretes succinate from the cell naturally. This is a mutant exclusive reaction for obvious reasons, since in the wild type no succinate is excreted. There are other reactions that only occur in the mutants, but the only one that appears in a significant number of mutants is SUCOAS. This reaction is also present in the wild type in its reversed form.

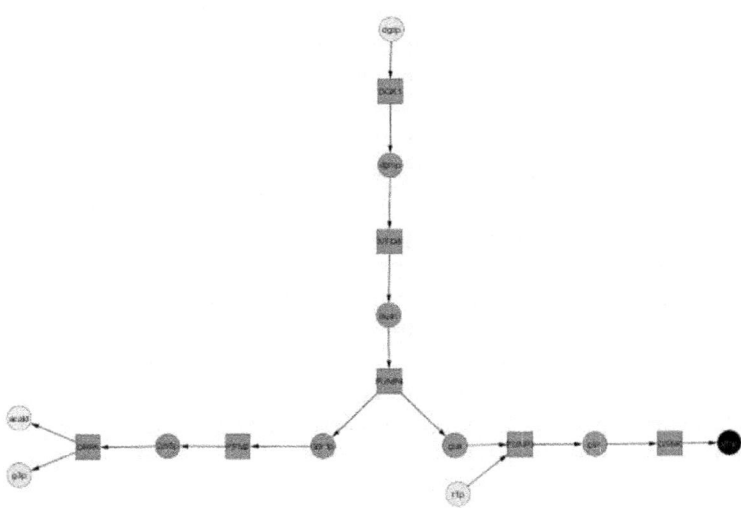

Fig. 3. Sub-network for dGDP consumption

dGDP Consumption. Originally, this module (Figure 3) appeared as two independent reaction chains exclusive to the wild type:

Guanine (gua) + alpha-D-Ribose 1-phosphate (r1p) → PUNP3 → Guanosine (gsn) → GSNK

2-Deoxy-D-ribose 1-phosphate (2dr1p) → PPM2 → 2-Deoxy-D-ribose 5-phosphate (2dr5p) → DRPA

The analysis of the variation network revealed that these two chains were actually a consequence of the wild exclusive chain:

dGDP (dgdp) → DGK1 → dGMP (dgmp) → NTD8 → Deoxyguanosine (dgsn) → PUNP4 → Guanine (gua) + Deoxy-D-ribose 1-phosphate (2dr1p)

This chain was not initially identified because its value of exclusivity was slightly below the defined threshold. It was determined that this module is wild type exclusive because the reaction DGK1 rarely appears on mutants. However, neither DGK1 or any of the reactions which produce the metabolites it consumes is a common knockout, which means that some alteration in the mutants' metabolism provokes the redirection of the flux of some of the compounds used by DGK1. Initial observations of the flux values of the reactions which compose this cycle and their immediate neighbors revealed that the reason for its wild type exclusivity is a consequence of the significant reduction of dGDP in the mutants. This module does not produce any essential metabolite that can not be obtained by other reactions. Also, dGDP is necessary for the production of dGTP which is a biomass precursor. The reduction of the concentration of dGDP leads to all dGDP being channelled to the production of dGTP to ensure the survival an growth of the organism.

A more thorough analysis of the reactions revealed that the reduction of the dGDP production in the mutants is ultimately due to the reduced production of a compound used in its synthesis: 3-phosphohydroxypyruvate. This compound is obtained from a reaction which uses 3-phospho-D-glycerate as a substrate, while the production of 3-phospho-D-glycerate is not being reduced in the mutants (in fact it is somewhat increasing). Most of it is being used to produce D-glycerate 2-phosphate, which in turn is used for the production of phosphoenolpyruvate.

Continuing the analysis of the reactions which use phosphoenolpyruvate, it was determined that the increased production of phosphoenolpyruvate in the mutants is a consequence of the change in the TCA cycle due to a reduction in the production of L-malate. This, in turn, leads to a need for an increase of the production of phosphoenolpyruvate in order to maintain the cycle.

The alterations in the production of L-malate are due to the reduction of the production of fumarate, a metabolite used by a reaction external to the TCA cycle which produces L-malate. Its reduction is a consequence of the flux reduction of the two major fumarate production reactions in the mutant:

1. The main fumarate production reaction is also the main succinate consuming reaction; since the objective is to maximize succinate production this means that this reactions is a frequent knockout and even when it is not, it tends to be inactive.

2. The other reactions which produce fumarate compete with for the consumption of L-Aspartate 4-semialdehyde with the alternative L-threonine production chain which how it will shown later is essential for the survival of most mutants,

It is interesting to note that the flux reduction of the fumarate production reactions is related with two other modules which indicates that the modules of the variation network are not as independent as the initial observations of the network implied.

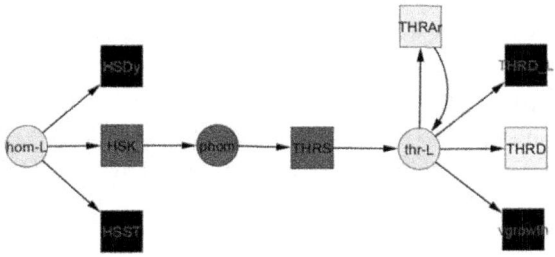

Fig. 4. Sub-network for Alternative L-threonine production

Alternative L-Threonine Production. This module, shown in Figure 4, is characterized by a mutant exclusive reaction chain:

HSK → O-Phospho-L-homoserine (phom) → THRS → L-Threonine (thr-L)

Our analysis revealed that the reason for this chain being mutant exclusive lies in the reaction THRAr which normally produces L-Threonine, which is inverted in most mutants. This chains is necessary to compensate this inversion. The inversion of reaction THRAr occurs because in the inverted form it produces glycine and the alternative reaction for the production of glycine (GHMT2) is wild type exclusive.

Initially, the fact that GHMT2 was wild type exclusive appeared strange. However, we eventually concluded that this reaction is a major producer of nadph and its removal forces the mutants to compensate by using a reaction which produces succinate as a byproduct, thus increasing the production of succinate. This fact was difficult to determine because nadph is a currency metabolite removed from the networks in the pre-processing stages.

It should be noted that the second fumarate production reaction mentioned in the dGDP Consumption Module is the producer of L-Homoserine which is at the beginning of the reaction chain central to this module. Again, this shows the unity of the metabolism and it gives further evidence to the idea that the variations are not distinct modules but a closely related group of metabolic changes.

4   Conclusions

The recent development of computational methods for strain optimization based on the use of genome-scale metabolic models has opened new avenues for

Metabolic Engineering. This work aims to contribute to this effort proposing computational methods for the analysis of the solutions of strain optimization metaheuristics, such as EAs and SA. The aim is to provide tools that allow to identify the most common patterns used by successful mutant strains and therefore understand the strategies used, prior to wet lab experiments that will ultimately validate the results.

Further work will address the full implementation of these features in the OptFlux platform, making them available to the community. Also, the work will be extended to other interesting case studies in ME, by considering other metrics of the network topology and by improving the analysis methodology so that it takes partial flux variations between the mutants and the wild type into account.

Acknowledgments. This work is supported by project PTDC/EIA-EIA/ 115176/2009, funded by Portuguese FCT and Programa COMPETE. JPP work is funded by a PhD grant from the Portuguese FCT (ref. SFRH/BD/41763/2007).

References

1. Feist, A.M., Herrgard, M.J., Thiele, I., Reed, J.L., Palsson, B.Ø.: Reconstruction of biochemical networks in microorganisms. Nature Reviews Microbiology 7(2), 129 (2008)
2. Feist, A.M., Palsson, B.Ø.: Nature Biotechnology 26(6), 659–667 (2008)
3. Ibarra, R.U., Edwards, J.S., Palsson, B.G.: Escherichia coli k-12 undergoes adaptive evolution to achieve in silico predicted optimal growth. Nature 420, 186–189 (2002)
4. Kauffman, K.J., Prakash, P., Edwards, J.S.: Advances in flux balance analysis. Curr. Opin. Biotechnol. 14, 491–496 (2003)
5. Lee, S.Y., Hong, S.H., Moon, S.Y.: In silico metabolic pathway analysis and design: succinic acid production by metabolically engineered escherichia coli as an example. Genome Informatics 13, 214–223 (2002)
6. Patil, K., Rocha, I., Forster, J., Nielsen, J.: Evolutionary programming as a platform for in silico metabolic engineering. BMC Bioinformatics 6(308) (2005)
7. Reed, J.L., Vo, T.D., Schilling, C.H., Palsson, B.O.: An expanded genome-scale model of escherichia coli k-12 (ijr904 gsm/gpr). Genome Biology 4(9), R54.1–R54.12 (2003)
8. Rocha, I., Maia, P., Evangelista, P., Vilaça, P., Soares, S., Pinto, J.P., Nielsen, J., Patil, K.R., Ferreira, E.C., Rocha, M.: Optflux: an open-source software platform for in silico metabolic engineering. BMC Systems Biology 4(45) (2010)
9. Rocha, M., Maia, P., Mendes, R., Pinto, J.P., Ferreira, E.C., Nielsen, J., Patil, K.R., Rocha, I.: Natural computation meta-heuristics for the in silico optimization of microbial strains. BMC Bioinformatics 9 (2008)

Wrapper- and Ensemble-Based Feature Subset Selection Methods for Biomarker Discovery in Targeted Metabolomics

Holger Franken[1], Rainer Lehmann[2,3], Hans-Ulrich Häring[2,3], Andreas Fritsche[2,3], Norbert Stefan[2,3], and Andreas Zell[1]

[1] Center for Bioinformatics (ZBIT), University of Tübingen,
D-72076 Tübingen, Germany
holger.franken@uni-tuebingen.de
[2] Division of Clinical Chemistry and Pathobiochemistry (Central Laboratory),
University Hospital Tübingen, D-72076 Tübingen, Germany
[3] Paul-Langerhans-Institute Tübingen,
Member of the German Centre for Diabetes Research (DZD),
Eberhard Karls University Tübingen, Tübingen, Germany

Abstract. The discovery of markers allowing for accurate classification of metabolically very similar proband groups constitutes a challenging problem. We apply several search heuristics combined with different classifier types to targeted metabolomics data to identify compound subsets that classify plasma samples of insulin sensitive and -resistant subjects, both suffering from non-alcoholic fatty liver disease. Additionally, we integrate these methods into an ensemble and screen selected subsets for common features. We investigate, which methods appear the most suitable for the task, and test feature subsets for robustness and reproducibility. Furthermore, we consider the predictive potential of different compound classes. We find that classifiers fail in discriminating the non-selected data accurately, but benefit considerably from feature subset selection. Especially, a Pareto-based multi-objective genetic algorithm detects highly discriminative subsets and outperforms widely used heuristics. When transferred to new data, feature sets assembled by the ensemble approach show greater robustness than those selected by single methods.

1 Introduction

1.1 Background

Non-alcoholic fatty liver disease (NAFLD) is associated with insulin resistance, but can also be detected in insulin sensitive subjects. Insulin resistant individuals with NAFLD have a very high risk of developing type 2 diabetes (T2D) at an early stage and therefore require a timely initiation of ongoing, preventive intervention and drug therapy. In contrast, it is currently assumed that insulin sensitive people with NAFLD have a low risk of developing T2D, i.e., metabolic

M. Loog et al. (Eds.): PRIB 2011, LNBI 7036, pp. 121–132, 2011.

control at longer intervals is sufficient [17,21]. In this work, we examine different methods for the discovery of novel metabolite biomarkers to discriminate benign versus malign fatty liver in prediabetic subjects.

Biomarkers are defined as a characteristic that is objectively measured and evaluated as an indicator of normal biological processes, pathogenic processes, or pharmacologic responses to a therapeutic intervention [1]. For the discovery of novel biomarkers, the young discipline of metabolomics has received increased attention in recent years. It measures small molecules or metabolites contained in cells, tissues, or fluids involved in metabolism to reveal information about physiological processes. These processes may be influenced by both, genetic predisposition and environmental factors such as nutrition, exercise or medication [5]. Modern high-throughput techniques are capable of performing great numbers of measurements to produce datasets which stand up to statistical scrutiny. At the same time, the amounts of data generated are too voluminous to be interpreted by hand and therefore require dimensionality reduction. Thus, data mining and bioinformatics techniques are essential to identify and verify markers that are biochemically interpretable and biologically relevant. In contrast to projection- or compression-based methods for dimensionality reduction, like Principal Component Analysis (PCA) or the use of information theory, feature selection methods select a subset of variables instead of altering them. Thus, they preserve the original meaning of the variables and facilitate interpretation by a domain expert [19].

Targeted profiling schemes are used to quantitatively screen for known compounds, which depict relevant metabolic pathways of the investigated conditions. In such an approach, features reflect calculated concentrations of predefined metabolites. Data mining techniques are affected by factors such as noise, redundancy, and relevance in the experimental data. Feature selection is therefore focused on the process of identifying and removing as much irrelevant or redundant information as possible [15]. In this context, feature selection techniques can be organized into filter and wrapper methods.

Filter methods, in most cases, compute a feature relevance score and discard low-scoring features. The remaining subset of features serves as input to a classification algorithm. Filter techniques easily scale to very high-dimensional datasets, are computationally fast and are independent of the classification algorithm. A common disadvantage of filter methods is that most proposed techniques are univariate; i.e., each feature is considered separately, ignoring feature dependencies [19].

In wrapper methods, a search procedure generates and evaluates various subsets of features. A classification algorithm is trained and tested to evaluate specific subsets. To search the space of all feature subsets, the search algorithm is *wrapped* around the classification model. This problem is NP-hard, as the space of feature subsets grows exponentially with the number of features. Heuristic methods are therefore used to search for an optimal subset. Advantages of wrapper approaches include the interaction between feature subset search and model selection and the ability to take feature dependencies into account. These methods have demonstrated their utility in various studies [19,12].

But the challenge in the search for biomarkers is not only to identify highly discriminative feature sets from a single data set. Subsets that are robust in the sense that they yield good results in the classification of new data sets, are particularly interesting. We assume that features that often contribute to highly predictive subsets independent of the method used for their selection can form a particularly robust pattern. Thus we applied a voting scheme (see Sec. 2.3) that integrates the single selection procedures into an ensemble.

1.2   State of the Art

Large numbers of features and limited sample sizes are typical drawbacks in the search for biomarkers in human studies. These give rise to hardly classifiable data sets that necessitate dimensionality reduction. Therefore, multivariate approaches such as PCA or partial-least-squares discriminant analysis (PLS-DA) are common. These techniques represent the extracted information as a set of new variables called components [7]. As these variables are not part of the original data set such methods lack in interpretability. Machine-learning algorithms are a more recent class of multivariate analysis techniques. Besides their desirable characteristic of preserving the original variables, they have demonstrated superior predictive accuracy than PLS-DA and PCA [14].

Several heuristic search algorithms have previously been applied as wrappers. Genetic algorithms have been shown to be a good choice for finding small feature subsets with high discriminatory power [23]. Furthermore, a modified form of SVMs was applied, and the unified maximum separability analysis (UMSA) algorithm was introduced for proteomic profiling [8,11]. Ressom et. al. proposed particle swarm optimization (PSO) for biomarker selection in mass spectrometric profiles of peptides and proteins [18].

In this work, we apply several search heuristics combined with various types of classifiers. The goal is to find feature subsets in different targeted metabolomics data sets that are able to classify plasma samples of insulin sensitive and insulin resistant subjects, both suffering from non-alcoholic fatty liver disease. This data set holds additional challenges as the group of sample donors was particularly singled out to be very homogeneous.

2   Methods

2.1   Plasma Samples

Plasma samples of 40 adults with NAFLD (20 insulin sensitive and 20 -resistant subjects) were analysed by a targeted metabolomics approach. All individuals were intensively phenotyped as part of the Tübingen Lifestyle Intervention Program (TULIP) and considered healthy according to physical examination and routine laboratory tests. Plasma samples were taken before and after a nine month lifestyle intervention including dietary counseling and increased physical activity. In this work, *baseline* and *follow-up* denote samples and data acquired before and after the lifestyle intervention, respectively.

2.2   Data Acquisition

Biocrates (Innsbruck, Austria) measured the concentrations of 247 compounds in EDTA-plasma by targeted IDQ. This targeted metabolomics analytical platform combines flow injection, liquid chromatography, and gas chromatography mass spectrometric approaches. The applied instruments were an API 4000 QTrap tandem MS, a 7890 GC, and a 5795 MSD (Agilent, Waldbronn, Germany).

We considered a measured concentration to be reliable if it exceeds the noise level by at least a factor of three. To ensure validity we only considered metabolites that contained at least 70 % reliable measurements. This restriction led to an exclusion of 69 metabolites from the data set, leaving 178 compounds for the data analyses (21 amino acids, 21 acylcarnitines, 5 bile acids, 37 free fatty acids, 15 sphingomyelins, 70 phosphatidylcholines, 9 lysophosphatidylcholines).

2.3   Feature Subset Selection

To perform feature selection we implemented a modular Java^{TM} software environment. This environment integrates the optimization framework EvA2 [9] and classification algorithms implemented in WEKA [4].

EvA2 is a comprehensive metaheuristic optimization framework with emphasis on Evolutionary Algorithms (EA) implemented in Java. It integrates several derivative-free, preferably population-based, optimization methods. From these methods, we applied the following algorithms as wrappers:

- a standard Genetic Algorithm (GA) [6]
- a Multi Objective Genetic Algorithm (MOGA) [3]
- the Cluster-Based Niching EA (CBN-EA) [22]
- Population-Based Incremental Learning (PBIL) [10]
- Hill Climbing (HC) [16]
- Monte Carlo Search (MCS)

The MOGA applies a Pareto-based ranking scheme to simultaneously optimize two objectives: the first is to maximize the classification performance, the second is to minimize the number of features contained in the subset.

From WEKA, we applied different classifier types:

- k-nearest-neighbor (kNN), an instance-based learner, which compares each new instance to existing ones using a distance metric
- K*NN, another instance-based method using an entropic distance measure [2]
- Naive Bayes, a classifier based on Bayes' formula for conditional probabilities, depending on the assumption that attributes are independent
- the J4.8 decision tree, a reimplementation of the C4.5 decision tree which has been shown to have a very good combination of error rate and speed [13]

- random forest, a metalearner which baggs ensembles of random trees
- linear SVM, a maximum-margin based classifier we chose due to its amenities concerning interpretability compared to nonlinear models [20]

While we are interested in features, that provide good classification performance, this is not the sole criterion of our search. We also want our selected features to be interpretable by a domain expert. Hence, we focused on classification methods that allow conclusions to be drawn about the involved features.

The modular structure of our software environment allows us to combine any of the aforementioned search strategies with an arbitrary classification method and thereby design a multitude of wrapper-based feature subset selection procedures. Each procedure performs a search for a good subset using the selected classifier to evaluate feature subsets.

For the evaluation of each feature subset, we performed a stratified nested cross validation consisting of an inner loop for model selection of optimal parameters and an outer loop for external validation. The optimal parameter combination was determined in the inner loop within a two-fold cross-validation. The performance of the selected parameters was then evaluated in the outer loop using three-fold cross-validation according to the area under the receiver operating characteristic curve (AUC). We carried out this validation scheme five times to avoid bias induced by the random number generator and computed the average AUC.

We combined each of the listed search strategies from EvA2 with each classifier to determine the feature subset that produces the best classification performance. Thereby we also assessed the question of the optimal wrapper-classifier combination for this task.

Additionally, we regarded the tested wrapper-classifier combinations as an ensemble and screened their selected subsets for common features. We implemented a voting scheme to extract features that were selected by many of the search procedures assuming that these form a particularly robust pattern. For this purpose, we compared all feature subsets selected by the individual procedures by their average AUC. Those with an AUC less than the median of all subsets were discarded. Thus, we only incorporate feature subsets that show a comparatively high predictive potential. Each of these subsets votes for the features it contains. We weight the votes by the average AUC of the voting subset and sum them into a voting score for every feature. Hence, votes from highly predictive subsets have a greater impact than others. From these voting scores we calculated the 90 % quantile as a lower bound. All features that received a voting score exceeding that bound were included in the consensus feature subset.

2.4   Experimental Design

From the measurements described in Sec. 2.2, we generated several different data sets. Firstly, we formed a data set from each of the listed compound classes individually. Secondly, we integrated all metabolites into one dataset to select subsets containing features from all groups under consideration. We performed

a)
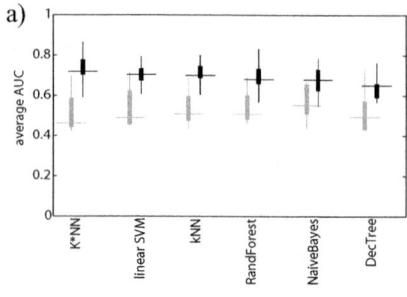

b)
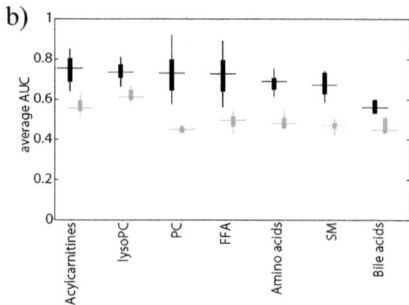

Fig. 1. Comparison of a) the classification performances of the applied wrappers and b) the predictive potential of the compound classes before (grey) and after (black) feature subset selection

this approach for the baseline as well as for the follow-up data. All data sets were mean-centered and scaled to unit variance. We applied each of the feature subset selection procedures described in Sec. 2.3 to these data sets to find subsets that accurately discriminate between insulin sensitive and insulin resistant subjects.

We analysed the results from different perspectives. The primary question is whether feature subset selection enhances classification performance at all. From this the question arises whether the applied procedures differ in their performance and which of them appears the most suitable for the task. Furthermore, we test our assumption that features that are selected by many different procedures form a particularly robust pattern. From the biological perspective we further consider the predicitve potential of the individual compound classes. We investigate whether these differ in their discriminative power and what impact the lifestyle intervention has on the different compound classes.

3   Results and Discussion

3.1   Benefit of Feature Subset Selection

First, we addressed the question whether the applied classification algorithms benefit from feature subset selection in terms of their classification accuracies. In Fig. 1a), grey boxes indicate the distribution of average AUCs each classifier achieved when applied to the complete data sets. Black boxes display the average AUCs when applied to the best selected feature subset, respectively. On each box, the central mark is the median, the edges of the box are the 25th and 75th percentiles, the whiskers extend to the most extreme data points not considered outliers, and outliers are plotted individually. Fig. 1b) shows the prediction accuracies based on the individual data sets. Again, the distributions of the classification results across all applied classifiers on the complete data sets are given by the grey boxes and black boxes illustrate the corresponding distributions of AUCs based on the best selected feature subsets from each data set. Both figures demonstrate that, in general, classification performances without

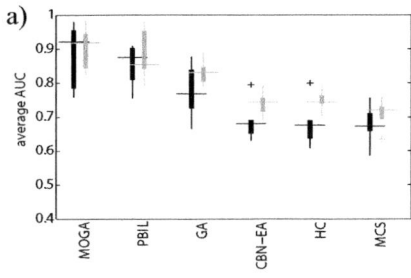 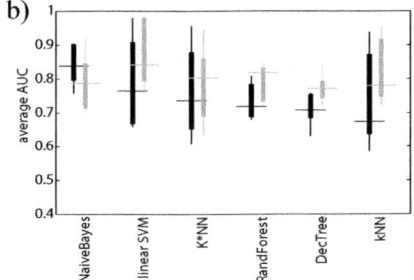

Fig. 2. Prediction accuracies achieved by the applied a) wrappers and b) classifiers on the baseline (black) and follow-up (grey) data sets

feature subset selection are close to an AUC of 0.5 and were significantly improved by the reduction of irrelevant information. All classifiers similarly benefit from the use of feature subset selection.

The individual compound classes differ in their gain in predictive potential. On the one hand, lysoPCs already classify the samples fairly well without feature subset selection and show a slight increase. On the other hand, phosphatidylcholines (PC), which show hardly any classification performance without feature subset selection, strongly benefit and even contain the most discriminative feature subsets afterwards.

3.2   Performance of Feature Subset Selection Procedures

We tested all feature subset selection procedures described in Sec. 2.3 on each individual compound class data set. The most discriminative subset selected from the baseline data consists of 10 phosphatidylcholines (PC). It yields an average AUC of 0.92 and was selected using MOGA and the K*NN classifier. Using the follow-up data sets, the best performing subset from a single compound class was again selected from the phosphatidylcholines. It also contains 10 metabolites and achieved an average AUC of 0.93. The employed classifier in this case was the linear SVM also wrapped by MOGA. The most discriminative subsets from the baseline and the follow-up data sets were selected from the same compound class, respectively, but they do not overlap with regard to individual compounds. Analogously, we analysed the data sets consisting of all considered metabolites. From the baseline data set, a subset of 19 compounds yields the best classification performance with an average AUC of 0.980. From the follow-up data, the most predicitve feature subset includes 23 metabolites and has an average AUC of 0.984. Both subsets were selected using MOGA as the wrapper and the linear SVM as the classifier. In this case, we found three compounds that were in both subsets.

In Fig. 2a), we compare the performance of the applied optimization heuristics. For each heuristic, it shows the distributions of average AUCs across all classifiers achieved by the best feature subset selected from all metabolites. Black

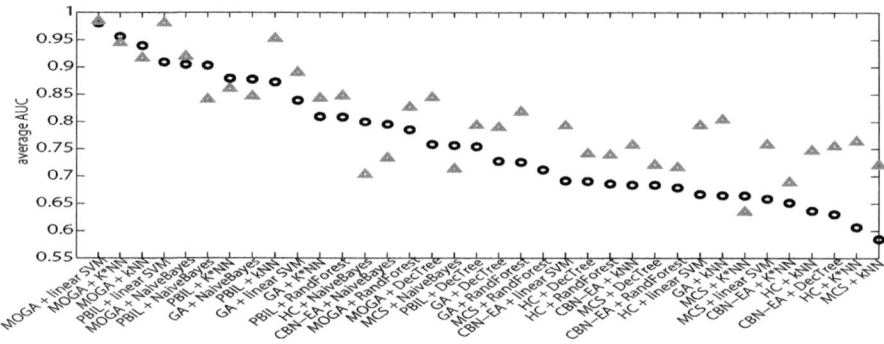

Fig. 3. Prediction accuracies achieved by the applied wrapper-classifier combinations on the baseline (black circles) and follow-up (grey triangles) datasets

boxes represent results obtained from the baseline data set and grey boxes depict results from the data acquired after lifestyle intervention. Entries are arranged in descending order of the median results per wrapper on the baseline data. MOGA and PBIL achieved the best results on both the baseline and the follow-up data sets. In both cases, they are followed by the GA. CBN-EA, HC and MCS achieve comparable results, which are considerably worse than those of MOGA and PBIL on both data sets.

Fig. 2b) displays a comparison of the classifiers applied to evaluate the tested feature subsets. For each classifier it shows the distribution of average AUCs across all wrappers achieved by the best feature subset selected from all metabolites. Again, black boxes depict results obtained from baseline data and grey boxes represent results from follow-up data. They are arranged in descending order of the median results per classifier on the baseline data. The results scatter strongly for most of the considered classifiers. The Naive Bayes classifier performs well on the baseline data, though the overall best feature set was selected using the linear SVM. Naive Bayes performs noticeably worse on the follow-up data sets. In this case the linear SVM ranks highest in terms of the median performance and overall best subset. The AUCs obtained by the J4.8 decision tree display the smallest variance, but no highly discriminative feature subsets could be selected using this classifier. The random forest shows a very similar behaviour, though its median performance is slightly better. However, a ranking of the performed classification algorithms is not revealing, due to the great variance in their results.

Fig. 3 gives an overview of the results achieved by each applied combination of classification algorithm and search heuristic. The presented average AUCs were obtained from the data sets including all considered compounds acquired before (black circles) and after lifestyle intervention (grey triangles). The data is sorted in descending order of the baseline results. The top three classification results were achieved using MOGA as the employed search heuristic. Among the top five classification results, MOGA can be found four times, followed by

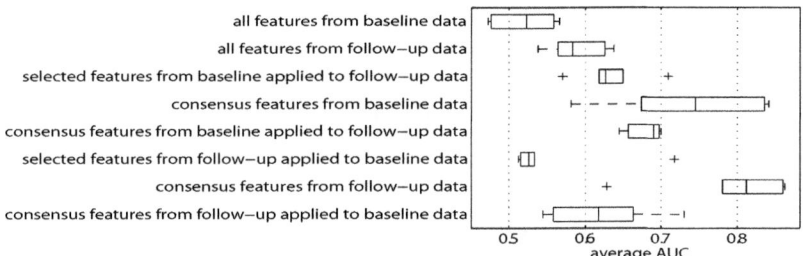

Fig. 4. Performance comparison of consensus feature sets with selected feature sets and complete data sets

PBIL, which appears four times among the top ten results. All feature sets selected using either of these heuristics are placed within the upper half of this arrangement. As previously mentioned, the results concerning the classifiers are not as clear. However, it can be observed that the achieved results of the applied combinations of wrappers and classifiers show significant correlation between baseline and follow-up data sets despite the scatter in the classifier performances (Spearman's $\rho = 0.73$, $p = 3.7e - 7$). Figures 2 and 3 show that, in the applied configurations, the choice of the search heuristic has a greater influence on the achieved results than the classifier applied for the evaluation of feature sets.

From the considered wrappers, MOGA delivered the most discriminative feature subsets and outperformed widely used heuristics such as the standard GA. This is an interesting finding because the MOGA has to deal with a tradeoff between the classifiaction accuracy and the size of the selected feature sets, whereas the GA is focused only on the accuracy. A possible explanation for this outcome might be that the said tradeoff prevents the MOGA from getting trapped in local optima and therefore it directs the search trajectory towards more generalizable solutions. This observation requires further investigation.

3.3   Ensemble Method

As described in Sec. 2.3, we assembled consensus feature sets from metabolites that were contained in many highly predictive subsets. From the baseline and follow-up data we obtained consensus sets that each contain 18 features. Seven features are in both of them. We tested the consensus sets for their predictive potential by using them as inputs for all of the applied classification algorithms. To measure their benefit, we compare the achieved classification results to those obtained using the data sets containing all metabolites. To evaluate the robustness of the consensus subsets, we applied the set that was assembled from the baseline analyses to the follow-up data and vice versa. Fig. 4 presents the results of these comparisons. In this figure, the term *selected features* refers to the best feature sets selected from the corresponding data as described in Sec. 3.2, *consensus features* denotes the consensus feature set as characterized in Sec. 2.3, respectively, and *all features* stands for the complete data sets.

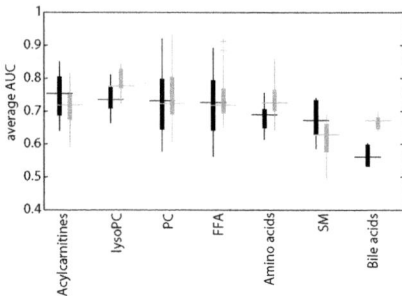

Fig. 5. Discriminative potential of compound classes at baseline (black) and follow-up (grey)

With regard to the benefit of the consensus feature sets, the results show that in both cases these sets have a significantly greater discriminative potential than the corresponding complete data sets. Furthermore, they also achieve significantly better results than the complete data sets, when applied to the data they were not derived from. In contrast, the *selected* feature sets did not perform considerably better than the complete data sets on the data, they were not selected from. The consensus feature sets do not quite attain the discriminative power of the best subsets selected by the single procedures. But when transferred from the baseline to the follow-up data sets or vice versa, the consensus sets prove to be more robust than those selected by single methods and reproducible results are obtained. This is also indicated by the finding, that the consensus feature sets obtained from baseline and follow-up data overlap in 7 out of 18 metabolites whereas the best subsets selected by single procedures, overlap by only 3 out of 19 and 23 compounds, respectively. The best subsets from the individual compound class data sets do not overlap at all.

3.4   Discriminative Potential of Compound Classes

Fig. 5 shows the distributions of classification performances that were achieved based on the individual compound class data sets (see Sec. 2.4) using all applied wrappers and classifiers. Black and grey boxes represent results from the baseline and the follow-up data, respectively. The entries are arranged in descending order by the median results from the baseline data sets. With the exception of the bile acids, no considerable differences exist in the results from the baseline data sets, with regard to the medians. In terms of the best selected feature subsets, however, the differences between the compound classes are more distinct. Highly discriminative subsets could be selected from the phosphatidylcholines (PC) and the free fatty acids (FFA) in the baseline and follow-up data. In general, the differences between the median results are greater at follow-up. LysoPCs, amino acids and bile acids show an increase in their discriminative power, whereas acylcarnitines and sphingomyelins decrease.

The selected feature subsets and actual metabolites will be analysed and discussed in more detail from the metabolic point of view in a separate publication.

4   Conclusions

The results demonstrate that highly discriminative subsets of interpretable features can be selected from otherwise hardly classifiable metabolomics data using a wrapper approach. Generally, in such an approach the choice of the wrapper has greater influence on the results than the applied classifier. It turned out that the J4.8 decision tree and random forests are not a good choice for the task, though. As a search strategy the Pareto-based MOGA proved to perform especially well and might be a useful tool in similar issues. With respect to transferability, however, the comparison and combination of several independently generated individual solutions revealed robust and reproducible subsets, which might become starting points for further investigation and test development for the early diagnosis of malign fatty liver in prediabetic subjects.

Acknowledgements. This investigation was supported in parts by the Kompetenznetz Diabetes mellitus (Competence Network for Diabetes mellitus) funded by the Federal Ministry of Education and Research (FKZ 01GI0803-04) and a grant from the German Federal Ministry of Education and Research to the German Center for Diabetes Research (DZD eV).

References

1. Atkinson, A., Colburn, W., DeGruttola, V., DeMets, D., Downing, G., Hoth, D., Oates, J., Peck, C., Schooley, R., Spilker, B., et al.: Biomarkers and surrogate endpoints: Preferred definitions and conceptual framework. Clinical Pharmacology & Therapeutics 69(3), 89–95 (2001)
2. Cleary, J., Trigg, L.: K*: An Instance-based Learner Using an Entropic Distance Measure. In: Proceedings of the 12th International Conference on Machine Learning (1995)
3. Deb, K., Pratap, A., Agarwal, S., Meyarivan, T.: A fast and elitist multiobjective genetic algorithm: NSGA-II. IEEE Transactions on Evolutionary Computation 6(2), 182–197 (2002)
4. Hall, M., Frank, E., Holmes, G., Pfahringer, B., Reutemann, P., Witten, I.: The WEKA data mining software: an update. ACM SIGKDD Explorations Newsletter 11(1), 10–18 (2009)
5. Harrigan, G., Goodacre, R.: Metabolic profiling: its role in biomarker discovery and gene function analysis. Springer, Netherlands (2003)
6. Holland, J.: Adaptation in natural and artificial systems. University of Michigan Press, Ann Arbor (1975)
7. Holmes, E., Nicholson, J., Nicholls, A., Lindon, J., Connor, S., Polley, S., Connelly, J.: The identification of novel biomarkers of renal toxicity using automatic data reduction techniques and PCA of proton NMR spectra of urine. Chemometrics and Intelligent Laboratory Systems 44(1), 245–255 (1998)

8. Koopmann, J., Zhang, Z., White, N., Rosenzweig, J., Fedarko, N., Jagannath, S., Canto, M., Yeo, C., Chan, D., Goggins, M.: Serum diagnosis of pancreatic adenocarcinoma using surface-enhanced laser desorption and ionization mass spectrometry. Clinical Cancer Research 10(3), 860 (2004)

9. Kronfeld, M., Planatscher, H., Zell, A.: The EvA2 optimization framework. In: Blum, C., Battiti, R. (eds.) LION 4. LNCS, vol. 6073, pp. 247–250. Springer, Heidelberg (2010)

10. Larranaga, P., Lozano, J.: Estimation of distribution algorithms: A new tool for evolutionary computation, vol. 2. Springer, Netherlands (2002)

11. Li, J., Zhang, Z., Rosenzweig, J., Wang, Y., Chan, D.: Proteomics and bioinformatics approaches for identification of serum biomarkers to detect breast cancer. Clinical Chemistry 48(8), 1296 (2002)

12. Li, T., Zhang, C., Ogihara, M.: A comparative study of feature selection and multiclass classification methods for tissue classification based on gene expression. Bioinformatics (Oxford, England) 20(15), 2429–2437 (2004)

13. Lim, T.: A Comparison of Prediction Accuracy, Complexity, and Training Time of Thirty-Three Old and New Classification Algorithms. Machine Learning 40, 203–228 (2000)

14. Mahadevan, S., Shah, S.L., Marrie, T.J., Slupsky, C.M.: Analysis of metabolomic data using support vector machines. Analytical Chemistry 80(19), 7562–7570 (2008)

15. Masseglia, F., Poncelet, P., Teisseire, M.: Successes and new directions in data mining. Information Science Publishing (2008)

16. Mitchell, M., Holland, J., Forrest, S.: When will a genetic algorithm outperform hill climbing? Ann Arbor 1001, 48109

17. Petersen, K., Dufour, S., Befroy, D., Lehrke, M., Hendler, R., Shulman, G.: Reversal of Nonalcoholic Hepatic Steatosis, Hepatic Insulin Resistance, and Hyperglycemia by Moderate Weight Reduction in Patients With Type 2 Diabetes. Metabolism 54, 603–608 (2005)

18. Ressom, H.W., Varghese, R.S., Abdel-Hamid, M., Eissa, S.A.L., Saha, D., Goldman, L., Petricoin, E.F., Conrads, T.P., Veenstra, T.D., Loffredo, C.A., Goldman, R.: Analysis of mass spectral serum profiles for biomarker selection. Bioinformatics (Oxford, England) 21(21), 4039–4045 (2005)

19. Saeys, Y., Inza, I.N., Larrañaga, P.: A review of feature selection techniques in bioinformatics. Bioinformatics (Oxford, England) 23(19), 2507–2517 (2007)

20. Schölkopf, B., Smola, A.: Learning with Kernels: Support Vector Machines, Regularization, Optimization, and Beyond (Adaptive Computation and Machine Learning), 1st edn. The MIT Press (2001)

21. Stefan, N., Kantartzis, K., Häring, H.U.: Causes and metabolic consequences of Fatty liver. Endocrine Reviews 29(7), 939–960 (2008)

22. Streichert, F., Stein, G., Ulmer, H., Zell, A.: A clustering based niching EA for multimodal search spaces. In: Liardet, P., Collet, P., Fonlupt, C., Lutton, E., Schoenauer, M. (eds.) EA 2003. LNCS, vol. 2936, pp. 293–304. Springer, Heidelberg (2004)

23. Zou, W., Tolstikov, V.: Probing genetic algorithms for feature selection in comprehensive metabolic profiling approach. Rapid Communications in Mass Spectrometry 22(8), 1312–1324 (2008)

Ensemble Logistic Regression
for Feature Selection

Roman Zakharov and Pierre Dupont

Machine Learning Group,
ICTEAM Institute,
Université catholique de Louvain,
B-1348 Louvain-la-Neuve, Belgium
{roman.zakharov,pierre.dupont}@uclouvain.be

Abstract. This paper describes a novel feature selection algorithm embedded into logistic regression. It specifically addresses high dimensional data with few observations, which are commonly found in the biomedical domain such as microarray data. The overall objective is to optimize the predictive performance of a classifier while favoring also sparse and stable models.

Feature relevance is first estimated according to a simple t-test ranking. This initial feature relevance is treated as a feature sampling probability and a multivariate logistic regression is iteratively reestimated on subsets of randomly and non-uniformly sampled features. At each iteration, the feature sampling probability is adapted according to the predictive performance and the weights of the logistic regression. Globally, the proposed selection method can be seen as an ensemble of logistic regression models voting jointly for the final relevance of features.

Practical experiments reported on several microarray datasets show that the proposed method offers a comparable or better stability and significantly better predictive performances than logistic regression regularized with Elastic Net. It also outperforms a selection based on Random Forests, another popular embedded feature selection from an ensemble of classifiers.

Keywords: stability of gene selection, microarray data classification, logistic regression.

1 Introduction

Logistic regression is a standard statistical technique addressing binary classification problems [5]. However logistic regression models tend to over-fit the learning sample when the number p of features, or input variables, largely exceeds the number n of samples. This is referred to as the *small n large p* setting, commonly found in biomedical problems such as gene selection from microarray data.

A typical solution to prevent over-fitting considers an l_2 norm penalty on the regression weight values, as in ridge regression [10], or an l_1 norm penalty for

M. Loog et al. (Eds.): PRIB 2011, LNBI 7036, pp. 133–144, 2011.

the (Generalized) LASSO [20,16], possibly a combination of both, as in Elastic Net [22]. The l_1 penalty has the additional advantage of forcing the solution to be *sparse*, hence performing feature selection jointly with the classifier estimation.

Feature selection aims at improving the interpretability of the classifiers, tends to reduce the computational complexity when predicting the class of new observations and may sometimes improve the predictive performances [8,17]. The feature selection obtained with a LASSO type penalty is however typically unstable in the sense that it can be largely affected by slight modifications of the learning sample (*e.g.* by adding or removing a few observations). The stability of feature selection has received a recent attention [12,1] and the interested reader is referred to a comparative study of various selection methods over a number of high-dimensional datasets [11].

In this paper, we propose a novel approach to perform feature (*e.g.* gene) selection jointly with the estimation of a binary classifier. The overall objective is to optimize the predictive performance of the classifier while favoring at the same time sparse and stable models. The proposed technique is essentially an embedded approach [8] relying on logistic regression. This classifier is chosen because, if well regularized, it tends to offer good predictive performances and its probabilistic output helps assigning a confidence level to the predicted class.

The proposed approach nonetheless starts from a t-test ranking method as a first guess on feature relevance. Such a simple univariate selection ignores the dependence between features [17] but generally offers a stable selection. This initial feature relevance is treated as a feature sampling probability and a multivariate logistic regression model is iteratively reestimated on subsets of randomly, and non-uniformly, sampled features. The number of features sampled at each iteration is constrained to be equal to the number of samples. Such a constraint enforces the desired sparsity of the model without resorting on a l_1 penalty. At each iteration, the sampling probability of any feature used is adapted according to the predictive performance of the current logistic regression. Such a procedure follows the spirit of wrapper methods, where the classifier performance drives the search of selected features. However it is used here in a smoother fashion by increasing or decreasing the probability of sampling a feature in subsequent iterations. The amplitude of the update of the sampling probability of a feature also depends on its absolute weight in the logistic regression. Globally, this feature selection approach can be seen as an ensemble learning made of a committee of logistic regression models voting jointly for the final relevance of each feature.

Regularized logistic regression methods are briefly reviewed in section 2.1. Section 2.2 further details our proposed method for feature selection. Practical experiments of gene selection from various microarrays datasets, described in section 3, illustrate the benefits of the proposed approach. In particular, our method offers significantly better predictive performances than logistic regression models regularized with Elastic Net. It also outperforms a selection with Random Forests, another popular ensemble learning approach, both in terms of predictive performance and stability. We conclude and present our future work in section 4.

2   Feature Selection Methods

Ensemble learning has been initially proposed to combine learner decisions, which aggregation produces a single regression value or class label [7]. The idea of ensemble learning has also been extended to feature selection [8]. Approaches along those lines include the definition of a feature relevance from Random Forests [3] or the aggregation of various feature rankings obtained from a SVM-based classifier [1]. Those approaches rely on various resamplings of the learning sample. Hence, the diversity of the ensemble is obtained by considering various subsets of training instances. We opt here for an alternative way of producing diversity, namely by sampling the feature space according to a probability distribution, which is iteratively refined to better model the relevance of each feature.

In section 2.1, we briefly review logistic regression, which serves here as the base classifier. Section 2.2 further details the proposed approach of ensemble logistic regression with feature resampling.

2.1   Regularized Logistic Regression

Let $\mathbf{x} \in \mathbf{R}^p$ denote an observation made of p feature values and let $y \in \{-1, +1\}$ denote the corresponding binary output or class label. A logistic regression models the conditional probability distribution of the class label y, given a feature vector $\mathbf{x}$ as follows.

$$\text{Prob}(y|\mathbf{x}) = \frac{1}{1 + \exp\left(-y(\mathbf{w}^T\mathbf{x} + v)\right)}, \tag{1}$$

where the weight vector $\mathbf{w} \in \mathbf{R}^p$ and intercept $v \in \mathbf{R}$ are the parameters of the logistic regression model. The equation $\mathbf{w}^T\mathbf{x} + v = 0$ defines an hyperplane in feature space, which is the decision boundary on which the conditional probability of each possible output value is equal to $\frac{1}{2}$.

We consider a supervised learning task where we have n i.i.d. training instances $\{(\mathbf{x}_i, y_i), i = 1, \ldots, n\}$. The likelihood function associated with the learning sample is $\prod_{i=1}^{n} \text{Prob}(y_i|\mathbf{x}_i)$, and the negative of the log-likelihood function divided by n, sometimes called the *average logistic loss*, is given by

$$l_{avg}(\mathbf{w}, v) = \frac{1}{n} \sum_{i=1}^{n} f\left(y_i(\mathbf{w}^T\mathbf{x}_i + v)\right), \tag{2}$$

where $f(z) = \log\left(1 + \exp(-z)\right)$ is the *logistic loss function*.

A maximum likelihood estimation of the model parameters $\mathbf{w}$ and v would be obtained by minimizing (2) with respect to the variables $\mathbf{w} \in \mathbf{R}^p$ and $v \in \mathbf{R}$. This minimization is called the *logistic regression* (LR) problem. When the number n of observations is small compared to the number p of features, a logistic regression model tends to over-fit the learning sample. When over-fitting occurs many features have large absolute weight values, and small changes of those values have a significant impact on the predicted output.

The most common way to reduce over-fitting is to add a penalty term to the loss function in order to prevent large weights. Such a penalization, also known as regularization, gives rise to the l_2-regularized LR problem:

$$\min_{\mathbf{w},v} l_{avg}(\mathbf{w}, v) + \lambda \|\mathbf{w}\|_2^2 = \min_{\mathbf{w},v} \frac{1}{n} \sum_{i=1}^{n} f\left(y_i(\mathbf{w}^T \mathbf{x}_i + v)\right) + \lambda \sum_{j=1}^{p} w_j^2. \quad (3)$$

Here $\lambda > 0$ is a regularization parameter which controls the trade-off between the loss function minimization and the size of the weight vector, measured by its l_2-norm.

As discussed in [15], the l_2-regularized LR worst case sample complexity grows at least linearly in the number of (possibly irrelevant) features. This result means that, to get good predictive performance, adding a feature to the model requires the inclusion of an additional learning example. For small n, large p problems the l_1-regularized LR is thus usually considered instead by replacing the l_2-norm $\|\cdot\|_2^2$ in (3) by the l_1-norm $\|\cdot\|_1$. This is a natural extension to the LASSO [20] for binary classification problems.

The benefit of the l_1-regularized LR is its logarithmic rather than linear sample complexity. It also produces *sparse* models, for which most weights are equal to 0, hence performing an implicit feature selection. However l_1-regularized LR is sometimes too sparse and tends to produce a highly unstable feature selection. A trade-off is to consider a mixed regularization relying on the Elastic Net penalty [22]:

$$\min_{\mathbf{w},v} l_{avg}(\mathbf{w}, v) + \lambda \sum_{j=1}^{p} \left[\frac{1}{2}(1 - \alpha)w_j^2 + \alpha|w_j| \right], \quad (4)$$

where $\alpha \in [0, 1]$ is a meta-parameter controlling the influence of each norm. For high-dimensional datasets the key control parameter is usually still the l_1 penalty, with the l_2 norm offering an additional smoothing.

We argue in this paper that there is an alternative way of obtaining sparse and stable logistic regression models. Rather than relying on a regularization including an l_1 penalty, the sparsity is obtained by constraining the model to be built on a number of features of the same order as the number of available samples. This constraint is implemented by sampling feature subsets of a prescribed size. The key ingredient of such an approach, further detailed in section 2.2, is a non-uniform sampling probability of each feature where such a probability is proportional to the estimated feature relevance.

2.2   Ensemble Logistic Regression with Feature Resampling

The proposed feature selection is essentially an embedded method relying on regularized logistic regression models. Those models are built on small subsets of the full feature space by sampling at random this space. The sampling probability is directly proportional to the estimated feature relevance. The initial relevance of each feature is estimated according to a t-test ranking. Such a simple univariate ranking does not consider the dependence between features but is observed to be stable with respect to variations of the learning sample. This initial relevance index is iteratively refined as a function of the predictive performance of regularized logistic regression models built on resampled features. This procedure iterates until convergence of the classifier performance.

Our method relies on the l_2-regularized LR as estimated by the optimization problem (3). The sparsity is not enforced here with an l_1 penalty but rather by explicitly limiting the number of features on which such a model is estimated. The sample complexity result from [15] gives us a reasonable default number of features to be equal to the number of training examples n. Those n features could be drawn uniformly from the full set of p features (with $p \gg n$) but we will show the benefits of using a non-uniform sampling probability. We propose here to relate the sampling probability of a given feature to its estimated relevance.

Since our primary application of interest is the classification of microarray data, a t-test relevance index looks to be a reasonable choice as a first guess [21,8]. This method ranks features by their normalized difference between mean expression values across classes:

$$t_j = \frac{\mu_{j+} - \mu_{j-}}{\sqrt{\sigma_{j+}^2/m_+ + \sigma_{j-}^2/m_-}}, \tag{5}$$

where μ_{j+} (respectively μ_{j-}) is the mean expression value of the feature j for the m_+ positively (respectively m_- negatively) labeled examples, and σ_{j+}, σ_{j-} are the associated standard deviations. The score vector $\mathbf{t}$ over the p features is normalized to produce a valid probability distribution vector **prob**. We note that there is no need here to correct for multiple testing since the t-test is not used to directly select features but to define an initial feature sampling probability.

At each iteration the learning sample is split into training (80%) and validation (20%) sets. Next, a subset of n features is drawn according to **prob** and a $l2$-regularized LR model is estimated on the training data restricted to those features. The resulting classifier is evaluated on the validation set according to its balanced classification rate BCR, which is the average between specificity and sensitivity (see section 3.2).

The BCR performance of the current model is compared to the average $\overline{BCR}$ (initially set to 0.5) obtained for all models built at previous iterations. The current model quality is estimated by $\log(1 + BCR - \overline{BCR})$. The relative quality of the current model is thus considered positive (resp. negative) if its performance is above (resp. below) average.

Finally, the probability vector **prob** controlling the sampling of features at the next iteration is updated according to the relative model quality and its respective weight vector $\mathbf{w}$. The objective is to favor the further sampling of important features (large weight values) whenever the current model looks good and disfavor the sampling of non-important features (small weight values) when the current model looks poor. This process is iterated until convergence of the classification performance. The net result of this algorithm summarized below is the vector **prob** which is interpreted as the final feature relevance vector.

Algorithm 1. Ensemble Logistic Regression with Feature Resampling

---

Algorithm ELR

Input: A learning sample $\mathbf{X} \in \mathbf{R}^{n \times p}$ and class labels $\mathbf{y} \in \{-1, 1\}^n$

Input: A regularization parameter λ for estimating a l2-LR model (3)

Output: A vector **prob** $\in [0, 1]^p$ of feature relevance

Initialize **prob** according to a t-test ranking
$\overline{BCR} \leftarrow 0.5$ // Default initialization of the average BCR

repeat
   | Randomly split $\mathbf{X}$ into TRAINING (80%) and VALIDATION (20%)
   | Draw n out of p features at random according to **prob**
   | $(\mathbf{w}, v) \leftarrow$ a l2-LR model M learned on TRAINING restricted to n features
   | Compute BCR of M on VALIDATION
   | $quality \leftarrow \log(1 + BCR - \overline{BCR})$
   | // Update the feature relevance vector
   | **foreach** j among the n sampled features **do**
   | | $\mathbf{prob}_j \leftarrow \frac{1}{Z} \left(\mathbf{prob}_j + quality \cdot \mathbf{w}_j{}^{2 \cdot \mathrm{sign}(quality)} \right)$
   | | // Z is the normalization constant to define a distribution
   | Update average $\overline{BCR}$
until *no significant change of* $\overline{BCR}$ *between consecutive iterations;*
return prob

---

3 Experiments

3.1 Microarray Datasets

We report here practical experiments of gene selection from 4 microarray datasets. Table 1 summarizes the main characteristics of those datasets: the number of samples, the number of features (genes) and the class ratios.

The classification task in DLBCL (diffuse large B-cells) is the prediction of the tissue type [18]. Chandran [4] and Singh [19] are two datasets related to prostate cancer and the task is to discriminate between tumor or normal samples. The Transbig dataset is part of a large breast cancer study [6]. The original data measures the time to metastasis after treatment. We approximate this task here (for the sake of considering an additional dataset) by considering a time threshold of 5 years after treatment. Such a threshold is commonly used as a critical value in breast cancer studies. The question of interest is to discriminate between patients with or without metastasis at this term, and hence reduces to a binary classification problem. We focus in particular on ER-positive/HER2-negative patients, which form the most significant sub-population as reported in [13].

Table 1. Characteristics of the microarray datasets

Dataset	Samples (n)	Features (p)	Class Priors
DLBCL	77	7129	25%/75%
Singh	102	12625	49%/51%
Chandran	104	12625	17%/83%
Transbig	116	22283	87%/13%

3.2 Evaluation Metrics

The main objective is to assess the predictive performance of a classifier built on the selected genes. The performance metric is estimated according to the *Balanced Classification Rate*:

$$BCR = \frac{1}{2}\left(\frac{TP}{P} + \frac{TN}{N}\right), \tag{6}$$

where $TP(TN)$ is the number of correctly predicted positive (negative) test examples among the P positive (N negative) test examples. BCR is preferred to the classification accuracy because microarray datasets often have unequal class prior, as illustrated in Table 1. BCR is the average between specificity and sensitivity and can be generalized to multi-class problems more easily than ROC analysis.

To further assess the quality of feature (= gene) selection methods, we evaluate the stability of the selection on k resamplings of the data. The Kuncheva index [12] measures to which extent k sets of s selected features share common features.

$$K\left(\{\mathbf{S}_1,\dots,\mathbf{S}_k\}\right) = \frac{2}{k\left(k-1\right)}\sum_{i=1}^{k-1}\sum_{j=i+1}^{k}\frac{\mid \mathbf{S}_i \cap \mathbf{S}_j \mid -\frac{s^2}{p}}{s - \frac{s^2}{p}}, \tag{7}$$

where p is the total number of features, and $\mathbf{S}_i$, $\mathbf{S}_j$ are two gene lists built from different resamplings of the data. The s^2/p term corrects a bias due to chance of selecting common features among two sets chosen at random. The Kuncheva index ranges within (-1,1] and the greater its value the larger the number of common features across the k gene lists.

3.3 Experimental Methodology

We report experimental performances of the gene selection approach introduced in section 2.2 and referred to as ELR. A simple variant, denoted by ELR_WOTT, uses a uniform distribution over the p genes to initialize the sampling probability distribution **prob**, instead of the t-test values. We also consider, as an additional baseline denoted TTEST, a direct ranking of the genes according to the t-test statistics without any further refinement (hence reducing ELR to its initialization).

A further competing approach is ENET: a gene selection based on the absolute values of the feature weights estimated from a logistic regression regularized with Elastic Net. In such an approach, the sparsity is controlled by the regularization constants λ and α (see equation (4)). We choose $\alpha = 0.2$ as in the original work on microarray classification from the authors [22], and let λ vary in the range $[10^{-6}, 10]$ to get more or fewer selected features. In contrast, in the ELR method, which uses a $l2$-regularized logistic loss, λ is fixed to a default value equal to 1. In this case, the sparsity results from the limited number of sampled features and the final result is a ranking of the full set of features according to the **prob** vector.

In contrast to the ELR method, which relies on various resamplings from the set of *features*, alternative methods use several bootstrap samples from the set of *training examples*. We report comparative results with BoRFE a bootstrap extension to RFE [1]. This method is similar to BoLASSO [2] but tailored to classification problems. Following [1], we rely on 40 bootstrap samples while discarding 20 % of features at each iteration of RFE.

Random Forests (RF) are another competing approach to define a relevance measure on genes [3]. Here, a key control parameter is the number of trees considered in the forest. Preliminary experiments (not reported here) have shown that the predictive performance obtained with RF is well stabilized with 1,000 trees. The stability of the gene selection itself can be improved by considering a larger number of trees ($\approx 5,000$) however resulting sometimes in a lower BCR. Hence we stick to $\approx 1,200$ trees on DLBCL, $\approx 2,000$ trees on Singh and Chandran, and $\approx 2,500$ trees on Transbig. Those numbers also happen to be of a similar order of magnitude as the number of iterations used by the ELR method to converge. We thus compare two different selection methods from approximately the same number of individual models. The second RF meta-parameter is the number of features selected at random when growing each node of the trees in the forest. To be consistent with the ELR method, we choose to sample n features at each node. We note that, given the characteristics of the datasets under study (see table 1), the n value tends to be close to $\sqrt{p}$, which is a common choice while estimating a RF [8].

All methods mentioned above produce ranked gene lists from expression values as measured by microarrays. Specific gene lists are obtained by thresholding such lists at a prescribed size. We report performances for various list sizes from 512 genes down to 4 genes. We consider 200 independent sub-samplings without replacement forming binary splits of each dataset into 90% training and 10% tests. The stability of the gene selection is evaluated according to the Kuncheva index over the 200 training sets. The predictive performance of classifiers built on the training sets from those genes is evaluated and averaged over the 200 test sets. To compare predictive results only influenced by the gene selection, we report the average BCR of $l2$-regularized LR classifiers, no matter which selection method is used.

3.4 Results

Figures 1 and 2 report the BCR and stability performances of the various methods tested on the 4 datasets described in table 1. The gene selection stability of the ELR method is clearly improved when using a t-test to initialize the feature sampling probability rather than a uniform distribution (ELR_WOTT). This result illustrates that a non-uniform feature sampling related to the estimated feature relevance is beneficial for the selection stability with no significant influence on the predictive performance. ELR is also more stable than RF on 3 out of 4 datasets while it offers results comparable to ENET and BoRFE in this regard.

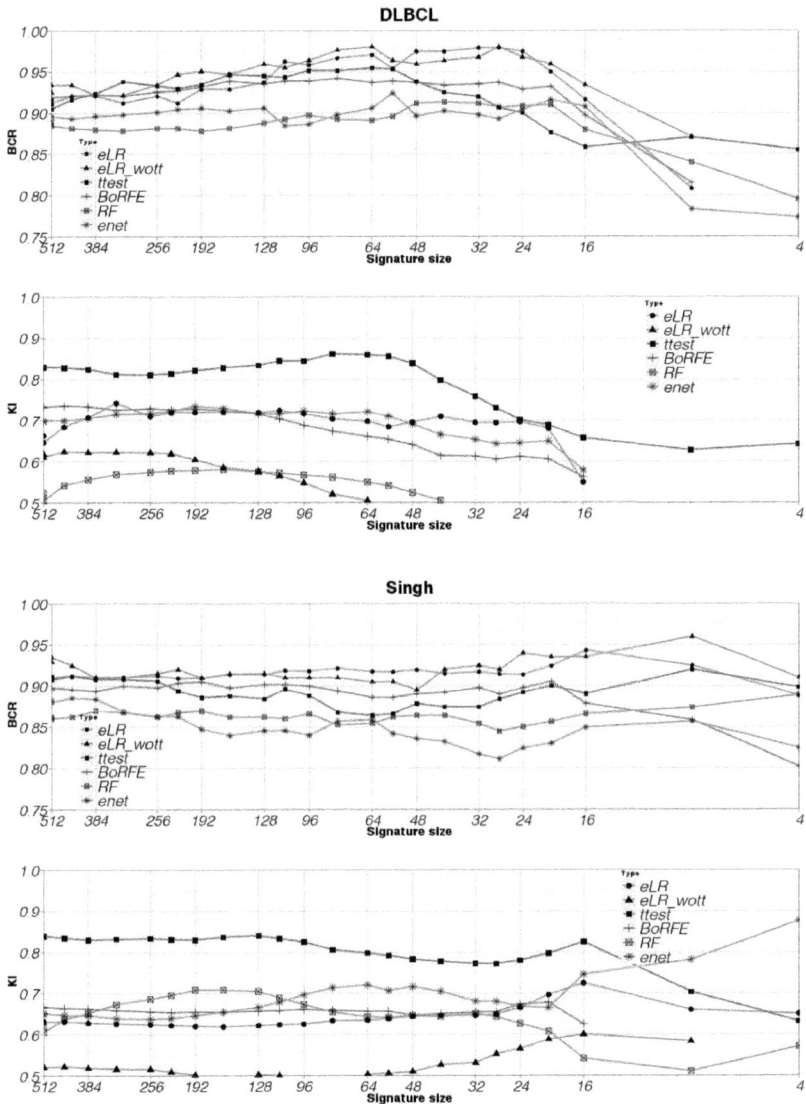

Fig. 1. Classification performance (BCR) and signature stability (Kuncheva index) of the competing methods on the **DLBCL** and **Singh** datasets

The ELR method outperforms, generally significantly, its competitors in terms of predictive performance. To support this claim, we assess the statistical significance of the differences of average BCR obtained with the ELR method and each of its competitors. We resort on the corrected resampled t-test proposed in [14] to take into account the fact that the various test folds do overlap. The significance is evaluated on the smallest signature size which show the highest BCR value: 24 genes for DLBCL, 16 genes for Singh, 28 genes for Chandran and 160 genes for Transbig. Table 2 reports the p-values of the pairwise comparisons between ELR and its competitors. Significant results (p-value < 0.05) are reported in bold in this table. ELR clearly outperforms ENET on all datasets and RF on 3 out of 4 datasets. It offers better performances than BoRFE and TTEST on all datasets, and significantly on DLBCL and Transbig respectively.

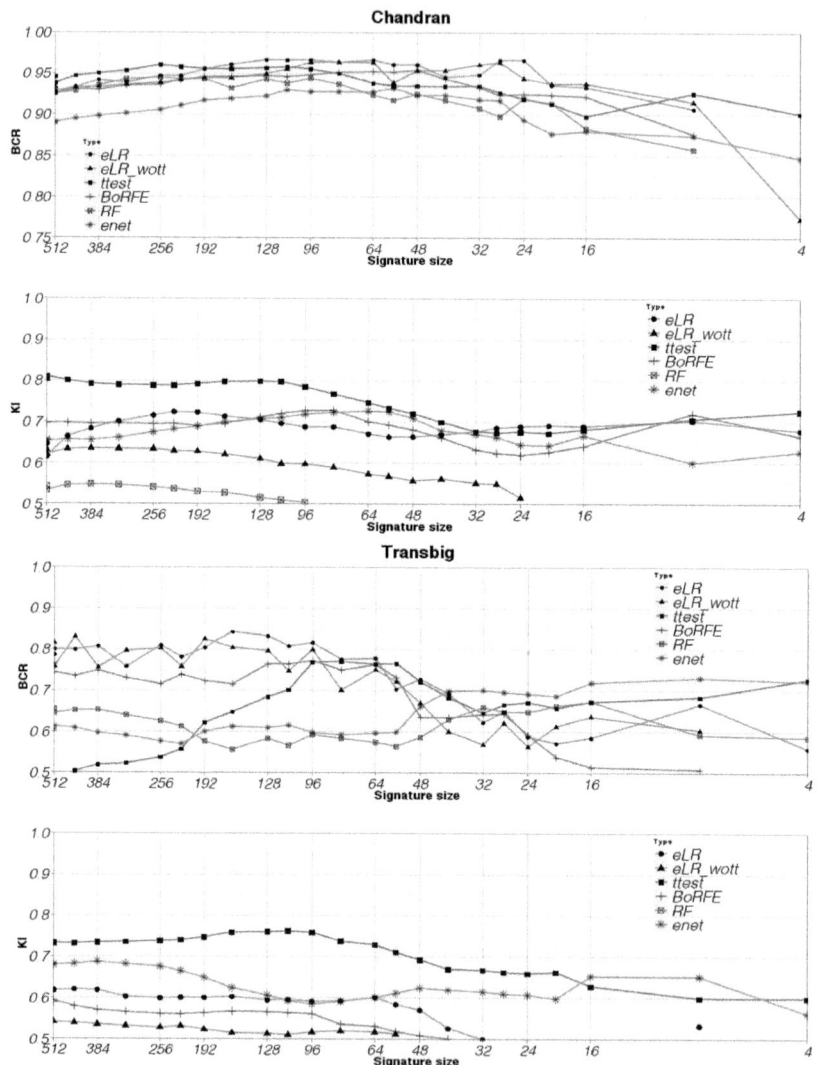

Fig. 2. Classification performance (BCR) and signature stability (Kuncheva index) of the competing methods on the **Chandran** and **Transbig** datasets

Table 2. Pairwise comparison of the average BCR obtained of ELR and its competitors. Reported results are p-values computed according to the corrected resampled t-test proposed in [14].

ELR vs.	DLBCL	Singh	Chandran	Transbig
Elastic Net	**0.039**	**0.001**	**0.043**	**0.033**
Random Forests	**0.023**	**0.007**	0.086	**0.005**
Boost. RFE	**0.042**	0.116	0.091	0.056
t-test	0.089	0.135	0.092	**0.046**
# genes	24	16	28	160

4 Conclusion and Future Work

We propose a novel feature selection method tailored to high dimensional datasets. The selection is embedded into logistic regression (LR) with non-uniform feature sampling. The sampling distribution of features is directly proportional to the estimated feature relevance. Such relevance is initialized with a standard t-test and further refined according to the predictive performance and weight values of LR models built on the sampled features. Experiments conducted on 4 microarray datasets related to the classification of tumor samples illustrate the benefits of the proposed approach in terms of predictive performance and stability of the gene selection.

Our future work includes a more formal analysis of the sampling probability update rule. On a practical viewpoint, the initial feature relevance could also be adapted according to some prior knowledge on some genes *a priori* believed to be more relevant. Such an approach would be an interesting alternative to the partially supervised gene selection method proposed in [9].

References

1. Abeel, T., Helleputte, T., Van de Peer, Y., Dupont, P., Saeys, Y.: Robust biomarker identification for cancer diagnosis with ensemble feature selection methods. Bioinformatics 26, 392–398 (2010)
2. Bach, F.R.: Bolasso: model consistent lasso estimation through the bootstrap. In: Proceedings of the 25th International Conference on Machine Learning, pp. 33–40. ACM (2008)
3. Breiman, L.: Random forests. Machine Learning 45, 5–32 (2001)
4. Chandran, U.R., Ma, C., Dhir, R., Bisceglia, M., Lyons-Weiler, M., Liang, W., Michalopoulos, G., Becich, M., Monzon, F.A.: Gene expression profiles of prostate cancer reveal involvement of multiple molecular pathways in the metastatic process. BMC Cancer 7(1), 64 (2007)
5. Cox, D.R., Snell, E.J.: Analysis of binary data. Monographs on statistics and applied probability. Chapman and Hall (1989)

6. Desmedt, C., Piette, F., Loi, S., Wang, Y., Lallemand, F., Haibe-Kains, B., Viale, G., Delorenzi, M., Zhang, Y., D'Assignies, M.S., Bergh, J., Lidereau, R., Ellis, P., Harris, A., Klijn, J., Foekens, J., Cardoso, F., Piccart, M., Buyse, M., Sotiriou, C.: Strong time dependence of the 76-gene prognostic signature for node-negative breast cancer patients in the transbig multicenter independent validation series. Clinical Cancer Research 13(11), 3207–3214 (2007)
7. Dietterich, T.G.: Ensemble methods in machine learning. In: Kittler, J., Roli, F. (eds.) MCS 2000. LNCS, vol. 1857, pp. 1–15. Springer, Heidelberg (2000)
8. Guyon, I., Gunn, S., Nikravesh, M., Zadeh, L. (eds.): Feature Extraction. Foundations and Applications. Studies in Fuzziness and Soft Computing. Physica-Verlag, Springer (2006)
9. Helleputte, T., Dupont, P.: Feature Selection by Transfer Learning with Linear Regularized Models. In: Buntine, W., Grobelnik, M., Mladenić, D., Shawe-Taylor, J. (eds.) ECML PKDD 2009. LNCS, vol. 5781, pp. 533–547. Springer, Heidelberg (2009)
10. Hoerl, A.E., Kennard, R.W.: Ridge regression: Biased estimation for nonorthogonal problems. Technometrics 12, 55–67 (1970)
11. Kalousis, A., Prados, J., Hilario, M.: Stability of feature selection algorithms: a study on high-dimensional spaces. Knowledge and Information Systems 12, 95–116 (2007), doi:10.1007/s10115-006-0040-8
12. Kuncheva, L.I.: A stability index for feature selection. In: Proceedings of the 25th International Multi-Conference Artificial Intelligence and Applications, pp. 390–395. ACTA Press, Anaheim (2007)
13. Li, Q., Eklund, A.C., Juul, N., Haibe-Kains, B., Workman, C.T., Richardson, A.L., Szallasi, Z., Swanton, C.: Minimising immunohistochemical false negative er classification using a complementary 23 gene expression signature of er status. PLoS ONE 5(12), e15031 (2010)
14. Nadeau, C., Bengio, Y.: Inference for the generalization error. Machine Learning 52, 239–281 (2003)
15. Ng, A.Y.: Feature selection, l_1 vs. l_2 regularization, and rotational invariance. In: Proceedings of the Twenty-First International Conference on Machine Learning (ICML), vol. 1, pp. 78–85 (2004)
16. Roth, V.: The generalized LASSO. IEEE Transactions on Neural Networks 15(1), 16–28 (2004)
17. Saeys, Y., Inza, I., Larrañaga, P.: A review of feature selection techniques in bioinformatics. Bioinformatics 23(19), 2507–2517 (2007)
18. Shipp, M., Ross, K., Tamayo, P., Weng, A., Kutok, J., Aguiar, R., Gaasenbeek, M., Angelo, M., Reich, M., Pinkus, G., Ray, T., Koval, M., Last, K., Norton, A., Lister, A., Mesirov, J.: Diffuse large b-cell lymphoma outcome prediction by gene-expression profiling and supervised machine learning. Nature Medicine 8, 68–74 (2002)
19. Singh, D., Febbo, P.G., Ross, K., Jackson, D.G., Manola, J., Ladd, C., Tamayo, P., Renshaw, A.A., D'Amico, A.V., Richie, J.P., Lander, E.S., Loda, M., Kantoff, P.W., Golub, T.R., Sellers, W.R.: Gene expression correlates of clinical prostate cancer behavior. Cancer Cell 1, 203–209 (2002)
20. Tibshirani, R.: Regression shrinkage and selection via the lasso. Journal of the Royal Statistical Society, Series B 58, 267–288 (1994)
21. Witten, D.M., Tibshirani, R.: A comparison of fold-change and the t-statistic for microarray data analysis. Stanford University. Technical report (2007)
22. Zou, H., Hastie, T.: Regularization and variable selection via the elastic net. Journal of the Royal Statistical Society, Series B 67, 301–320 (2005)

Gene Selection in Time-Series Gene Expression Data

Prem Raj Adhikari[1], Bimal Babu Upadhyaya[1],
Chen Meng[2], and Jaakko Hollmén[1]

[1] Aalto University School of Science,
Department of Information and Computer Science,
P.O. Box 15400, FI-00076 Aalto, Espoo, Finland
prem.adhikari@aalto.fi
[2] Royal Institute of Technology,
School of Computer Science and Communication,
Department of Computational Biology,
P.O. Box S-100 44, Stockholm, Sweden
chenm@kth.se

Abstract. The dimensionality of biological data is often very high. Feature selection can be used to tackle the problem of high dimensionality. However, majority of the work in feature selection consists of supervised feature selection methods which require class labels. The problem further escalates when the data is time–series gene expression measurements that measure the effect of external stimuli on biological system. In this paper we propose an unsupervised method for gene selection from time–series gene expression data founded on statistical significance testing and swap randomization. We perform experiments with a publicly available mouse gene expression dataset and also a human gene expression dataset describing the exposure to asbestos. The results in both datasets show a considerable decrease in number of genes.

Keywords: Feature Selection, Statistical Significance, Time–series, Randomization.

1 Introduction

Biological datasets are often high-dimensional in terms of the number of measurements whereas the number of samples is often very low. However, performance of learning algorithms is poor when data is high–dimensional and sample size is small because of curse of dimensionality. The problem of high dimensionality has inspired several concentrated research in feature selection and feature extraction [1–3]. Furthermore, biological systems are very complex systems and measurements in biology are performed using high–resolution and high–throughput techniques thus producing high–dimensional data. Hence, bioinformatics community is also highly interested in feature selection and feature extraction methods to be used in their learning algorithms [3]. Moreover, data incorporating

M. Loog et al. (Eds.): PRIB 2011, LNBI 7036, pp. 145–156, 2011.

temporal information, often referred to as time–series data in machine learning literature, increases complexity of feature selection. Each subsequent feature consists of temporal information in that time-stamp thus making feature selection a difficult task. Hence, several researchers have extensively investigated the problem of feature selection in time–series data [4]. In [5], authors study the problem of feature selection in a time-series prediction problem.

In biology, recent advancement in technology has also provided means to study the complex biological system. Micro–array technology, such as CGH (Comparative Genomic Hybridization) and aCGH (Array Comparative Genomic Hybridization) offer the facilities to study the genomes and the genes in high resolution [6]. Hence, in most biological experiments that measure genes of a sample (for e.g. a cancer patient), data generated would consist of dimensions from tens of thousands to hundreds of thousands. However, number of samples are severely restricted in biological experiments. For example, only small number of measurement samples are available from cancer patients. Hence, feature selection is inevitable in biological problems such as the analysis of genes. Considering its importance, research community has coined a new term called gene selection to describe the feature selection procedure in genes [3, 7]. Additionally, gene selection also helps in determining relevant genes for future detailed studies of gene expression by extensive and expensive methods such as qRT-PCR or further detailed experiments with specific focus on selected genes leading to bio–marker identification. In some cases, problem of bio-marker identification could be solved by feature selection alone.

Time–series data in biology usually originates from the effects of external stimuli such as drug treatment on certain biological entity such as genes in different time intervals. Difficulty of gene selection in time–series is further compounded by presence of few data points in the time–series because it is difficult to capture a trend from small number of data samples. Available gene selection methods in bioinformatics such as [7] and [8] would not be applicable in time–series data. Gene selection methods in time–series data has received considerable interest. Authors in [9] use average–periodogram to select the most discriminative genes. However, there is a graphical component in their pipeline that needs to be inspected to determine the periodic components in data thus involving human intuition. Additionally, Fisher's test is used only on the peaks but we are interested in overall changes in data not only the peaks. In [10], authors define candidate temporal profiles in terms of inequalities among mean expression levels at the time points and select the genes that meet a bootstrap-based criterion for statistical significance and assign each selected gene to the best fitting candidate profile. However, the algorithm requires specifying the candidate profile which is often an arduous task. Similarly, authors in [11] use ranking of genes, HMMs and cross–validation to select genes. However, their method uses backward selection which is computationally very inefficient for large number of genes. Moreover, all of these methods are supervised and require a class label to select the features. Similar to other application areas, biology also has difficulty in getting the class label.

In wrapper settings, gene selection requires a modelling technique before the genes are selected and if the model selection procedure is problematic, the gene selection results are not reliable. Moreover, traditional feature selection algorithms such as forward/backward selection requires modelling of each gene and their combination, which is computationally expensive and inefficient. Feature extraction methods such as Principal Component Analysis (PCA) and Independent Component Analysis (ICA) [12] are not suitable because these methods can not determine most discriminative set of genes instead they transform the entire data and extract new set of features. Determining set of most discriminative genes is important because often these analysis of micro–array data is used as a guide to further precise studies of gene expression by extensive and expensive methods such as qRT-PCR. Additionally, the available gene selection methods are complicated and difficult to implement. They also require considerable amount of prior knowledge and a number of parameters.

This paper proposes an unsupervised method founded on statistical significance of gene expression and swap randomization for gene selection. The proposed algorithm is based on existing methods but novelty in this contribution is the construction of the pipeline combining the available methods with application to gene selection. The selection of biologically motivated special test statistic of longest run of consecutive 1s improves the selection of discriminative genes which is also a novel contribution of this paper. We consider the problem of gene selection as a series of hypothesis testing problems; first in determining if the gene is expressed at any given time instance and second when we select the final set of genes using swap randomization. The methods used in the pipeline are easy to implement individually but the implementations are also freely available. We experiment our method on two time-series gene expression datasets and obtain considerable reduction in number of genes.

2   Time–Series Gene Expression Dataset

We performed experiments on two different time–series gene expression datasets. One of them on asbestos exposure was the data presented in [13] and the other dataset on T cell differentiation is publicly available from NCBI[1] database [14]. The T cell differentiation dataset is an Affymetrix mouse genome array and mouse T cell line CTLL-2 data. It describes the expression from Murine T cell in response to Interleukin-2 at different time points in 24 hours after Interleukin-2 treatment. This dataset is used in [14] to analyze the gene expression profiles of Interleukin-2 regulated genes over ten different time points. There are 45101 genes measured in ten different time points. Control for every time point in time-series is not available. Hence, a random sample not treated with Interleukin-2 is taken as the control group.

Asbestos exposure dataset [13] is also a micro–array data that describes the effect of drug crocidolite exposed through an asbestos on the gene expression of humans. It represents the evolution of gene expression levels over time for both

---

[1] http://www.ncbi.nlm.nih.gov/geo/query/acc.cgi?acc=GSE6085

asbestos exposed and non–exposed samples of A549 Caucasian lung carcinoma. There were 27 micro–arrays and each array has 54675 probe sets; 38500 of which corresponds to 38500 well established genes. Hence, there are 54675 samples of data measured in 5 different time intervals. Unlike T cell differentiation dataset, control is available for every five data points in the time–series.

3   Methodology of Pipeline

We had access to two datasets as discussed in Section 2 which were already pre-processed using image processing techniques. Quality control was also already performed on the dataset. Our pipeline proceeds using the methodology as shown in Figure 1 starting with normalization of data and ending with the swap randomization to determine the statistically significant genes. These statistically significant genes are the final result of gene selection.

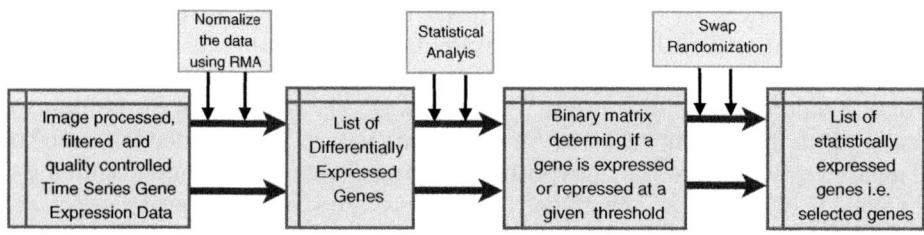

Fig. 1. The pipeline of the methodology. It starts with generic micro–array processing techniques and completes with swap randomization which enlists the statistically significant genes in the gene expression time-series. These statistically significant genes are the selected genes by our pipeline.

3.1   Normalization of Micro–Array Data

Normalization is the first transformation method applied to the image processed data from the micro–array experiments. Normalization adjusts the hybridization intensities such that they are suitable for biological comparisons [15]. It is important to adjust unequal amounts of initial RNA, to account for differences in labeling or detection efficiencies of fluorescence dyes used, and remove the bias of the system arising during the measurement [15]. There are different methods of normalization, such as lo(w)ess normalization, and intensity based filtering of array elements. The choice of normalization method also depends upon the source micro–array platform used. For example, methods such as MAS5.0, RMA or GC-RMA are suitable for Affymetrix micro–array data. For spotted arrays the normalization method depends on the experimental setup. For example, for within array normalization, lo(w)ess normalization is one of the most suitable option while methods such as median centering and scale normalization is suitable in other situations. In our experiments we used RMA (Robust Multi-chip

Average) normalization [15] method to normalize the time-series gene expression data. RMA normalization comprises of three distinct steps: background adjustment, quantile normalization, and summarization [15].

3.2   Gene Expression Analysis

Our test of statistical significance is based on simple comparison of two gene expression measurements of experimental and control patients [16]. Since we had access to two sets of measurements control and experimental in T cell differentiation dataset, we use the statistical significance to determine the differentially expressed genes. Asbestos exposure data at our disposal had no replicate copies required for determining the statistical significance. Hence, we generated the random samples from within the 95% confidence interval of the original gene expression measurements as the new replicate measurements. Our hypothesis is that "statistically significant genes differentially expressed whereas statistically insignificant genes are not differentially expressed" [16]. We calculate the statistical significance using paired t-test from the logarithm of the expression values to assess the significance of the measured relative expression or repression of the genes at different time instances. Multiple testing corrections were performed using bootstrapping with replacement. Confidence interval is chosen as discussed in Section 4 which is directly correlated with the number of genes required for the final output. If fewer number of genes are required; confidence interval is higher and if more number of genes are required confidence interval is lower.

3.3   Swap Randomization

After statistical significance testing is performed as discussed in Section 3.2, we can determine the list of genes that are differentially expressed at different time-points. It is important to note that a single gene can be expressed at certain time instance whereas it can be unexpressed at other time instances. Hence, we generate a binary matrix where 1 denotes that the gene is differentially expressed at that time instance where as 0 denotes the absence of differential expression of gene at that time instance. Now, the gene expression dataset is a 0-1 dataset where each row represents one gene at different time instances. Now the problem that arises is that the normal statistical significance testing can not be used to select the statistically significant genes because data belongs to a class of empirical distributions thus integrating over the PDF (Probability Density Function) is not feasible. Additionally, for any given data set $\mathcal{D}$, its PDF i.e. true generating model is often unknown. However, we can fix a null distribution and sample the data from the null distribution [17]. Swap randomization can be used to sample the data from null distribution [18].

Test Statistic. We introduce a new test statistic in this paper i.e. the longest run of consecutive 1s and 0s. It measures how often 1s and 0s comes together (adjacently) in each row i.e. for a gene in all time instances. Table 1 gives an example of longest run of consecutive 1s and consecutive 0s. To accommodate

this special test statistic, we only consider more extreme values while calculating
p–values instead of the original definition which also considers the results that
are as extreme as the original result i.e. equal to the original result.

Table 1 shows an example of calculation of longest run of consecutive 1s and
0s test statistic in a 5 dimensional dataset. In our experiments, if the longest
run of consecutive 0s is statistically significant and occurs more than three times
then Gene E is not selected.

Table 1. A toy data explaining the procedure of counting consecutive 1s and
consecutive 0s

Data Points						Consecutive	
Gene	x_1	x_2	x_3	x_4	x_5	1s	0s
A	1	1	1	1	0	4	1
B	1	1	0	0	1	2	2
C	1	1	1	0	1	3	1
D	1	0	1	0	1	1	1
E	0	1	0	0	0	1	3

Null Distribution

The null distribution considered throughout this paper is similar to [18] and
satisfies the following properties for a 0-1 dataset, $\mathcal{D}$:

1. The size of the random datasets, $\mathcal{D}_1, \mathcal{D}_2, \ldots \mathcal{D}_n$ is same as the original
   dataset, $\mathcal{D}$. If the original dataset, $\mathcal{D}$, has m rows and k columns, each
   of the sampled datasets $\mathcal{D}_1, \mathcal{D}_2, \ldots \mathcal{D}_n$ also has m rows and k columns.
2. The random datasets, $\mathcal{D}_1, \mathcal{D}_2, \ldots \mathcal{D}_n$ has the same column and row margins
   (sums) as the original dataset, $\mathcal{D}$. This constraint preserves the number of
   times the genes are expressed at any time instance and also the number of
   times a gene is expressed different time instance. It also automatically preserves the number of 1s in the dataset i.e. total number of genes differentially
   expressed in the data.

If we add more properties to this list, the randomization is more conservative.
This is advantageous because we want to compare our results to the closely
related datasets. If the randomization is unconstrained, then most of the results
obtained by an algorithm will be statistically significant. For example, if we
compare the height of grade 10 students with grade 5 students, almost everyone
will be significantly tall. However, we want to state that a student in grade 10 is
significantly tall than we need to compare his height with closely related students
such as other students in grade 10. This constrained randomization is one of the
main advantages offered by swap randomization.

Swap randomization [18] algorithm proceeds as follows: Let us consider a 0-1
dataset, $\mathcal{D}$ and let $\mathcal{D}_1, \mathcal{D}_2, \ldots \mathcal{D}_n$ be the randomized datasets. Also consider an
algorithm, $\mathcal{A}$, run on the dataset, $\mathcal{D}$, resulting in $\mathcal{A}(\mathcal{D})$ which determines the
structural measure of the dataset $\mathcal{D}$. The structural measure used in our case is

the longest run of consecutive 1s and 0s in each row i.e. number of times genes are expressed in successive time intervals (1s) or not expressed in successive time intervals (0s). The algorithm $\mathcal{A}$ is also applied to the randomized datasets producing results $\mathcal{A}(\mathcal{D})_1, \mathcal{A}(\mathcal{D})_2, \ldots \mathcal{A}(\mathcal{D})_n$. Now, p–values can be calculated from these results. The calculated p–values can be used to compare if the result on original data is significantly different from random data. The null hypothesis, $\mathcal{H}_0$, considered in this paper is that for all datasets, $\mathcal{D}$, satisfying the above constraints, test statistic follows the same distribution.

p−Values

p-value is the probability of attaining a test statistic in randomized data as extreme as the original data, assuming that the null hypothesis, $\mathcal{H}_0$, is true [17, 19]. For a dataset, $\mathcal{D}$, let us sample i.i.d. datasets $\hat{\mathcal{D}} = \{\mathcal{D}_1, \mathcal{D}_2, \mathcal{D}_n\}$ from the null distribution, $\mathcal{H}_0$. Now, one-tailed *empirical p−value* of $\mathcal{A}(\mathcal{D})$ for $\mathcal{A}(\mathcal{D})$ being as extreme can be defined by

$$\tilde{p} = \frac{1}{n+1} \left(\sum_{i=1}^{n} I(\mathcal{A}(\mathcal{D}_i) \geq \mathcal{A}(\mathcal{D})) + 1 \right), \tag{1}$$

where $i \in \{1, 2 \ldots n\}$

Equation 1 defines the fraction of randomized dataset, $\hat{\mathcal{D}}$, whose structural measure $\mathcal{A}(\hat{\mathcal{D}})$ is as extreme as the original data $\mathcal{A}(\mathcal{D})$. Since our test statistic is special, we consider only those results that are more extreme than original results. Swap randomization [18] produces 10^4 samples of randomized dataset data using Markov Chain Monte Carlo (MCMC) using forward backward approach. Furthermore, Holm-Bonferroni correction was used to address the problem of multiple hypothesis testing [20].

4   Experiments and Results

We ran the proposed gene selection pipeline depicted in Figure 1 on two time–series gene expression datasets which were presented in Section 2. The CEL files was extracted and relevant preprocessing was performed necessary for data from Affymetrix micro–array data using Affy package in R. The dataset was also normalized to account for experimental variations using RMA [15] as discussed in Section 3.1. The normalized data was subjected to statistical significance analysis as discussed in Section 3.2. We implemented a paired t–test to the logarithm of the expression levels to assess the significance of the measured relative expression or repression of the genes at various time points in the time–series. After the application of statistical significance, the resulting matrix is a matrix of p–values where each elements, p_{ij}, take the value between 0 and 1 such that $0 \leq p_{ij} \leq 1$. The lesser the value, the more statistically significant the gene is at that time instance.

The p–values were obtained by bootstrapping strategy with replacement to correct for the multiple tests. We arbitrarily selected 90% as the confidence interval to determine if a gene is statistically expressed ($\alpha = 0.1$). In 90% confidence interval more than two-thirds of the genes were significantly expressed at different time points. We created a 0-1 matrix where 1 represents that the gene at that time instance is differentially expressed while 0 denotes that the gene in that time instance is not differentially expressed at 90% confidence interval. A number of genes were not statistically expressed at any time instance so we excluded them from further experiments i.e. swap randomization as these genes are not interesting and their expression is not affected by external stimuli. Thus, achieving reduction in number of genes from 45101 to 21478 genes in T cell differentiation dataset and from 54675 to 38510 genes in asbestos exposure dataset.

4.1   Swap Randomization and Convergence Analysis

One of the important parameters to consider in the swap randomization process is number of swaps to be performed on the dataset i.e. the parameter $\mathcal{J}$. In our experiments and as suggested in [18], we consider the swap randomization procedure to converge when the distance, Frobenius norm, between the original data and the randomized data has converged.

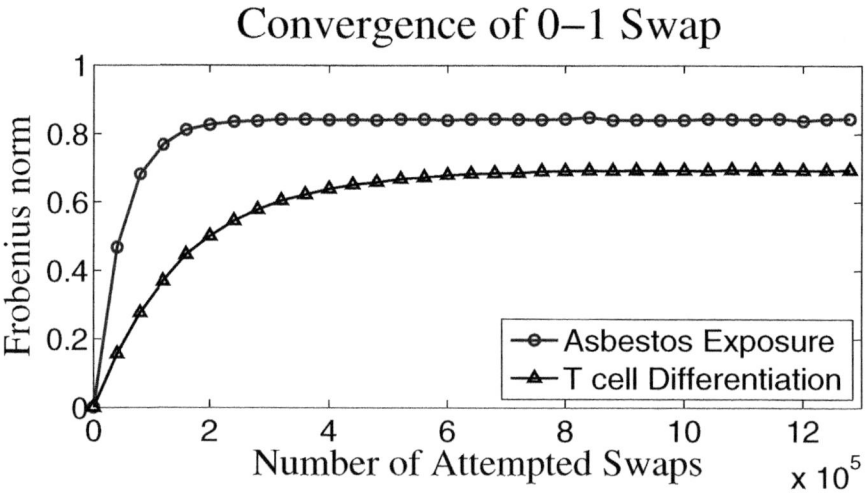

Fig. 2. Convergence analysis of 0–1 Swap. Y–axis shows the Frobenius norm between original data and swapped data after the number of performed swaps specified in the X-axis.

First we initialize the number of attempted swaps to 1 and increase it in the step-size of 1 which is different from the strategy used in [18] where the number of attempted swaps is fixed to $\mathcal{K}$ equal to the number of 1s in the dataset. Figure 2 shows that Frobenius norm initially increases speedily as the number of attempted swaps increases but tends to reach steady state after certain number of attempted swaps. Reaching the steady state is considered as the point of convergence. We repeat this experiment 10 times and report the mean of the results. For example, in the case of T cell differentiation dataset as shown in Figure 2, the convergence is reached when the number of swaps is approximately 6×10^5. Similarly, for the Asbestos exposure dataset convergence is reached when the number of swaps is approximately 2×10^5. However, taking a conservative approach we use 9×10^5 and 3×10^5 swaps for T cell differentiation and asbestos exposure datasets, respectively.

4.2   Selection of Genes

We generate 10^4 samples of random datasets after performing selected number of swaps for each random dataset. Now, we calculate the longest run of consecutive 1s in original dataset and also in each of the randomized datasets for each gene. As shown in the Figure 3, consecutive occurrence of 1s is mostly rare in the original data. However, among these rare occurrence of 1s; we find if these rare occurrence of 1s are statistically significant. Hence, we calculate the p–values comparing two p–values from original and randomized data as discussed in Section 3.3. The p–values were corrected for multiple corrections using Holm-Bonferroni [20]. The crux of the problem we are trying to solve is the selection of a set of genes. Finally, we select the genes that are significant at 99% confidence interval. In T cell differentiation dataset, number of statistically significant genes at 99% confidence interval were 12814. However, these 12814 sets of genes also includes those genes which have only one as the longest run of consecutive occurrence of 1s.

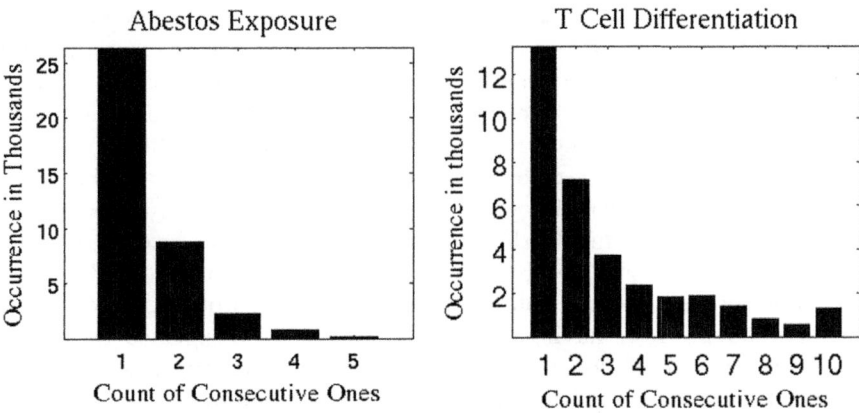

Fig. 3. Histogram showing number of longest run of consecutive 1s in different genes in original dataset. X-axis shows the number of longest run of consecutive 1s in both the figures. Consecutive occurrence of more than two 1s is very rare. We are interested in those genes that have significantly high number of consecutive occurrence of 1s and significantly less number of consecutive 0s.

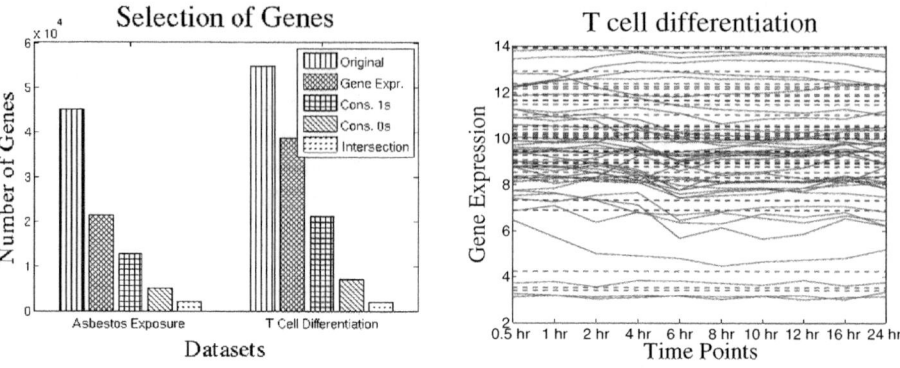

Fig. 4. Left panel shows the reduction in number of genes during different phases of analysis pipeline. Initially the number of genes are greater than 45000 in both datasets but finally the number of selected genes is around 2000. Right panel shows the gene expression profiles of first 40 selected genes at ten different time points. Solid line shows the expression profiles of treatment group and dashed lines represents the expression profiles of control group. Figure in the right panel shows that expression of selected genes are comparatively different from the control group over a period of time.

It is also to be noted that we are also not interested in genes having significantly high longest run of consecutive occurrence of 0s. Hence, we also calculated the genes that have significantly high occurrence of longest run of consecutive 0s with a threshold of 3; i.e. we calculate if a gene shows occurrence of 3 or more longest run of consecutive 0s. Unlike the longest run of consecutive 0s longest run of consecutive 1s were not subjected to a threshold. Since, we've not thresholded the longest run of consecutive 1s property, it allows for some incorrect assignment of differential expression at a time instance. Finally, as shown in the Figure 4, we take the intersection of genes that have significantly high longest run of consecutive 1s (12814 genes) and genes that do not have significantly high occurrence of 3 or more longest run of consecutive 0s (4980 genes). The intersection results in 2015 genes which computationally and even biologically seems appropriate number of genes for any experiment. Instead of longest run of consecutive 1s other thresholding methods would not produce useful results. Row sums are definitely one of them but we can not test for their statistical significance because row and column sums are preserved during randomization process. Removing rows having column sums below the specified threshold as with zero vectors would decrease the size of data significantly thus affecting the randomization process. Similarly, in asbestos exposure dataset, number of genes having significant number of consecutive 1s is 21174 and that of consecutive 0s with threshold of 3 is 7104. As shown in the Figure 4, the intersection of these two sets results in 1811 genes which is the final set of selected genes.

Figure 4 shows that gene expression profiles of selected genes differ from their control group over a longer timespan showing that selected genes are relevant set of genes and most responsive to the applied external stimuli. From the numbers seen in the results, we can perceive that the number of genes selected from the total set of genes is small. These selected set of genes are statistically significant and are more likely to be affected by the external stimuli such as the application of drugs. Since each micro–array is a snapshot whereas the biological process is dynamic, the time-series selection helps us to trace the dynamics of biology from snapshots. These selected genes can be used for diagnostics using them as features for different classification and regression models.

5   Summary and Conclusions

In this paper, we propose a novel unsupervised method for gene selection in time-series gene expression data. The pipeline is based on the combination of statistical testing founded on a paired t–test and statistical significance testing founded on swap randomization [18]. Experiments are performed on two datasets: one of them on a publicly available mouse T cell differentiation and the other one on asbestos exposure. The experiments showed that number of genes are considerably reduced in both the experimented datasets. One of the major strength of the proposed methodology is its unsupervised nature thus it does not require expensive and difficult–to–get class labels and results in large reduction in number of genes.

References

1. Kira, K., Rendell, L.A.: A practical approach to feature selection. In: Proceedings of the ninth international workshop on Machine Learning, ML 1992, pp. 249–256. Morgan Kaufmann Publishers Inc., San Francisco (1992)
2. Blum, A.L., Langley, P.: Selection of relevant features and examples in machine learning. Artificial Intelligence 97(1-2), 245–271 (1997)
3. Saeys, Y., Inza, I., Larrañaga, P.: A review of feature selection techniques in bioinformatics. Bioinformatics 23(19), 2507–2517 (2007)
4. Mörchen, F.: Time series feature extraction for data mining using DWT and DFT. Technical Report 33, Department of Mathematics and Computer Science, University of Marburg, Germany (2003)
5. Tikka, J., Hollmén, J.: A Sequential Input Selection Algorithm for Long-term prediction of Time Series. Neurocomputing 71(13-15), 2604–2615 (2008)
6. Heller, M.J.: DNA microarray technology: Devices, systems, and applications. Annual Review Of Biomedical Engineering 4, 129–153 (2002)
7. Guyon, I., Weston, J., Barnhill, S., Vapnik, V.: Gene selection for Cancer Classification using Support Vector Machines. Machine Learning 46, 389–422 (2002)
8. Diaz-Uriarte, R., Alvarez de Andres, S.: Gene selection and classification of microarray data using random forest. BMC Bioinformatics 7(1), 3 (2006)
9. Wichert, S., Fokianos, K., Strimmer, K.: Identifying periodically expressed transcripts in microarray time series data. Bioinformatics 20, 5–20 (2004)
10. Peddada, S.D., Lobenhofer, E.K., Li, L., Afshari, C.A., Weinberg, C.R., Umbach, D.M.: Gene selection and clustering for time–course and dose–response microarray experiments using order–restricted inference. Bioinformatics 19(7), 834–841 (2003)

11. Lin, T., Kaminski, N., Bar-Joseph, Z.: Alignment and classification of time series gene expression in clinical studies. Bioinformatics 24(13), i147–i155 (2008)
12. Hyvärinen, A., Karhunen, J., Oja, E.: Independent component analysis. Adaptive and learning systems for signal processing, communications, and control. John Wiley and Sons (2001)
13. Nymark, N., Lindholm, P.M., Korpela, M.V., Lahti, L., Ruosaari, S., Kaski, S., Hollmén, J., Anttila, S., Kinnula, V.L., Knuutila, S.: Gene Expression Profiles in Asbestos-exposed Epithelial and Mesothelial Lung Cell Lines. BMC Genomics 8(1), 62 (2007)
14. Zhang, Z., Martino, A., Faulon, J.: Identification of expression patterns of IL-2-responsive genes in the murine T cell line CTLL-2. Jounal of Interferon & Cytokine Research 27(12), 991–995 (2007)
15. Bolstad, B.M., Irizarry, R.A., Astrand, M., Speed, T.P.: A comparison of normalization methods for high density oligonucleotide array data based on variance and bias. Bioinformatics 19(2), 185–193 (2003)
16. Parmigiani, G.: The analysis of gene expression data: methods and software. Springer, Heidelberg (2003)
17. Good, P.: Permutation Tests: A Practical Guide to Resampling Methods for Testing Hypotheses, 2nd edn. Springer, Heidelberg (2000)
18. Gionis, A., Mannila, H., Mielikäinen, T., Tsaparas, P.: Assessing data mining results via swap randomization. ACM Transactions on Knowledge Discovery from Data 1(3), 14 (2007)
19. Schervish, M.J.: P Values: What They Are and What They Are Not. American Statistician 50(3), 203–206 (1996)
20. Holm, S.: A simple sequentially rejective multiple test procedure. Scandinavian Journal of Statistics 6, 65–70 (1979)

Multi-task Drug Bioactivity Classification with Graph Labeling Ensembles

Hongyu Su and Juho Rousu

Department of Computer Science,
P.O. Box 68, 00014 University of Helsinki, Finland
{hongyu.su,juho.rousu}@cs.helsinki.fi

Abstract. We present a new method for drug bioactivity classification based on learning an ensemble of multi-task classifiers. As the base classifiers of the ensemble we use Max-Margin Conditional Random Field (MMCRF) models, which have previously obtained the state-of-the-art accuracy in this problem. MMCRF relies on a graph structure coupling the set of tasks together, and thus turns the multi-task learning problem into a graph labeling problem. In our ensemble method the graphs of the base classifiers are random, constructed by random pairing or random spanning tree extraction over the set of tasks.

We compare the ensemble approaches on datasets containing the cancer inhibition potential of drug-like molecules against 60 cancer cell lines. In our experiments we find that ensembles based on random graphs surpass the accuracy of single SVM as well as a single MMCRF model relying on a graph built from auxiliary data.

Keywords: drug bioactivity prediction; multi-task learning; ensemble methods; kernel methods.

1 Introduction

Molecular classification, the task of predicting the presence or absence of the bioactivity of interest, has been benefited from variety of methods in statistics and machine learning [7]. In particular, kernel methods [9,16,2,7] have emerged as an effective way to handle the non-linear properties of chemicals. However, classification methods focusing on a single target variable are probably not optimally suited to drug screening applications where large number of target cell lines are to be handled.

In [15] a multi-task (or multilabel) learning approach was proposed to classify molecules according to their activity against a set of cancer cell lines. It was shown that the multilabel learning setup improves predictive performance over a set of support vector machine based single target classifiers. The multilabel classifier applied, Max-Margin Conditional Random Field (MMCRF) [11] relies on a graph structure coupling the outputs together. In [15] the graph was extracted

M. Loog et al. (Eds.): PRIB 2011, LNBI 7036, pp. 157–167, 2011.

from auxiliary data, concerning sets of experiments conducted on the cancer cell lines, by simple techniques such as correlation thresholding and maximum weight spanning tree extraction.

In this paper, we develop ensemble learning methods for the multi-task learning setup. In our method, MMCRF models are used as the ensemble components. Unlike other ensemble learners for multi-task setups, our method does not require bootstrapping of the training data or changing instance weights to induce diversity among the ensemble components. In our case, the diversity is provided by the randomization of the output graphs, which combined with discriminative training of the base MMCRF classifiers, realizes the benefits typically seen from ensemble approaches. The random graph approach is compared against single classifiers and ensembles on graphs built from auxiliary data with different graph extraction methods, including inverse covariance learning [5] that is theoretically superior to correlation thresholding for extracting statistical dependencies.

Ensembles of multi-task or multilabel classifiers have been proposed in a few papers prior to ours, but with important differences both in the methods and the applications. In general, the previous approaches can be divided into two groups based on the source of diversity among the base classifiers of the ensemble. The first group of methods, boosting type, relies on changing the weights of the training instances so that difficult-to-classify instances gradually receive more and more weight. The Boostexter method [12] by the inventors of boosting has a multilabel learning variant. Later, Esuli et al. [4] developed a hierarchical multilabel variant of AdaBoost. Neither method explicitly considers label dependencies but the multilabel is considered essentially a flat vector. The second group of methods, bagging, is based on bootstrap sampling the training set several times and building the base classifiers from the bootstrap samples. Averaging over the ensemble gives the final predictions. Schietgat et al [13] concentrate a bagging in multilabel gene function prediction. They build ensembles of predictive clustering trees (PCT) by bagging, that is, bootstrap sampling of the data several times to arrive at a set of different models. In their approach, there is also no structure defined for the tasks, but the multilabel is essentially treated as a flat vector. Finally, Yan et al. [18] select different random subsets of input features and examples to induce the base classifiers.

The remainder of the article is structured as follows. In section 2 we present the base classifier MMCRF and the multi-task ensemble learning approach. In section 3 we validate the methods empirically, in particular we show that the ensemble approach exceeds the accuracy of MMCRF, which to our knowledge currently has the state-of-the-art predictive performance. In section 4 we aim to provide intuition of the hows and whys of the behaviour of the new method by relating the new ensemble approach to other multi-task and multilabel ensemble approaches. In section 5 finish with concluding remarks.

2   Ensemble Learning with Max-Margin Conditional Random Field Models on Random Graphs

Ensemble learners [3,8], such as boosting [12] and bagging [1] are based on the notion that a set of *weak leaners*, those that have accuracy higher than coin tossing, may produce a strong learner with high accuracy when appropriately combined. It has been found that the diversity among the base models is the key property. The diversity may arise from re-weighting of examples, bootstrap resampling of examples, from the different inductive biases of the base learners, or in multiclass setup, or by generating a set of derived binary classification tasks (one-vs-the rest, one vs. one, and error-correcting output codes [3]).

In this section we present our ensemble learning approach where the diversity among the base learners comes from a different source, namely randomized graph structures that are used to couple the tasks. We use a majority voting approach over the predictions of the base classifiers, namely labelings of the randomized graphs. Two basic types of graphs are used, random spanning trees and random pairings of targets (Figure 1). As the base learner, we use the MMCRF algorithm [11].

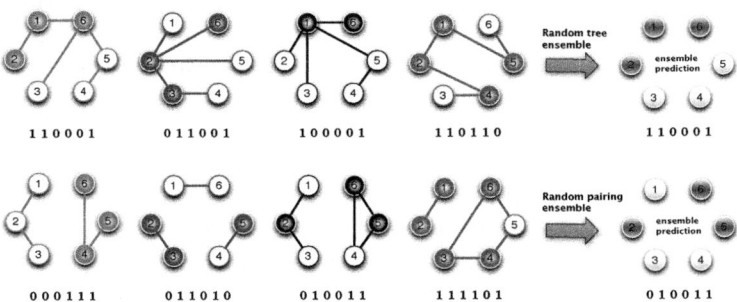

Fig. 1. Ensemble prediction from a set of random spanning trees (top) and a set of random pairings of tasks (bottom). The varying graph structures provide the required diversity among the ensemble components. Majority vote decides the final predicted label for each task.

The method for generating the ensemble is depicted in Algorithm 1. The algorithm receives a training sample of molecules x_i, computes the input kernel K and embarks on the ensemble learning phase. For each base model, a random graph G_t of the type specified by the user is drawn to couple the outputs $\mathbf{y}_i$ which are the inhibition potentials of molecule x_i against 60 cancer cell lines. The input kernel, label data and the graph are given as input to the MMCRF (see Section 2.1) that learns the graph labeling. After the ensemble has been generated, the ensemble prediction is extracted in post-processing: we extract the majority vote over the graph labelings from the sign of the mean of the base classifier predictions.

Algorithm 1. Ensemble learning algorithm with random graph multi-task classifiers

---

Input: Training sample $S = \{(x_i, \mathbf{y}_i)\}_{i=1}^m$, ensemble size T, type of the graph generated $graphType$, n the number of nodes in the graph, type of input kernel applied $kernelType$

Output: Multi-task classification ensemble $\left(f^{(1)}, \ldots, f^{(T)}\right)$

1: $K = buildKernel(\{x_i\}_{i=1}^m, kernelType)$
2: $t = 0$
3: **while** $t < T$ **do**
4:      $t = t + 1$
5:      $G_t = randomGraph(n, graphType)$
6:      $f^{(t)} = learnMMCRF(K, (\mathbf{y}_i)_{i=1}^m, G_t)$
7: **end while**
8: **return** $f = \left(f^{(1)}, \ldots, f^{(T)}\right)$

2.1   Learning Graph Labeling with MMCRF

The MMCRF method used as the base learner in the multi-task ensembles is an instantiation of the structured output prediction framework MMCRF [11] for associative Markov networks and can also be seen as a sibling method to HM^3[10], which is designed for hierarchies. We give a brief outline here, the interested reader may check the details from the above references.

The MMCRF learning algorithm takes as input a matrix $K = (k(x_i, x_j))_{i,j=1}^m$ of kernel values $k(x_i, x_j) = \phi(x_i)^T \phi(x_j)$ between the training patterns, where $\phi(x)$ denotes a feature description of an input pattern (in our case a potential drug molecule), and a label matrix $Y = (\mathbf{y}_i)_{i=1}^m$ containing the multilabels $\mathbf{y}_i = (y_1, \ldots, y_k)$ of the training patterns. The components $y_j \in \{-1, +1\}$ of the multilabel are called microlabels, which in multi-task learning setup, correspond to labels of different tasks. In addition, the algorithm assumes an associative network $G = (V, E)$ to be given, where node $j \in V$ corresponds to the j'th component of the multilabel and the edges $e = (j, j') \in E$ correspond to a microlabel dependency structure.

The model learned by MMCRF takes the form of a conditional random field with exponential edge-potentials,

$$P(\mathbf{y}|x) \propto \prod_{e \in E} \exp\left(\mathbf{w}_e^T \varphi_e(x, \mathbf{y}_e)\right) = \exp\left(\mathbf{w}^T \varphi(x, \mathbf{y})\right),$$

where $\mathbf{y}_e = (y_j, y_{j'})$ denotes the pair of microlabels of the edge $e = (j, j')$. A joint feature map $\varphi(x, \mathbf{y}) = \phi(x) \otimes \psi(\mathbf{y})$ is composed via tensor product of input $\phi(x)$ and output feature map $\psi(\mathbf{y})$, thus including all pairs of input and output features. The output feature map is composed of indicator functions $\psi_e^u(\mathbf{y}) = [\![\mathbf{y}_e = u]\!]$ where u ranges over the four possible labelings of an edge given binary node labels. The corresponding weights are denoted by $\mathbf{w} = (\mathbf{w}_e)_e$. The benefit of the tensor product representation is that context (edge-labeling) sensitive weights can be learned for input features and no prior alignment of input and output features needs to be assumed.

The parameters are learned by maximizing the minimum loss-scaled margin between the correct training examples $(x_i, \mathbf{y}_i)$ and incorrect pseudo-examples $(x_i, \mathbf{y}), \mathbf{y} \neq \mathbf{y}_i$, while controlling the norm of the weight vector. The dual soft-margin optimization problem takes the form

$$\min_{\alpha \geq 0} \sum_{i,\mathbf{y}} \alpha(i, \mathbf{y})\ell(\mathbf{y}_i, \mathbf{y}) - \frac{1}{2} \sum_{i,\mathbf{y}} \sum_{j,\mathbf{y}'} \alpha(i, \mathbf{y})K(x_i, \mathbf{y}; x_j, \mathbf{y}')\alpha(i, \mathbf{y}')$$

$$\text{s.t.} \sum_{\mathbf{y}} \alpha(i, \mathbf{y}) \leq C, \forall i, \tag{1}$$

where $K(x_i, \mathbf{y}; x_j, \mathbf{y}') = \Delta\varphi(i, \mathbf{y})^T \Delta\varphi(j, \mathbf{y}') = K_X(x_i, x_j) \odot K_{\Delta Y}(\mathbf{y}, \mathbf{y}')$ is the joint kernel composed of the input $K_X(x_i, x_j)$ and output $K_{\Delta Y}(\mathbf{y}_i, \mathbf{y}')$ kernels. The underlying joint feature map is expressed by

$$\Delta\varphi(i, \mathbf{y}) = (\varphi(x_i, \mathbf{y}_i) - \varphi(x, \mathbf{y})) = \phi(x_i) \otimes (\psi(\mathbf{y}_i) - \psi(\mathbf{y})),$$

that is, joint feature difference vectors between the true $(\mathbf{y}_i)$ and a competing output $(\mathbf{y})$.

As the input kernel we use the hash fingerprint Tanimoto kernel [9] that was previously shown [15] to be a well performing kernel in this task. Hash fingerprints enumerate all linear fragments of length n in a molecule. A hash function assigns each fragment a hash value that determines its position in descriptor space. Given two fingerprint vectors x and z, Tanimoto kernel is the way to measure their similarity defined as

$$K_X(x, z) = \frac{|I(x) \cap I(z)|}{|I(x) \cup I(z)|},$$

where $I(x)$ denotes the set of indices of 1-bits in x .

As the loss function we use *Hamming loss*

$$\ell_\Delta(\mathbf{y}, \mathbf{u}) = \sum_j [\![y_j \neq u_j]\!]$$

that is gradually increasing in the number of incorrect microlabels so that we can make a difference between 'nearly correct' and 'clearly incorrect' multilabel predictions.

2.2   Graph Generation for Cancer Cell Lines

In the anti-cancer bioactivity prediction problem, a single task entails classification of drug molecules according to whether they are active or inactive against one of the 60 cancer cell lines. The nodes of the graph G to be labeled thus correspond to cancer cell lines. The edges of the graph depict coupling of the tasks, denoting a potential statistical dependency that is to be utilized in predicting the graph labels (Figure 2).

To generate random graphs G_t we use two approaches.

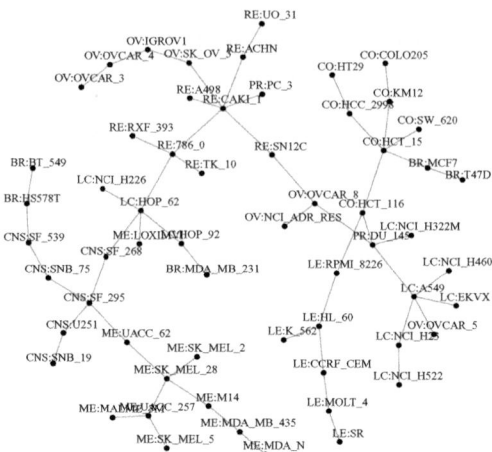

Fig. 2. Example of a cell line graph

- In the random spanning tree approach, we first generate a random correlation matrix and extract the spanning tree out of the matrix with the above described approach.
- In the random pairing approach, one takes each vertex in turn, randomly draws another vertex and couples the two with an edge.

We note that the random graph approach lets us build ensembles whose size is not limited in practice.

We compare the random graphs against the approach used by [15], namely, graphs built from Radiation RNA Array data, available for the cancer cell lines from NCI database[1]. To extract a graph out of the correlation matrix we use the graphical lasso [5] which estimates a sparse graph model by using L_1 (lasso) regularization on inverse covariance matrix, and is theoretically a better method than the simple thresholding of the covariance matrix, applied in [15]. Graphical lasso assumes multivariate Gaussian distribution over cell lines with mean μ and covariance matrix Σ. The inverse covariance matrix Σ^{-1} is a good indicator for conditional independencies [6], where variable i and j are conditional independent given other variables if the $ij$the entry of Σ^{-1} is zero. It imposes L_1 penalty during the estimation of Σ^{-1} to increase the sparsity of the resulted graph. The objective is to maximize the penalized log-likelihood

$$\log \det \Sigma^{-1} - \mathrm{tr}(S\Sigma^{-1}) - \rho ||\Sigma^{-1}||_1,$$

where tr is the trace of the matrix, S is empirical covariance matrix, and $||\Sigma^{-1}||_1$ is the L_1 norm of Σ^{-1}. Particularly in our application, we post processed the estimated sparse graph to be a tree-liked one.

---

[1] http://discover.nci.nih.gov/cellminer/home.do

3   Experiments

3.1   Data and Preprocessing

In this paper we use the NCI-Cancer dataset obtained through PubChem Bioassay[2] [17] data repository. The dataset, initiated by National Cancer Institute and National Institutes of Health (NCI/NIH), contains bioactivity information of large number of molecules against several human cancer cell lines in nine different tissue types including leukemia, melanoma and cancers of the lung, colon, brain, ovary, breast, prostate, and kidney. For each molecule tested against a certain cell line, the dataset provide a bioactivity outcome that we use as the classes (active, inactive).

Currently, there are 43197 molecules in the PubChem Bioassay database together with their activities information in 73 cancer cell lines. 60 cell lines have screen experimental results for most molecules and 4547 molecules have no missing data in these cell lines. Therefore these cell lines and molecules are selected and employed in our experiments. However, molecular activity data are highly biased over the 60 cell lines: Around 60% of molecules are inactive in all cell lines, while still a relatively large proportion of molecules are active against all cell lines. These molecules are less likely to be potential drug candidates than the ones in the middle part of the histogram.

To tackle the skewness problem, Su et al. [15] prepared three different versions of the datasets, which approach is also followed here:

Full Data. This dataset contains all 4547 molecules in the NCI-Cancer dataset that have their activity class (active vs. inactive) recorded against all 60 cancer cell lines.

No-Zero-Active. From full data, we removed all molecules that are not active towards any of the cell lines. The remaining 2303 molecules are all active against at least one cell line.

Middle-Active. Here, we followed the preprocessing suggested in [14], and selected molecules that are active in more than 10 cell lines and inactive in more than 10 cell lines. As a result, 545 molecules remained and were employed in our experiments.

3.2   Compared Methods

Three kinds of multi-task classifier ensembles are compared:

- SVM: Support vector machines (SVM) are used as the single-task non-ensemble baseline classifier.
- MMCRF-Glasso: An MMCRF model where the underlying graph connecting the tasks is built by graphical lasso from auxiliary data.
- MMCRF-EnsRT: An ensemble of 1-500 MMCRF models, where the graph connecting the tasks is built by a random spanning tree.

---

[2] http://pubchem.ncbi.nlm.nih.gov

- MMCRF-EnsRP: An ensemble of 1-500 MMCRF models, where the graph connecting the tasks is built by random pairing of the tasks.

In the tests by [15], a relatively hard margin ($C = 100$) emerged as the most favorable for SVM, while MMCRF proved to be quite insensitive as regarding margin softness. Here we used the same value for all compared classifiers.

3.3   Experiment Setup and Performance Measures

Because of the skewness of the multilabel distribution we used the following *stratified 5-fold cross-validation* scheme in all experiments reported such that we group the molecules in equivalence classes based on the number of cell lines they are active against. Then each group is randomly split among the five folds. The procedure ensures that also the smaller groups have representation in all folds. Besides overall classification accuracy, we also report microlabel F_1 score, the harmonic mean of precision and recall

$$F_1 = 2 \times \frac{Precision \times Recall}{Precision + Recall}.$$

In particular, we pool together individual microlabel predictions over all examples and all cell lines, and count accuracy and F_1 from the pool.

We generated hash fragments features from OpenBabel[3] which is a chemical toolbox available in public domain. We used default value for enumerating all linear structures up to length seven. Then Tanimoto kernel was built based on hash fingerprints features and normalized.

3.4   Results

Figure 3 illustrates the performance of the compared methods on the three versions of the datasets. All models based on MMCRF are clearly more accurate than SVM. Among single models and small ensembles, MMCRF-Glasso is the most competitive method, showing that the auxiliary data contains information that can be successfully used to improve predictive performance.

Both the random pairing and random tree based ensembles steadily improve accuracy and F_1 score as the number of base models increases. SVM falls behind the random graph ensembles even on small ensemble sizes ($T < 5$). With larger ensemble sizes, both types of ensembles end up superior to MMCRF-Glasso in terms of classification accuracy. In terms of F_1 score, the best method depends on the dataset: on the Middle-Active dataset, the random tree ensemble outperform random pairing one, and MMCRF-Glasso is slightly behind. On No-zero-Active and Full data, random pairing ensemble ends up the best method. This result might reflect the sizes of the datasets: the Middle-Active dataset is significantly smaller than the other two, and perhaps the random pairing ensemble requires more data for best results.

---

[3] http://openbabel.org

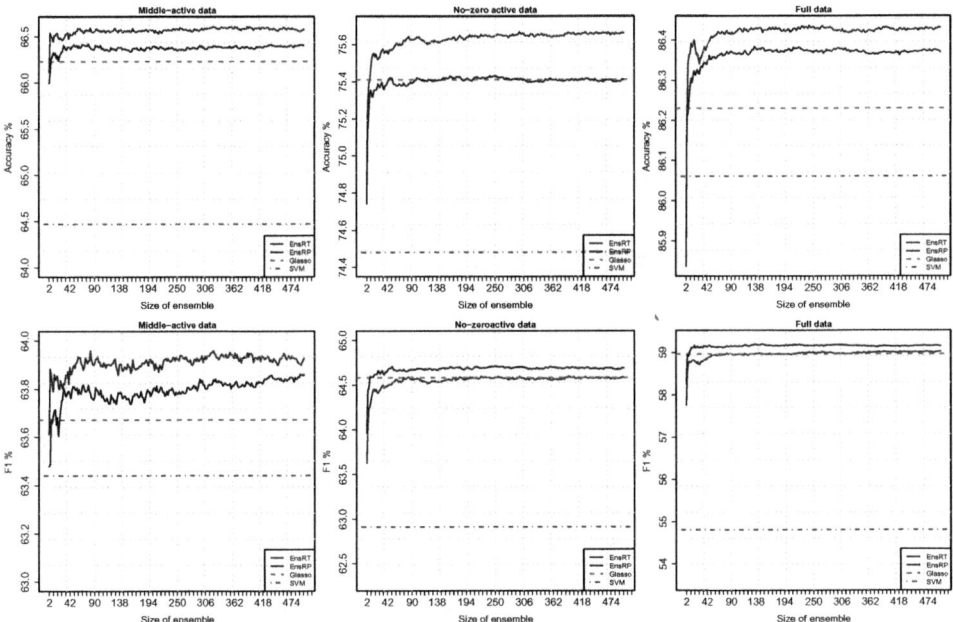

Fig. 3. Accuracy against number of individual classifiers in ensemble methods from different version of datasets. The red line corresponds to random tree ensemble, and blue line is random pairing ensemble. The performance of single models (SVM and MMCRF-Glasso) are depicted by the horizontal lines.

Table 1 shows the prediction performance from SVM, Glasso, EnsRP and EnsRT from three versions of the dataset. We performed two-tailed sign test to identify whether the differences in accuracy and F_1 score in individual cell lines are statistically significant. P-values for the difference over the worst classifier and the ones towards the best classifier are shown as asterisks and crosses. The result shows that, in terms of accuracy and F_1 the multi-task methods outperform SVM in all versions of datasets in a statistically significant manner. EnsRT outperforms Glasso in terms of accuracy in statistically very significant manner.

4   Discussion

The results of this paper show that ensemble methods can be effectively combined with a graph-based multi-task learner such as MMCRF. From machine learning point of view, perhaps the most surprising result obtained here is that in an ensemble, the base graph labeling models can be successfully learnt on random graphs, as opposed to using some auxiliary data or prior knowledge to extract graphs that aim to reflect statistical dependencies.

The present ensemble method differs from previous approaches in that the diversity among the base classifiers arises from the different random output

Table 1. Overall accuracy and microlabel F_1 scores. P-values for the differences over the worst classifier in each version of the dataset are marked with asterisks. P-values for the differences towards the best classifier are marked with crosses. Single, double and triple symbols correspond to p-value below 0.05, 0.01 and 0.001.

Dataset	Accuracy				F_1			
	SVM	Glasso	EnsRP	EnsRT	SVM	Glasso	EnsRP	EnsRT
Middle-Active	$64.5\%_{\dagger\dagger\dagger}$	$66.2\%^{***}_{\dagger\dagger}$	$66.5\%^{***}_{\dagger}$	$\mathbf{66.6\%}^{***}$	$63.4\%_{\dagger}$	63.7%	$63.9\%^{*}$	$\mathbf{63.9\%}^{*}$
No-Zero-Active	$74.5\%_{\dagger\dagger\dagger}$	$75.4\%^{***}_{\dagger\dagger\dagger}$	$75.4\%^{***}_{\dagger\dagger\dagger}$	$\mathbf{75.7\%}^{***}$	$62.9\%_{\dagger\dagger\dagger}$	$64.6\%^{***}$	$\mathbf{64.7\%}^{***}$	$64.6\%^{***}$
Full	$86.1\%_{\dagger\dagger\dagger}$	$86.2\%^{***}_{\dagger\dagger\dagger}$	$86.3\%^{***}$	$\mathbf{86.4\%}^{***}$	$54.8\%_{\dagger\dagger\dagger}$	$59.0\%^{***}$	$\mathbf{59.2\%}^{***}$	$59.0\%^{***}_{\dagger\dagger\dagger}$

structures, we do not reweight training examples as in boosting and we do not resample the data like in bagging methods. At the same time, each weak learner is trained to discriminate between different multilabels as well as possible.

Another way to understand the phenomenon is to see the edges of the task network as 'experts', and the collection of edges adjacent to a node as a 'expert committee' voting on the node label, each from a different context. The random pairing of tasks then induces a random set of experts. Random tree of tasks, in addition, makes the experts to negotiate on all node labels in order to keep the tree labeled consistently. Our experiments suggest that enforcing this consistency also may be beneficial.

5   Conclusions

We presented an ensemble approach for multi-task classification of drug bioactivity. The base classifiers of the ensemble, learned by Max-Margin Conditional Random Field algorithm (MMCRF), predict a labeling of a graph coupling the tasks together. The predictive performance of two types of ensembles, one based on random pairing of tasks, another based on a random spanning tree of tasks, surpasses that of SVM as well as single MMCRF model where the underlying graph has been built from auxiliary data using graphical lasso.

Acknowledgements. The work was financially supported by Helsinki Doctoral Programme in Computer Science (Hecse), Academy of Finland grant 118653 (ALGODAN), and in part by the IST Programme of the European Community, under the PASCAL2 Network of Excellence, ICT-2007-216886. This publication only reflects the authors' views.

References

1. Breiman, L.: Bagging predictors. Machine Learning 24, 123–140 (1996)
2. Ceroni, A., Costa, F., Frasconi, P.: Classification of small molecules by two- and three-dimensional decomposition kernels. Bioinformatics 23, 2038–2045 (2007)
3. Dietterich, T.: Ensemble methods in machine learning. Multiple classifier systems, 1–15 (2000)

4. Esuli, A., Fagni, T., Sebastiani, F.: Boosting multi-label hierarchical text categorization. Information Retrieval 11(4), 287–313 (2008)
5. Hastie, T., Tibshirani, R.: Sparse inverse covariance estimation with the graphical lasso. Biostatistics 9(3), 432–441 (2008)
6. Meinshausen, N., Bühlmann, P., Zürich, E.: High dimensional graphs and variable selection with the lasso. Annals of Statistics 34, 1436–1462 (2006)
7. Obrezanova, O., Segall, M.D.: Gaussian processes for classification: Qsar modeling of admet and target activity. Journal of Chemical Information and Modeling 50(6), 1053–1061 (2010)
8. Opitz, D., Maclin, R.: Popular ensemble methods: an empirical study. Journal of Artificial Intelligence Research 11, 169–198 (1999)
9. Ralaivola, L., Swamidass, S., Saigo, H., Baldi, P.: Graph kernels for chemical informatics. Neural Networks 18, 1093–1110 (2005)
10. Rousu, J., Saunders, C., Szedmak, S., Shawe-Taylor, J.: Kernel-Based Learning of Hierarchical Multilabel Classification Models. The Journal of Machine Learning Research 7, 1601–1626 (2006)
11. Rousu, J., Saunders, C., Szedmak, S., Shawe-Taylor, J.: Efficient algorithms for max-margin structured classification. Predicting Structured Data, 105–129 (2007)
12. Schapire, R.E., Singer, Y.: Boostexter: A boosting-based system for text categorization. Machine Learning 39(2/3), 135–168 (2000)
13. Schietgat, L., Vens, C., Struyf, J., Blockeel, H., Kocev, D., Džeroski, S.: Predicting gene function using hierarchical multi-label decision tree ensembles. BMC bioinformatics 11(1), 2 (2010)
14. Shivakumar, P., Krauthammer, M.: Structural similarity assessment for drug sensitivity prediction in cancer. Bioinformatics 10, S17 (2009)
15. Su, H., Heinonen, M., Rousu, J.: Structured Output Prediction of Anti-cancer Drug Activity. In: Dijkstra, T.M.H., Tsivtsivadze, E., Marchiori, E., Heskes, T. (eds.) PRIB 2010. LNCS, vol. 6282, pp. 38–49. Springer, Heidelberg (2010)
16. Trotter, M., Buxton, M., Holden, S.: Drug design by machine learning: support vector machines for pharmaceutical data analysis. Comp. and Chem. 26, 1–20 (2001)
17. Wang, Y., Bolton, E., Dracheva, S., Karapetyan, K., Shoemaker, B., Suzek, T., Wang, J., Xiao, J., Zhang, J., Bryant, S.: An overview of the pubchem bioassay resource. Nucleic Acids Research 38, D255–D266 (2009)
18. Yan, R., Tesic, J., Smith, J.: Model-shared subspace boosting for multi-label classification. In: Proceedings of the 13th ACM SIGKDD International Conference on Knowledge Discovery and Data Mining, pp. 834–843. ACM (2007)

A Bilinear Interpolation Based Approach for Optimizing Hematoxylin and Eosin Stained Microscopical Images

Kaya Kuru[1] and Sertan Girgin[2]

[1] Gülhane Military Medical Academy
kkuru@gata.edu.tr
[2] INRIA Lille Nord Europe, Villeneuve d'Ascq, France
sertan.girgin@inria.fr

Abstract. Hematoxylin & Eosin (H&E) is a widely used staining technique in medical pathology for distinguishing nuclei and cytoplasm in tissues by dying them in different colors; this helps to ease the diagnosis process. However, usually the microscopic digital images obtained using this technique suffer from uneven lighting, i.e. poor Koehler illumination. The existing ad-hoc methods for correcting this problem generally work in RGB color model, and may result in both an unwanted color shift and loosing essential details in terms of the diagnosis. The aim of this study is to present an alternative method that remedies these deficiencies. We first identify the characteristics of uneven lighting in pathological images produced by using the H&E technique, and then show how the quality of these images can be improved by applying an interpolation based approach in the Lab color model without losing any important detail. The effectiveness of the proposed method is demonstrated on sample microscopic images.

1 Introduction

Hematoxylin & Eosin (H&E) technique is one of the most common methods in medical pathology. The method allows to distinguish between nuclei and cytoplasm in tissues by dying them in different colors, namely blue and red. Routinely processed tissue sections are put into hematoxylin (a bluish dye) for 3-5 minutes. After removal of the excess dye, sections are then treated with a bicarbonate solution for about 2 minutes until nuclei stand out sharply blue. This is followed by rinsing and putting the slides into ethanol; sections are stained with eosin (a reddish dye) for 1-2 minutes. The staining procedure is finished by consecutive steps using ethanol, xylene and a mounting medium before coverslipping. This bi-coloring process helps to better depict the microscopic morphology of diseases and eases the diagnosis[1].

---

[1] Coloring process for blue is between yellow and blue and coloring process for red is between red/magenta and green respectively. Thus, sometimes it is not possible to see distinct blue or red colors on images.

M. Loog et al. (Eds.): PRIB 2011, LNBI 7036, pp. 168–178, 2011.
© Springer-Verlag Berlin Heidelberg 2011

With the recent advances in digital imaging, the latest digital cameras coupled with powerful computer software now offer image quality that is comparable with traditional silver halide film photography and greater flexibility for image manipulation and storage; hence, it is increasingly being used for image capture for microscopy – an area that demands high resolution, color fidelity and careful management [1].

Perhaps one of the most misunderstood and often neglected concepts in optical microscopy is the proper configuration of the microscope with regards to illumination, which is a critical parameter that must be fulfilled in order to achieve optimum performance [2]. The intensity and wavelength spectrum of light emitted by the illumination source is of significant importance, but even more essential is that light emitted from various locations on the lamp filament be collected and focused at the plane of the condenser aperture diaphragm [2]. Both the filament and condenser alignment procedures are necessary to achieve good Koehler illumination. Moreover, when working with living cells one must avoid relatively high light intensities and long exposure times that are typically employed in recording images of fixed cells and tissues (where photo bleaching is the major consideration) [3]. The misalignment of the filament and the condenser, low light situations and short exposure time lead to illumination problems; this reduces the quality of the digital images, in particular those obtained by using the H&E technique.

In order to remedy defects on digital H&E images, pathologists rely on various image processing applications to enhance them using the common RGB color model. The RGB color model is an additive color model in which red, green, and blue light is added together in various ways to reproduce other colors; it is optimized for display on computer monitors and peripherals, and is device dependent, meaning that the output may vary from one device to another. Enhancing images is usually a time consuming and mostly ad-hoc process; sometimes even experienced pathologists may not be able to get images in correct appearance. For example, Fig. 1 shows two sample pathological images in

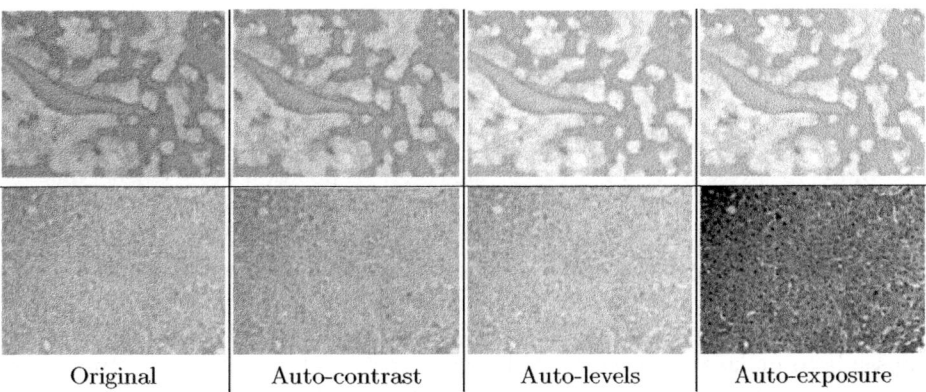

| Original | Auto-contrast | Auto-levels | Auto-exposure |

Fig. 1. Two sample H&E images before and after enhancements in RGB color model

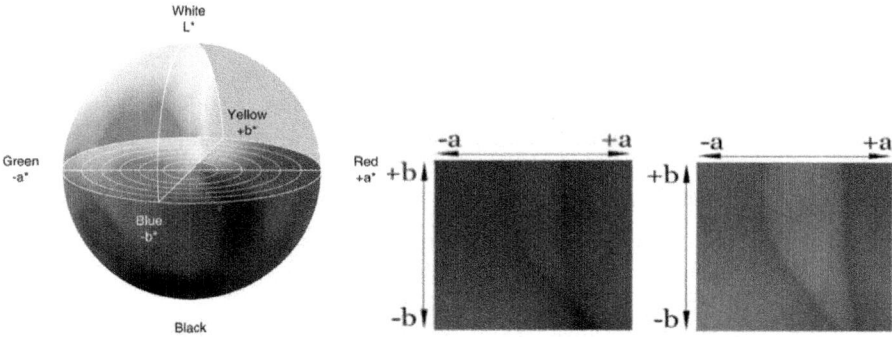

Fig. 2. The CIE LAB color model. The rectangles on the right show the colors in a* and b* dimensions with L* values of 25 and 50.

their original form and after employing three different enhancement procedures. Auto-contrast builds a histogram of the image and spreads out the values so the whole tonal-range is being used, auto-levels adjusts brightness and contrast to produce a balanced image with a good range of color intensities, and auto-exposure measures the darkest and lightest points in the image and adjusts the brightness accordingly. Note that, the effect of auto-contrast is minimal, and a color shift is apparent in the other enhanced images. The appearance of the image with color shift not only disturbs the appearance, but also causes the loss of essential details in the hematoxylin and eosin sections.

The difficulty stems from the fact that the RGB color model does not allow to modify the illumination of an image without altering the information stored in the red and blue channels of the image. Since, the nuclei of cells are colored in blue and the cytoplasms are colored in red in the H&E dying technique, illumination corrections in RGB color model inherently affect the data that the pathologists are interested in. In particular, some delicate details become barely perceptible impair the elaborateness of the images. This brings the need for a solution in which the lightness and the chromaticity of pixels in an image can be changed independent of each other. One possible option, is to use the CIE 1976 (L*, a*, b*), or CIE LAB, color space. The CIE LAB color space aims to conform to human vision and can describe all the colors visible to the human eye. It is a device-independent model established by the International Commission on Illumination [4]. The coordinates of CIE LAB color model, L*, a* and b*, represent the lightness of the color, its position between red/magenta and green, and its position between yellow and blue, respectively (Fig. 2). L* values range from 0 to 100; a value of 0 yields black and a value of 100 indicated diffuse white. Negative a* values indicate green while positive values indicate magenta; similarly, negative b* values indicate blue and positive values indicate yellow. L* component closely matches human perception of lightness. Thus, it is possible to make accurate color balance corrections by modifying output curves in the a* and b* components, or to adjust the lightness contrast using the L* component. As an example, the distribution of colors in a* and b* dimensions for

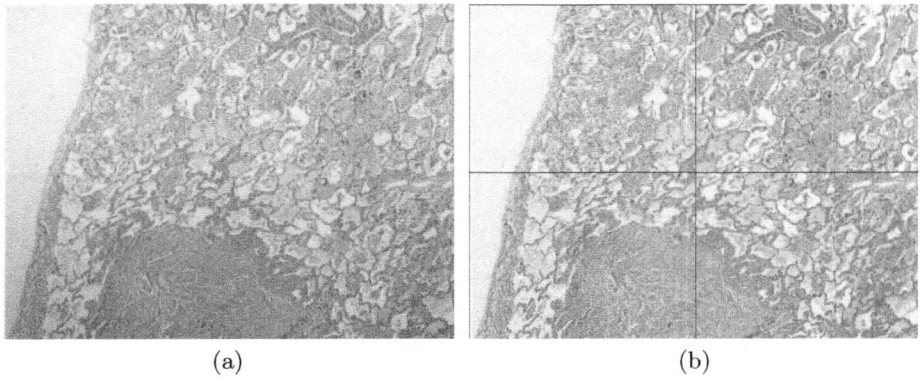

(a)   (b)

Fig. 3. Two images of the same specimen with light coming from different angles

two different L* values, 25 and 50, are shown in Fig. 2. Despite the change in the lightness, the color information is stable. By taking advantage of this property and the characteristics of the uneven lighting in H&E images, we propose a simple method to correct the illumination of the images and optimize them without losing any important detail which is essential for diagnosis.

The method will be described in detail and its effectiveness will be demonstrated on sample microscopic images in the next section.

2   The Method

As mentioned in the previous section, the CIE LAB color model provides a suitable and convenient way to represent and manipulate H&E images. In order to correct the illumination problems in these images, the remaining missing piece of the puzzle is the characteristics of uneven lighting that is observed in such images. For this purpose, we analyzed a large set of pathological images obtained using the H&E technique. Figure 3 shows two images of a sample specimen with light emitting from different angles. These images have a white area on the left side; because, there is no cell stained by the H&E technique at that area. But, in the first image, one can observe a gradient of increasing intensity (getting darker toward the bottom) which is caused by uneven lighting. The second image, on the other hand, has a relatively uniform illumination and does not suffer from such a problem; however, it still has an illumination problem spreading uniformly all over the image. The L*, a*,and b* and red, green, and blue components of both images in CIE LAB and RGB color models are presented in Fig. 4.

When we examine the lightness component, i.e. L*, of the images more carefully, we see that the L* values of pixels on the left side of the first image decrease linearly along the edge of the image (Fig. 5a); this trend spreads equally all over the image. As expected, the L* values of the pixels in the same region of the second image stay almost constant (Fig. 5b). Each pathological digital image with uneven lighting that we have examined had a similar linear pattern depending

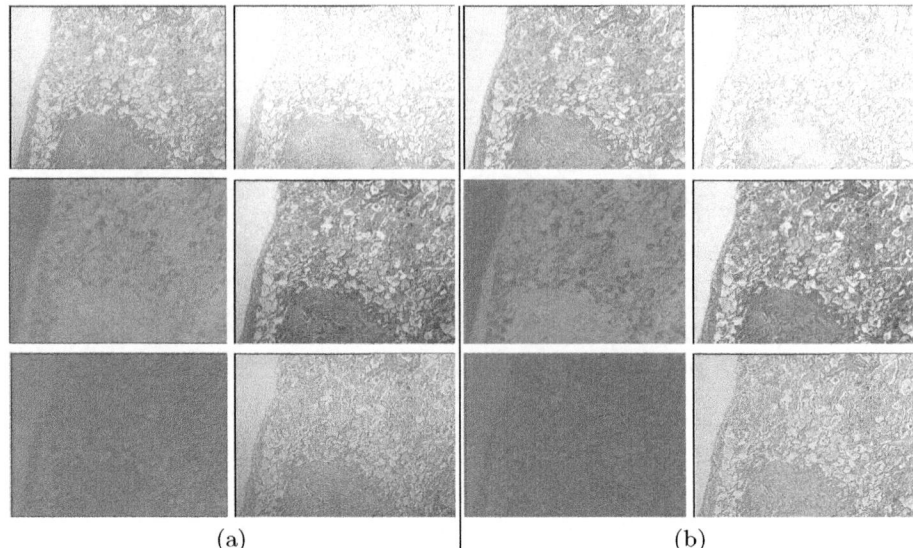

Fig. 4. L*, a*, and b* (first column, from top to bottom) and red, green, and blue (second column, from top to bottom) components of the images in Fig. 3 (a) and (b)

on the distinct positions of either the illumination apparatus or the microscope. From a global point of view, this common pattern tells us that the L* values of pixels in the CIE LAB color model and the relationship between them, for example, near the corners, indicate how the uneven lighting spreads through the image, either increasingly or decreasingly with a different angle with the XY plane. Given an image and a set of points on the image that are supposed to be white, i.e. the L* value equal to 100, by first converting the image to the CIE LAB color model, and then interpolating and exterpolating the L* values of these points subtracted from a reference value, in most cases 100, we can create a mask that approximates the distribution of the uneven lighting; subtracting this mask from the L* channel of the image and converting the image back to

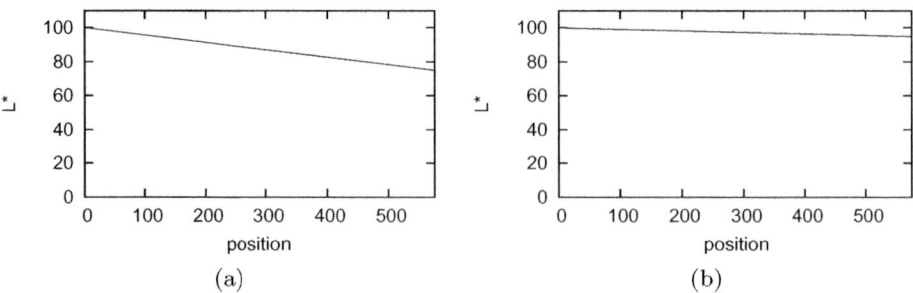

Fig. 5. The L* values of the pixels in the left-most area of the images in Fig. 3

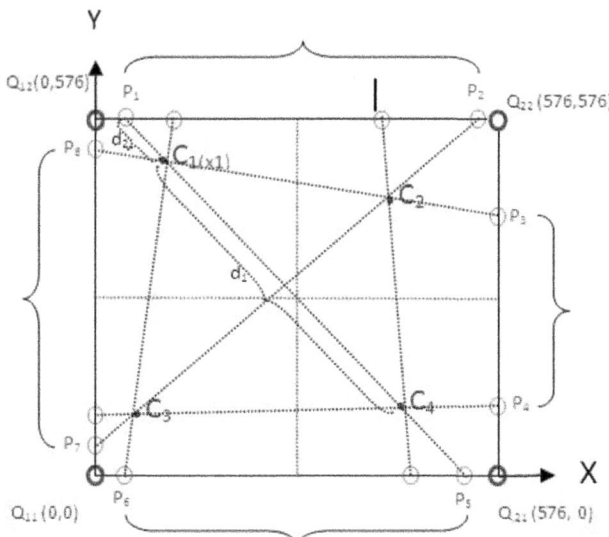

Fig. 6. The four red dots show the data click points and the red circled dots are the points at which we want to interpolate by using the calculated values (L values and xy coordinates values) of blue circled dots

the RGB color model allows us to reduce the uneven illumination, and in most cases eliminate it. The procedures for converting an image from the RGB color model to the CIE LAB color model and vice versa are provided in the appendix. Note that, during this process a* and b* channels of the image, which contain information about nuclei and cytoplasm in tissues, are not altered, and detail enhancement and sharpening can also be applied to this channels before converting the image back to the RGB color model to further improve the quality of the resulting image.

By taking into consideration the observed linear distribution, we opted for a bilinear interpolation scheme using four source points; this scheme allows to process images efficiently and produces good results. Let C_1, C_2, C_3 and C_4 be the four chosen points such that any three of them are not colinear. For each pair of points, C_i and C_j, $i < j$, we first determine the intersection points of the line passing from these two points and the image boundaries; if both points are on the same border of the image, then the intersection points are assumed to be the points themselves. The mask values of the intersection points are then calculated by linear extrapolation of the L* values of the corresponding source points subtracted from the reference value. Once the mask values of intersection points are known, the mask value of any point on a border of the image can be found by calculating a set of projected values and taking their mean; each projected value is calculated by interpolating or extrapolating linearly the mask values of a pair of intersection points on that border (see Fig. 6). Finally, in

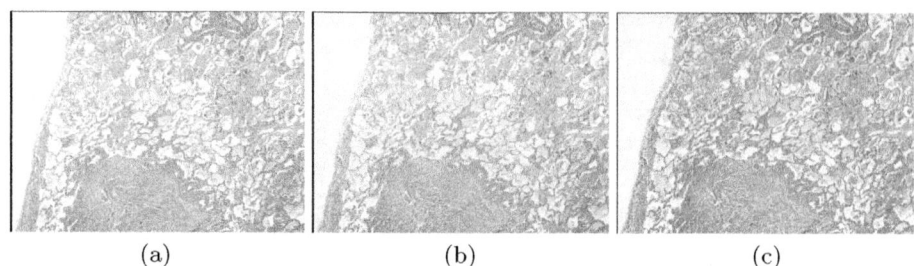

(a)   (b)   (c)

Fig. 7. (a, b) The resulting images after applying the proposed method to the sample images in Fig. 3; (c) the resulting image when correction is applied in the RGB color model

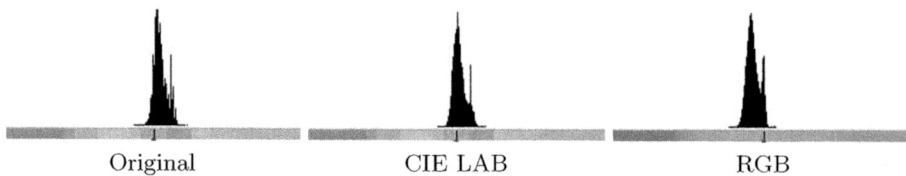

Original   CIE LAB   RGB

Fig. 8. The histograms of the b* channel of the sample image in CIE LAB color model (a) before and after correction in (b) CIE LAB and (c) RGB color models

order to find the mask value of any point (x, y) in the image, we apply linear interpolation in x and y directions separately using the mask values of the border points (i.e. interpolate the mask values of the two border points on column x and similarly on row y). Although each step is linear in the sampled values and in the position, the estimation as a whole is not linear but quadratic in the sample location.

Figure 7 shows the resulting images after applying the proposed method to the sample images in Fig. 3 with illumination problems. One can observe that the uneven lighting (poor Koehler illumination) is removed in both images. Correcting them in RGB color model with similar algorithm on all channels causes a color shift (Fig. 7c), demonstrating the advantage of using the CIE LAB color model. The histograms of the b* channel of the sample image with uneven illumination (Fig. 3a) before and after correction in CIE LAB and RGB color models are presented in Fig. 8. There is a nonsignificant difference between the histograms of the original image and the one processed in the LAB color model; the mean of the b* values is 133.32 in the first one (standard deviation 5.53) and 130.65 in the second (standard deviation 5.48). On the other hand, there is a significant difference between the histograms of the original image and the one processed in the RGB color mode with a shift to the left (the mean of the b* values is 118.75 with standard deviation 5.62); that is to say, there is a significant loss of essential data in terms of diagnosis.

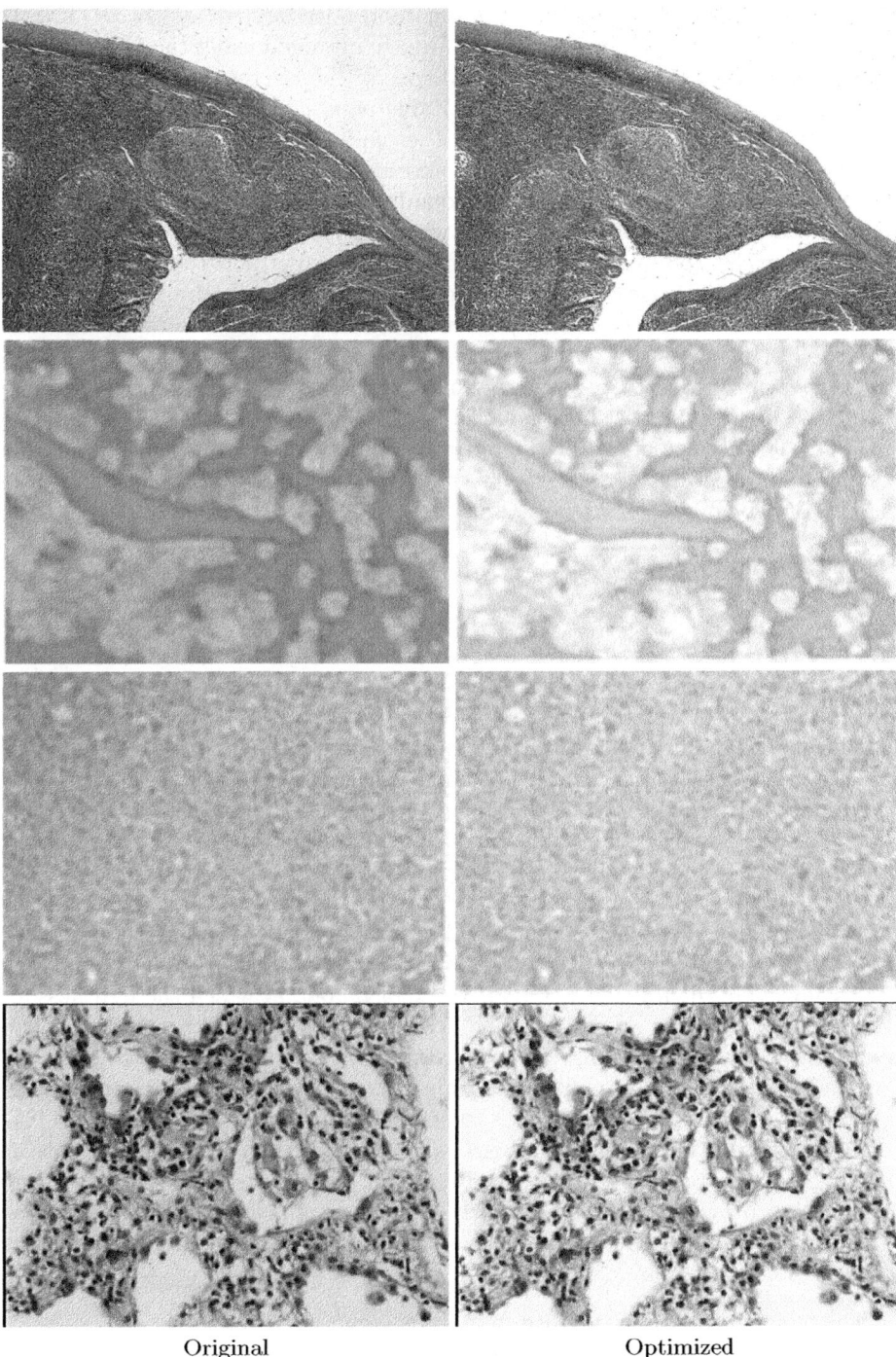

Original   Optimized

Fig. 9. Sample images before and after optimization in the CIE LAB color model

We have implemented the proposed method both in Java and in DOTNET[2]. The program allows the user to load an image obtained using the H&E staining technique, choose the source points and specify the reference L* value. In order to facilitate the process of choosing suitable points, the program can highlight the probable white areas of the image (i.e. with L* values in CIE LAB model close to 100 and not smaller than 90) and automatically determine the source points by dividing the image in four quadrants and picking in each quadrant the point among possible white points which is closest to the centroid of those points; if such a point does not exist in a quadrant, then the centroid of the points that are symmetric to the found points in other quadrants with respect to the XY axis is chosen as the source point and its mask value is set to the average of the mask values of the source points in other quadrants[3].

Some resulting corrected images that are produced by the application are presented in Fig. 9. Distinctive features of H&E technique in the images are made noticeably evident after the optimization of the illumination problem. No essential detail of nuclei and cytoplasm, in terms of diagnosis is lost, thanks to the LAB color model.

3   Conclusion

The digital camera systems have immediate advantages in the world of medicine - images can be distributed easily and without delay for second opinions or for educational purposes. However, no matter how good the camera is, it cannot produce outstanding images from a poorly configured microscope. Both the filament and condenser alignment procedures are necessary to achieve good Koehler illumination in microscopic images; furthermore, low light situations arise frequently and short exposure times may be compulsory. These affect the quality of the resulting images, and often they suffer from illumination problems.

In this paper, we proposed a simple but effective method for correcting uneven lighting in pathological images obtained using the Hematoxylin & Eosin staining technique. The method exploits the characteristics of such images by first converting them into the CIE LAB color model, in which the lightness and the chromaticy of pixels can be modified independent of each other, and then employing bilinear interpolation to mask out the uneven lighting that spreads through an image. The empirical studies on a diverse set of sample images show that the proposed method is effective in improving the quality of H&E images; unlike other ad-hoc methods in RGB color space, a color shift is not observed in the resulting images and essential details are kept. The developed application provides pathologists with little or no image processing experience an effective means of optimizing the H&E images for printing, image analysis, or

---

[2] The program can be downloaded from http://goo.gl/UGQR3

[3] Possible white areas can be seen easily in four quadrants of the images after they are highlighted, although they are not evident by naked eye. Choosing points other than white areas may cause unexpected results.

telepathology. We may explore some other enhancement for H&E images to improve diagnosis by adopting some basic image processing algorithms to other channels, "a" and "b", of LAB color mode as well as "L" channel as a future work.

References

1. Drent, P.: Digital imaging - new opportunities for microscopy. Microscopy, Nikon (2004),
   http://www.microscopyu.com/articles/digitalimaging/drentdigital.html
2. Parry-Hill, M., Sutter, R.T., Davidson, M.W.: Microscope alignment for kohler illumination. Microscopy, Nikon (2004),
   http://www.microscopyu.com/tutorials/java/kohler/
3. Hazelwood, K.L., Olenych, S.G., Griffin, J.D., Murphy, C.S., Cathcart, J.A., Davidson, M.W.: Introduction to digital imaging in microscopy. Hamamatsu (2004),
   http://learn.hamamatsu.com/articles/microscopyimaging.html
4. Fairchild, M.: Color appearance models, 2nd edn. Wiley - IS & T series in imaging science and technology. J. Wiley (2005); ISBN: 978-0-470-01216-1
5. Ford, A., Roberts, A.: Colour space conversions. Mendeley (1998),
   http://www.poynton.com/PDFs/coloureq.pdf

Appendix

Conversion between RGB and CIE LAB Color Models

In the study of color perception, one of the first mathematically defined color spaces was the CIE XYZ color space, created by the International Commission on Illumination (CIE) in 1931. CIE XYZ may be thought of as derived parameters from CIE RGB color space, the red, green, blue colors. CIE LAB color space is based directly on the CIE XYZ color space as an attempt to linearize the perceptibility of color differences. The non-linear relations for L^*, a^*, and b^* are intended to mimic the logarithmic response of the eye. In order to convert an image from RGB color space to CIE LAB color space (or vice versa), the CIE XYZ color space is used as an intermediate color space at transformation phases from one color space into other [5]. The conversion formulas between CIE XYZ and RGB color spaces are as follows:

$$X = 0.412453 * R + 0.357580 * G + 0.180423 * B$$
$$Y = 0.212671 * R + 0.715160 * G + 0.072169 * B$$
$$Z = 0.019334 * R + 0.119193 * G + 0.950227 * B$$
$$R = 3.240479 * X - 1.537150 * Y - 0.498535 * Z$$
$$G = -0.969256 * X + 1.875992 * Y + 0.041556 * Z$$
$$B = 0.055648 * X - 0.204043 * Y + 1.057311 * Z$$

The conversion formula from the CIE XYZ color space to the CIE LAB color space is defined as

$$L^* = 116f(\frac{Y}{Y_n}) - 16, a^* = 500[f(\frac{X}{X_n}) - f(\frac{Y}{Y_n})], b^* = 200[f(\frac{Y}{Y_n}) - f(\frac{Z}{Z_n})]$$

where $f(t) = t^{\frac{1}{3}}$ for $t > 0.008856$, otherwise $f(t) = 7.787t + \frac{16}{116}$ and X_n, Y_n and Z_n denote the CIE XYZ tristimulus values of the reference white point. The division of the $f(t)$ function into two domains is done to prevent an infinite slope at $t = 0$. $f(t)$ is assumed to be linear below some $t = t_0$, and is assumed to match the $t^{1/3}$ part of the function at t_0 in both value and slope.

$$t_0^{1/3} = t_0 + b \text{ (match in value) and } 1/(3t_0^{2/3}) = a \text{ (match in slope)}$$

In other words, the value of b was chosen to be 16/116. The above two equations can be solved for a and t_0 to obtain

$$a = \frac{1}{3\delta} = 7.787037 \text{ and } t_0 = \delta^3 = 0.008856$$

where $\delta = \frac{6}{29}$. The reverse transformation can then be calculated by applying the following rules:

1. Define $f_y = (L^* + 16)/116$
2. Define $f_x = f_y + a^*/500$
3. Define $f_z = f_x - b^*/200$
4. If $f_y > \delta$ then $Y = Y_n f_y^3$, else $Y = (f_y - 16/116)3\delta^2 Y_n$
5. If $f_x > \delta$ then $X = X_n f_x^3$, else $X = (f_x - 16/116)3\delta^2 X_n$
6. If $f_z > \delta$ then $Z = Z_n f_z^3$, else $Z = (f_z - 16/116)3\delta^2 Z_n$

Automatic Localization of Interest Points in Zebrafish Images with Tree-Based Methods

Olivier Stern[1], Raphaël Marée[1,2], Jessica Aceto[3], Nathalie Jeanray[3],
Marc Muller[3], Louis Wehenkel[1], and Pierre Geurts[1]

[1] GIGA-Systems Biology and Chemical Biology, Dept. of EE and CS,
University of Liège, Belgium
[2] GIGA Bioinformatics Core Facility, University of Liège, Belgium
[3] GIGA-Development, Stem Cells and Regenerative Medicine,
Molecular Biology and Genetic Engineering,
University of Liège, Belgium

Abstract. In many biological studies, scientists assess effects of experimental conditions by visual inspection of microscopy images. They are
able to observe whether a protein is expressed or not, if cells are going
through normal cell cycles, how organisms evolve in different experimental conditions, etc. But, with the large number of images acquired in
high-throughput experiments, this manual inspection becomes lengthy,
tedious and error-prone. In this paper, we propose to automatically detect specific *interest points* in microscopy images using machine learning methods with the aim of performing automatic morphometric measurements in the context of Zebrafish studies. We systematically evaluate variants of ensembles of classification and regression trees on four
datasets corresponding to different imaging modalities and experimental conditions. Our results show that all variants are effective, with a
slight advantage for multiple output methods, which are more robust to
parameter choices.

1 Context, Motivation, and Strategy

The zebrafish is a well-known model organism increasingly used for biological
studies on development, gene function, toxicology, and pharmacology. In addition
to its major biological advantages (ease of reproduction, quick growth, genome
close to the human's), the fact that embryos are transparent eases microscopic
observations. More specifically, in bone research studies, the skeleton can be
observed at different stages of development combined with appropriate staining
[1,13]. From these images, one seeks to perform morphometric measurements of
the cartilage skeleton to describe the effects of different experimental conditions
such as chemical treatments or gene knock-downs. Interesting measurements
include the length of cartilages or angles defined by specific interest points.

Traditionally, effects of biological experiments on zebrafish embryos are evaluated manually through microscopic observation. However, due to the large number of experimental protocols, chemical substances, acquisition modalities, and

M. Loog et al. (Eds.): PRIB 2011, LNBI 7036, pp. 179–190, 2011.
© Springer-Verlag Berlin Heidelberg 2011

the recent availability of high-throughput imaging equipments, visual inspection of zebrafish images by experts is becoming a limiting factor in terms of time and cost. Moreover, for humans it is often hard to distinguish visually subtle changes and in particular to perform measurements in a reproducible way. Using traditional low-level image processing methods (e.g. those based on thresholding and mathematical morphology) would also be limiting because a significant number of factors make images quite different from an experiment to another hence it would require tuning image processing operations for each and every experiment. Indeed, factors such as biological preparation protocols, imaging acquisition procedures, and experimental conditions produce very different types of images, such as those considered in this paper and illustrated by Figure 1.

These observations motivated us to consider generic machine learning methods to speed-up the reproducible extraction of quantitative information from these images. In our approach, experts first encode manually the localization of interest points within a few training images, for each batch of images. These annotations are then used to train either classification or regression models that are used in order to locate in a fully automatic way these interest points in the remaining images of the current batch.

2   Methods

As stated before, we follow a supervised learning approach, ie. we exploit manually annotated images (see Figure 1 for a few examples) where interest points coordinates have been localized by experts to train models able to predict those interest points in new, unseen images. The approach first extracts subwindows (or patches) around points of interest and at other randomly chosen positions within images, describe these patches by various visual features, and then either a classification or a regression model is built. In the classification scheme, the model is trained to predict whether the central pixel of a subwindow is an interest point or not (a binary classification problem). In the regression scheme, the model predicts the distance between the central pixel of a subwindow and the interest point. These models are built using either single output (one model per interest point) or multiple outputs (one model predicts simultanously all the interest points).

Table 1 describes the overall algorithmic approach within the single output setting. The different steps of this procedure are further explained in the following subsections.

2.1   Extraction and Description of Subwindows

The input of our learning algorithms is a learning set of subwindows of size l x l extracted within the training images in the following way: (i) for each pixel located within a circular region of radius r around an interest point a subwindow centered on this pixel is extracted; (ii) a certain number of subwindows are

Table 1. Training and testing algorithms (single output setting)

Parameters: radius of the interest region r, subwindows size l, *method*: either 'classification' or 'regression', a subwindow feature extractors $x(.;l)$, Extra-trees parameters T and K.

Train(LS)

Input: a learning sample of N images with the interest point position:

$$LS = \{\langle I_i, (p_{x,i}, p_{y,i}) \rangle | i = 1, \ldots, N\},$$

Output: an ensemble of trees defined on subwindows features
- $LS_{sw} = \emptyset$
- For each pair $\langle I, (p_x, p_y) \rangle \in LS$:

  – $S_p = \emptyset$
  – add in S_p all positions (p'_x, p'_y) such that $(p'_x - p_x)^2 + (p'_y - p_y)^2 < r^2$.
  – Let $P = |S_p|$ the size of S_p. Add in S_p $2P$ positions (p'_x, p'_y) randomly drawn in the image such that $(p'_x, p'_y) \notin S_p$
  – For each (p_x, p_y) in S_p, add $\langle x(p_x, p_y; l), y \rangle$ in LS_{sw} where:
    • $x(I, p_x, p_y; l) \in I\!R^m$ is the feature descriptors (of size m) of the $l \times l$ subwindow centered at (p_x, p_y) in I
    • y is either $1((p'_x - p_x)^2 + (p'_y - p_y)^2 < r^2)$ if *method*='classification' or $(p'_x - p_x)^2 + (p'_y - p_y)^2$ if *method*='regression'.

- Return the ensemble of trees obtained by the Extra-trees algorithm applied on LS_{sw}

Predict(I,ens)

Input: an image I, an ensemble of trees *ens* returned by the function **Train**
Output: a prediction $(\hat{p}_x, \hat{p}_y)$ of the position of the interest point in I
- For all pixels (p_x, p_y) in I, compute the predictions $\hat{y}(p_x, p_y)$, by propagating the feature vectors $x(I, p_x, p_y; l)$ into *ens*.
(*NB:* $\hat{y}(p_x, p_y)$ *is the predicted probability of class j in classification and the predicted distance to the interest point in regression.*)
- Compute as a prediction for the position of the interest point:

$$(\hat{p}^j_x, \hat{p}^j_y) = median(\{(p_x, p_y) \in I | \hat{y}^j(p_x, p_y) = \bar{y}\},$$

where:

  – $\bar{y} = \arg\max_{(p_x, p_y)} \hat{y}(p_x, p_y)$ if *method*='classification',
  – $\bar{y} = \arg\min_{(p_x, p_y)} \hat{y}(p_x, p_y)$ if *method*='regression'.

---

randomly extracted from the rest of the image. In our experiments we always used twice as many "other" subwindows as the number of interest point specific ones, which increases proportionally to the radius r.

Input Features. Each subwindow is then processed to compute the input features using three visual cues:

- Color and grayscale: Each pixel of a subwindow is decomposed in the Red-Green-Blue (RGB) color space (3 features per pixel), in Hue-Saturation-Value (HSV) color space (3 features per pixel), and in grayscale (luminance, 1 feature per pixel).
- Edges: The gradient of the Sobel operator is applied on each pixel and its direct neighbours. Considering A as a 3×3 matrix of a pixel and its eight neighbours in grayscale, we define the gradient G (1 feature per pixel) as:

$$G = \sqrt{G_x^2 + G_y^2} \text{ with } G_x = \begin{bmatrix} -1 & 0 & +1 \\ -2 & 0 & +2 \\ -1 & 0 & +1 \end{bmatrix} \times A \text{ and } G_y = \begin{bmatrix} -1 & -2 & -1 \\ 0 & 0 & 0 \\ +1 & +2 & +1 \end{bmatrix} \times A,$$

where $\times$ denotes the scalar product.
- Texture: We use the basic version of the local binary pattern (LBP) [16] to describe the texture of subwindows. Each pixel of a subwindow is compared to its 8 neighbors. If the intensity of the center pixel is greater than the compared neighbor, we encode it by 1, and otherwise by 0. This gives an 8-digit binary number per pixel, that we convert in decimal. We compute the histogram of these numbers over the subwindow, yielding a 256-dimensional feature vector.

Overall, we thus use $m = (3 + 3 + 1 + 1) \times l \times l + 256$ numerical features to describe the visual content of the subwindows.

Outputs. In the single output classification approach, the output is binary and equal to 1 if the central pixel is an interest point, 0 otherwise. In the multiple-output approach, the output is a vector of $N_p + 1$ class-indicator variables, where the N_p first components correspond to the N_p types of interest points whereas the last component corresponds to the background.

In the single output regression approach, the output is a number reflecting the distance of the center pixel to the interest point. In the multiple output approach, the output is a vector of N_p numbers corresponding to the distances to the N_p interest points.

2.2   Model Construction Using Extremely Randomized Tree Ensembles

Starting with the learning set of subwindows at the top-node, the Extra-Trees algorithm [10] builds an ensemble of T fully developed decision trees. At each node, it generates tests on input variables (features) in order to progressively partition the input space into hyper-rectangular regions where the output is constant. In order to select relevant tests, k features are chosen at random at each node, where the filtering parameter k can take all possible values from 1 to the total number m of features. For each of these k features, a numerical value is randomly chosen within the range of variation of that feature in the subset of subwindows available in the current tree node. The score of each binary test is then computed on the current subwindow subset, and the best test among the k tests is chosen to split the current node into two child nodes. For single output classification and regression trees, we use CART's standard score measures, i.e., Gini index reduction in classification and output variance reduction in the case of regression [5]. In the case of multiple classification or regression outputs, prediction at leaf nodes of the trees are extended to be vectorial and we use as a score measure the sum of the scores for each individual output (see e.g. [4,7] for a treatment of multiple output trees).

2.3   Prediction in Test Images

To localize interest points in a new, unseen image, we extract a subwindow centered at every pixel position and propagate it into the trees to predict one or more value(s) according to the model used. Indeed, as explained in Section 2, the output of classification and regression models will be different.

In the classification scheme, a model outputs probability estimates to determine if the central pixel of a subwindow is or not an interest point. The multiple output approach will consider all the interest points at once, while in the single output approach each model is applied separately to predict probability estimates for each interest point. In order to produce the coordinates of one interest point within a test image, we then compute the median of all pixel positions which obtained the highest predicted probability estimate.

In the regression scheme, we predict the euclidian distance from the central pixel of a subwindow to our interest points. In the multiple output variant, we consider as ouputs the distances to all the interest points at once, while in the single output scheme each model predicts the distance to a specific interest point. To obtain the predicted coordinates of one particular interest point within a test image, we compute the median of all pixel positions which obtained the smallest predicted distance.

2.4   Related Work

Various studies use computational techniques for zebrafish image quantification but none addresses the problem of skeleton/cartilage morphometric measurements using machine learning methods, to the best of our knowledge. In [2], a study of embryo images submitted to toxicological treatments is proposed. They aim at observing the mortality rate depending on the toxicological concentrations, by extracting image features (e.g. variance of pixel values) to distinguish dead and alive embryos. Images are then labeled by experts and classified thanks to the Matlab Gait-CAD toolbox in only two classes. [3] describes a way to automatically obtain images of zebrafish embryos thanks to a motorized microscope, and classifies them manually into phenotypes. [19] developed an automatic system of data acquisition and embryos' analysis in multi-well plates, where manual intervention is still needed to produce analysis routines based on several image segmentations. More recently, another approach to classify embryos of zebrafish depending on several phenotypes has been proposed in [12]. Manually acquired images of zebrafish embryos are first pre-processed to standardize images which are then submitted to a phenotypic classification. The supervised learning algorithm used is based on random subwindows extraction in images [15], their description by raw pixel values, and the use of ensembles of extremely randomized trees [10] to classify these subwindows hence images.

Beyond studies involving Zebrafish images, one can see similarities between our work and some applications in the broader pattern recognition literature. In the field of face recognition, multi-stage classification and regression approaches have been proposed (e.g. [6,8]) to localize facial features (e.g. eyes) as a preliminary step for face recognition. In scene and object recognition tasks, [18] emulates two generic interest point detectors (Hessian-Laplace and Kadir-Brady) using boosting approaches. [9,14] use randomized trees for object detection and real-time tracking. In medical imaging, [17] uses regression trees to detect bounding boxes of organs in CT images.

In protein bioinformatics, [11] compares classification and regression approaches for protein binding site prediction using patch based predictors similarly as our visual interest point localization with subwindow predictors.

3   Experiments

3.1   Datasets

We apply our method on four different image datasets illustrated in Figure 1. The image size in all datasets is 400 pixels × 300 pixels.

- CTL: a batch of 15 wild type zebrafish images stained with alcian blue. The four interest points are located on the skeleton.
- DRUG: a second batch of 20 zebrafish images from a toxicology experiment (also stained with alcian blue but with slightly different imaging settings). From the biological point of view, the goal is to compare these to the CTL database so as to quantify morphometric changes due to drug treatment.
- RED: a batch of 24 zebrafish control images stained with alizarin red to detect the bone skeleton.
- EMBRYO: a batch of 20 zebrafish embryo images where the four interest points characterize body parts.

3.2   Evaluation Protocols

Considering the experts' annotations in all images for each database, we want to evaluate the performance of the classification and the regression methods for single and multiple outputs. To do so, we will perform a leave-one-out cross validation for different values of method parameters: the number of trees in the ensemble (T), the value of the filtering parameter (k), the radius of the circular region around the interest points (r), and the size of the subwindows (l). The default values of these parameters are $T = 20$, $k = 10$, $r = 5$, and $l = 21$, leading to a good compromize between accuracy and computational complexity.

For each experiment (i.e. each leave-one-out computation on one dataset and with a given method setting and parameter values), we have constructed a distance accuracy graph which represents the percentage of predicted points within a distance d to the interest points. Figure 2 shows two such accuracy graphs for one of the variants, illustrating the effect of the parameter l. The faster the curves reach the upper bound of 100%, the better the approach. Hence, we will use the area under these curves (AUC) to assess and compare the different settings and parameter values.

For our four datasets, the results of our empirical assessment are in Figures 5, 6, 7, and 8. Each graph shows the AUC of the four methods as a function of one of the four parameters, the other parameters being kept constant and set to their default values. Notice that overall, these experiments involved roughly 2500 computer jobs (each one corresponding to one experiment).

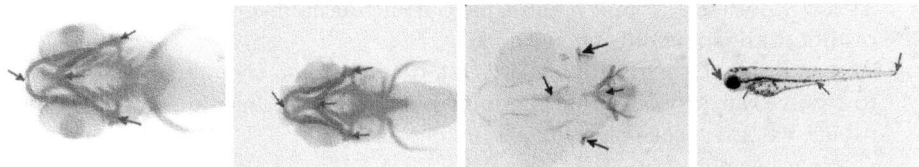

Fig. 1. An image from each database (CTL, Drug, Red, Embryo) where their manually annotated interest points are pointed by an arrow

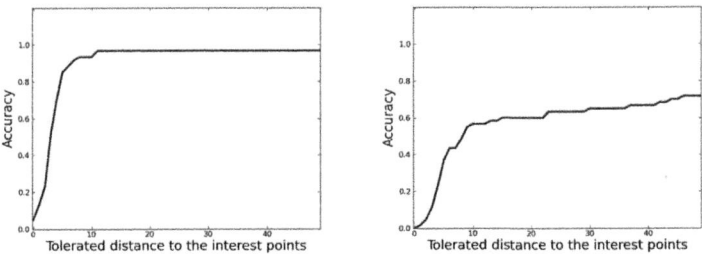

Fig. 2. Distance accuracy graphs for the CTL database (multiple-output classification setting; $T = 20, k = 10, r = 5$): left $l = 21$; right $l = 5$

3.3   Results and Observations

At first, we can see on the prediction example in Figure 3 that the results are visually very satisfying. We observed such results in each database for each method setting, provided that the parameter values are well chosen.

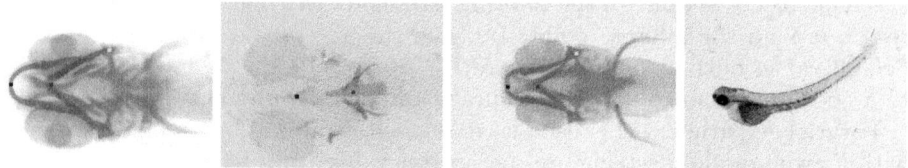

Fig. 3. Prediction of interest points in each database: multiple output regression approach with default parameter settings ($T = 20, k = 10, r = 5, l = 21$)

Let us further investigate the results according to the different AUC produced as a function of the parameters and method settings. According to figures 5, 6, 7 and 8 we can make the following observations:

– The parameter k has little impact on the accuracy
– For higher values of r, the single output regression method obtains the best results
– The size of subwindows is very important, whatever the method. Only high values of l obtain good performances
– At low values of l, T and r, multiple output methods obtain generally better results than single output methods
– At the light of these graphs, it's not obvious that there is a best method to resolve our problem, but the multiple output settings appear to be more robust with respect to parameter values.

A last remark concerns the difference between the output of the classification and the regression methods. Figure 4 illustrates the fact that the classification setting often finds several 'best predictions' while in the regression setting we observed that in general only one position is predicted as the most likely one.

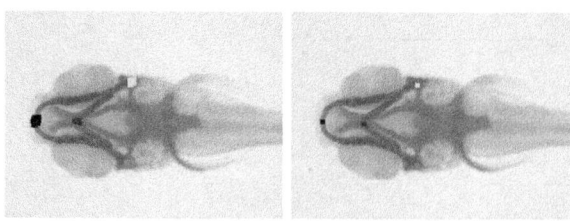

Fig. 4. Output of the best predictions in classification (left) and regression (right) methods for a same image, before evaluating the median

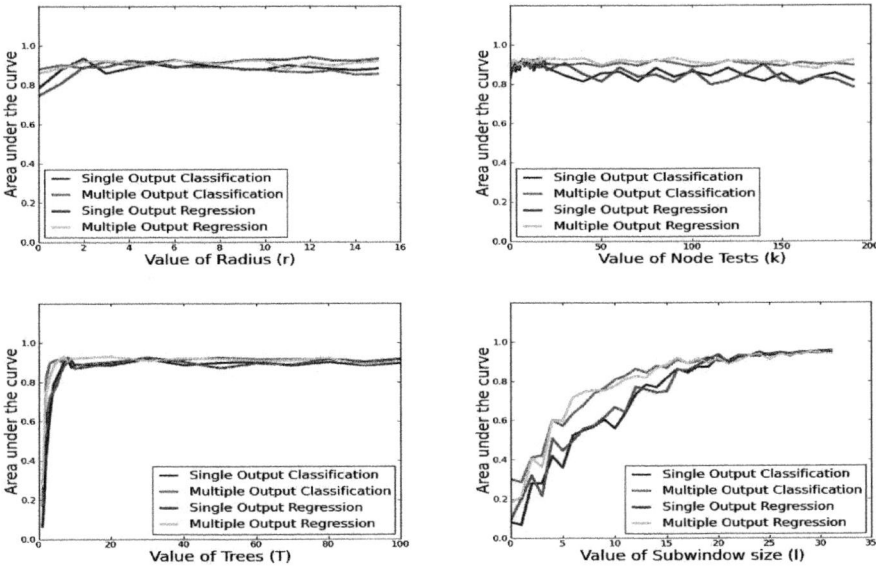

Fig. 5. [CTL database] Area under the curve computed from the distance accuracy graphs for the different values of the parameters r, k, T, l

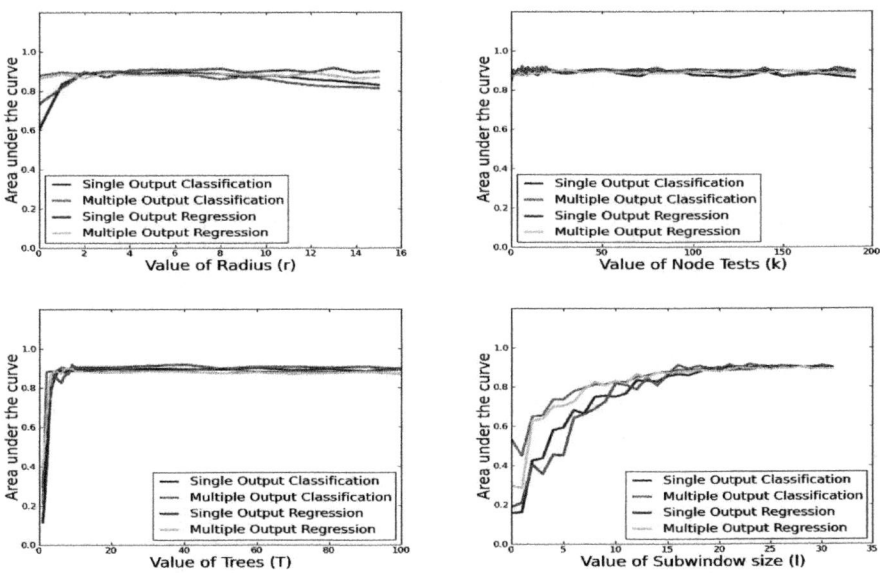

Fig. 6. [DRUG database] Area under the curve computed from the distance accuracy graphs for the different values of the parameters r, k, T, l

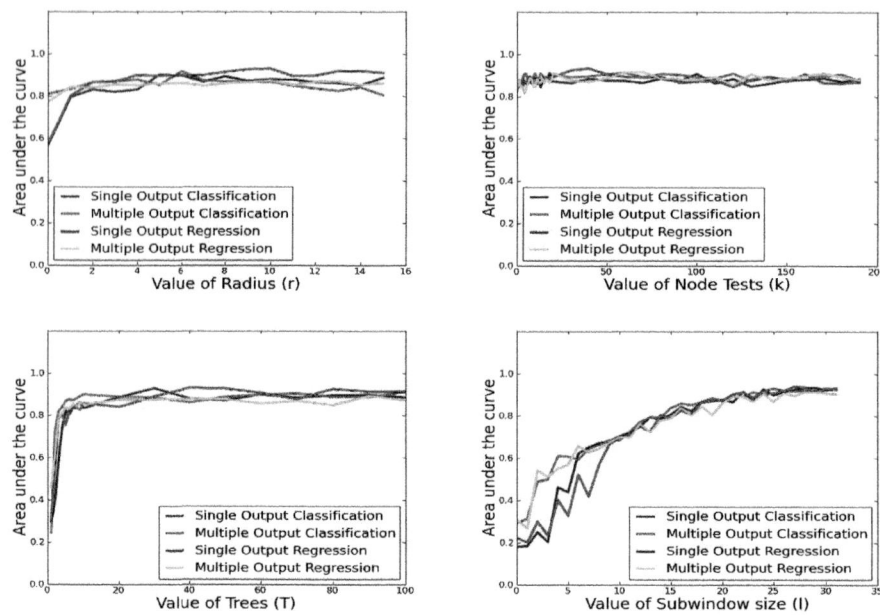

Fig. 7. [RED database] Area under the curve computed from the distance accuracy graphs for the different values of the parameters r, k, T, l

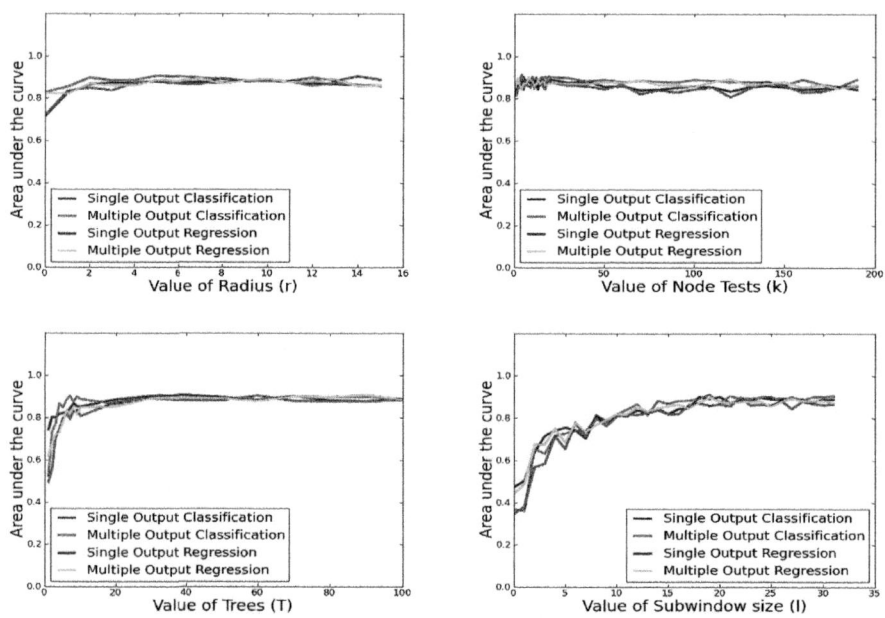

Fig. 8. [EMBRYO database] Area under the curve computed from the distance accuracy graphs for the different values of the parameters r, k, T, l

4   Conclusion and Future Work

In this paper, we tackled the task of specific interest point detection in zebrafish images using machine learning methods based on ensembles of randomized regression and classification trees. We compared different settings (multiple vs single output, regression vs classification) on four imaging datasets and have studied the effect of various parameters on accuracy.

Our study shows that all approaches give good results provided that parameters are well chosen. We also found that the parameter which has the strongest influence is the window size, and that the multiple output setting is less sensitive to parameter choices than the single output setting.

Although this work did not focus on the computational aspects, we notice that training and prediction with tree-based ensemble methods is highly scalable with respect to dataset size and feature space dimensionality. As a matter of fact, the main computational burden of the approach is related to the extraction of subwindows from the original images and the subsequent feature computations. These are specially demanding at the stage of prediction, since one subwindow for each pixel has to be extracted, represented and then classified.

Future work will thus look at computational optimizations and further algorithmic optimizations and more extensive large scale validation studies, specially in the context of toxicology studies. In terms of accuracy, we note that in this paper we used the precision of interest point localization as a criterion, while in practical biological studies these are used in order to compute more complex geometrical features or phenotype classifications, generally involving the computation of several interest points. Hence, accuracy evaluations should also be made on these end-outcomes used by biologists to assess statistical significance of the impact of the considered experimental conditions.

Acknowledgments. This paper presents research results of the ARC BIOMOD, the Interuniversity Attraction Poles Programme (IAP P6/25 BIOMAGNET), initiated by the Belgian State, Science Policy Office, and by the European Network of Excellence, PASCAL2. RM is supported by the GIGA interdisciplinary cluster of Genoproteomics of the University of Liège with the help of the Walloon Region and the European Regional Development fund, and by the CYTOMINE research grant n°1017072 of the Walloon Region (DGO6). PG is a research associate of the FNRS, Belgium.

References

1. Adkins, K.F.: Alizarin red s as an intravital fluorochrome in mineralizing tissues. Stain Technol. 40, 69–70 (1965)
2. Alshut, R., Legradi, J., Liebel, U., Yang, L., Van Wezel, J., Strähle, U., Mikut, R., Reischl, M.: Methods for automated high-throughput toxicity testing using zebrafish embryos. In: Dillmann, R., et al. (eds.), pp. 219–226 (2010)

3. Arslanova, D., Yang, T., Xu, X., Wong, S., Augelli-Szafran, C., Xia, W.: Phenotypic analysis of images of zebrafish treated with alzheimer's γ-secretase inhibitors. BMC Biotechnology, 10–24 (2010)
4. Blockeel, H., De Raedt, L., Ramon, J.: Top-down induction of clustering trees. In: Proceedings of ICML 1998, pp. 55–63 (1998)
5. Breiman, L., Friedman, J., Stone, C.J., Olshen, R.A.: Classification and Regression Trees. Wadsworth and Brooks, Monterey (1986)
6. Campadelli, P., Lanzarotti, R., Lipori, G.: Eye localization: a survey. The Fundamentals of Verbal and Non-verbal Communication and the Biometrical Issue (2007)
7. Dumont, M., Marée, R., Wehenkel, L., Geurts, P.: Fast multi-class image annotation with random subwindows and multiple output randomized trees. In: Proc. of the International Conference on Computer Vision Theory and Applications, vol. 2 (2009)
8. Everingham, M., Zisserman, A.: Regression and classification approaches to eye localization in face images. In: Proc. of the 7th Int. Conf. on Automatic Face and Gesture Recognition, pp. 441–448 (2006)
9. Gall, J., Yao, A., Razavi, N., Van Gool, L., Lempitsky, V.: Hough forests for object detection, tracking, and action recognition. IEEE Transactions on Pattern Analysis and Machine Intelligence, 1022–1029 (2011)
10. Geurts, P., Ernst, D., Wehenkel, L.: Extremely randomized trees. Machine Learning 36, 3–42 (2006)
11. Giard, J., Ambroise, J., Gala, J.-L., Macq, B.: Regression applied to protein binding site prediction and comparison with classification. BMC Bioinformatics 10(276) (2009)
12. Jeanray, N., Marée, R., Pruvot, B., Stern, O., Geurts, P., Wehenkel, L., Muller, M.: Phenotype classification of zebrafish embryos by supervised learning. Poster at Belgian Dutch Conference on Machine Learning, Benelearn (2011)
13. Kimmel, C.B., Miller, C.T., Kruze, G., Ullmann, B., BreMiller, R.A., Larison, K.D., Snyder, H.C.: The shaping of pharyngeal cartilages during early development of the zebrafish. Dev. Biol. 203, 245–263 (1998)
14. Lepetit, V., Fua, P.: Keypoint recognition using randomized trees. IEEE Transactions on Pattern Analysis and Machine Intelligence 28, 1465–1479 (2006)
15. Marée, R., Geurts, P., Piater, J., Wehenkel, L.: Random subwindows for robust image classification. In: Schmid, C., Soatto, S., Tomasi, C. (eds.) Proceedings of the IEEE International Conference on Computer Vision and Pattern Recognition (CVPR 2005), vol. 1, pp. 34–40. IEEE (2005)
16. Ojala, T., Pietikainen, M., Harwood, D.: A comparative study of texture measures with classification based on feature distributions. Pattern Recognition 29, 51–59 (1996)
17. Pathak, S.D., Criminisi, A., Shotton, J., White, S., Robertson, D., Sparks, B., Munasinghe, I., Siddiqui, K.: Validating automatic semantic annotation of anatomy in dicom ct images. In: Proceedings of the Medical Imaging 2011: Advanced PACS-based Imaging Informatics and Therapeutic Applications (2011)
18. Sochman, J., Matas, J.: Learning fast emulators of binary decision processes. International Journal of Computer Vision 83, 149–163 (2009)
19. Vogt, A., Cholewinski, A., Shen, X., Nelson, S., Lazo, J., Tsang, M., Hukriede, N.: Automated image-based phenotypic analysis in zebrafish enbryos. Developmental Dynamics 238, 656–663 (2009)

A New Gene Selection Method Based on Random Subspace Ensemble for Microarray Cancer Classification

Giuliano Armano[1], Camelia Chira[2], and Nima Hatami[1]

[1] DIEE-Department of Electrical and Electronic Engineering,
University of Cagliari,
Piazza D'Armi, I-09123 Cagliari, Italy
[2] Department of Computer Science,
Babes-Bolyai University,
Kogalniceanu 1, Cluj-Napoca 400084, Romania

Abstract. Gene expression microarray data provides simultaneous activity measurement of thousands of features facilitating a potential effective and reliable cancer diagnosis. An important and challenging task in microarray analysis refers to selecting the most relevant and significant genes for data (cancer) classification. A random subspace ensemble based method is proposed to address feature selection in gene expression cancer diagnosis. The introduced Diverse Accurate Feature Selection method relies on multiple individual classifiers built based on random feature subspaces. Each feature is assigned a score computed based on the pairwise diversity among individual classifiers and the ratio between individual and ensemble accuracies. This triggers the creation of a ranked list of features for which a final classifier is applied with an increased performance using minimum possible number of genes. Experimental results focus on the problem of gene expression cancer diagnosis based on microarray datasets publicly available. Numerical results show that the proposed method is competitive with related models from literature.

Keywords: random subspace ensembles, multiple classifier systems, multivariate feature selection, gene expression data analysis, pairwise diversity.

1 Introduction

High-throughput technologies are able nowadays to produce huge amount of valuable information which can be used in the identification and classification of various diagnostic areas. The required analysis of this information creates a real challenge for machine learning and new robust models are still required to efficiently tackle this task.

Cancer diagnosis based on gene expression data is an important emerging medical application domain of microarray analysis tools [14]. Clinical-based cancer classification has been shown to have limited diagnostic ability [11]. On the other

M. Loog et al. (Eds.): PRIB 2011, LNBI 7036, pp. 191–201, 2011.

hand, the classification of different tumor types using gene expression data is able to generate valuable knowledge for important tasks such as cancer diagnosis and drug discovery [11].

Classifying microarray samples (for example, cancer versus normal cells) according to their gene expression profiles represents an important and challenging task. The complexity of the problem rises from the huge number of features that contribute to a profile compared to the very low number of samples normally available in microarray analysis. Another challenge for classification is the presence of noise (biological or technical) in the dataset which further affects the classifier accuracies. Moreover, the inherent presence of a large number of irrelevant genes increases the difficulty of the classification task influencing the discrimination power of relevant features [11].

The problem of extracting significant knowledge from microarray data requires the development of robust methods able to address this task [9,12]. Lu and Han [11] emphasize that the most important aspects of classification and gene selection methods are their computation time, classification accuracy and ability to reveal biologically meaningful gene information. Many studies consider the selection of genes to be an important integral preprocessing step for the classification problem [11] able to reduce dimensionality, remove irrelevant or noisy genes and improve the learning accuracy [11,8,1]. Furthermore, gene selection can reduce the computational cost of the classifier and lead to more compact results easily interpretable in the diagnostics task [10].

Multiple classifier systems use base classifiers with complementary behaviour resulting in an efficient alternative to a complex and hard-to-manage single classifier. Among many well-known ensemble methods, Random Subspace Ensemble (RSE) [4] is an efficient model which obtained good results particularly for high-dimensional classification problems. RSE uses a number of base classifiers, each of them considering only a subset (randomly determined) of the original feature space.

In this paper, the *Diverse Accurate Feature Selection (DAFS)* method is introduced to deal with feature selection in gene expression cancer diagnosis. The main idea behind DAFS is to adapt RSE for feature selection by efficiently exploring accuracy and diversity information. To be more specific, the proposed DAFS method uses individual and ensemble accuracies of base classifiers specialized on random subspaces as well as pairwise diversity to rank features. This is achieved by assigning to each feature a score calculated by a metric which takes into account the mean ratio between individual and ensemble accuracies weighted by their diversity. This way, the introduced approach takes advantage of RSE to deal with the high-dimensionality of the addressed problem while in the same time building many different classifiers on the same samples to overcome the sample-size limitation.

The proposed method is evaluated for 11 cancer gene expression datasets [13] (from which nine are multiclass and two are binary classification problems). Computational experiments and comparissons to related methods indicate a good performance of the proposed DAFS method and emphasize the potential of RSE to deliver a fast effective feature selection method.

The structure of the paper is as follows: section 2 presents the DAFS method detailing the feature-space coverage approach, the feature evaluation and ranking phase and the DAFS algorithm; section 3 presents the computational experiments discussing the obtained results and comparing them with related methods and section 4 contains the conclusions of the paper.

2   Diverse Accurate Feature Selection

In this section, we introduce the *Diverse Accurate Feature Selection (DAFS)* method to facilitate multivariate feature selection for gene expression classification. The proposed DAFS relies on the RSE method [4] to select features triggering a high classification accuracy and furthermore on certain diversity measures to minimize the redundancy among selected features. It should be emphasized that the proposed method is intended for feature selection and not for classification, which means that DAFS should be used rather as a preprocessing step of the classification task (approach suggested in many studies [11,8,1,10]).

Let us consider a sample set $X_{M \times N}$ of size N and dimensionality M. Instead of using all features for each classifier in the ensemble, RSE [4] samples the feature set. The ensemble assigns a class label by majority or average voting of L classifiers built on different feature subsets of size m (where $m << M$) sampled randomly and uniformly from the original feature set. Each base classifier is invariant to points that are different from the training points only in the unselected features, thus encouraging diversity. This approach results in each classifier generalizing in a different way. Hence, while most other classification methods suffer from the curse of dimensionality, the RSE method can take advantage of high dimensionality to build an ensemble of "diverse" classifiers, each specialized on an small subspace [4].

2.1   Feature-Space Coverage

Let us denote by γ the total number of relevant features returned by the selection algorithm. The *selection ratio* is defined as $\frac{\gamma}{M}$. As indicated in many research reports [3,11], the number of useful genes resulted from feature selection is very small (without loss of generality, we can consider this number to be as small as 10 out of thousands of genes). Considering the high number of features present in microarray gene expression datasets (normally more than 5000), we can state that *selection ratio < 0.002* carying a large amount of information by a small number of genes. Therefore, any feature selection strategy should evaluate each gene at least once for an effective selection.

The probability that a particular feature f_i is hit in m trials of RSE is m/M (a measure called *SR - Subspace Ratio*). Therefore, the probability of not selecting a feature in any of the L classifiers of the ensemble is $P(\bar{f}_i) = (1 - SR)^L$. The probability of f_i being *at least* in one of the L selections is $1 - P(\bar{f}_i)$. Furthermore, the probability of all features to appear at least once in one of the L classifiers (called P_{cov} - feature coverage probability) is defined as follows:

$$P_{cov} = (1 - (1 - SR)^L)^\gamma \tag{1}$$

The assumption here is that the features within the selected subset of size m are sampled independently. The probability P_{cov} is considered in the proposed DAFS method to ensure the feature-space coverage by informing the selection of L and m parameters.

2.2 Feature Evaluation and Ranking

The core of any feature selection method is the process of evaluating each feature and selecting the most relevant ones for the problem at hand. Furthermore, this process should avoid redundant features because of their negative effect on the classification task (since redundant features do not bring any additional necessary information and further increase dimensionality).

The individual accuracy and the diversity among ensemble members are two key characteristics affecting ensemble accuracy [7]. The relation of the individual accuracy and diversity with the ensemble accuracy is still not clear but underlies the only way of designing an ensemble. The idea behind the DAFS method is to use the accuracy and diversity driven information not for designing an ensemble but for selecting the most relevant features from a very large set.

To minimize the redundancy between the selected features, the feature evaluation and ranking phase of DAFS makes use of an important ensemble characteristic i.e. diversity. As opposed to many standard methods which directly compare the similarity within a set of features (e.g. by Euclidean distance) to measure their redundancy, we propose to evaluate features by comparing the labels diversity of the base classifiers using a specific feature. The more diverse the classifiers to which a specific features contributes to, the less redundant the feature is.

The accuracy of a base classifier built on a specific feature subset is able to consider the contribution of those features to the classification task. Thanks to this important ensemble characteristic, the importance of a feature is *indirectly* measured by comparing the performances of the corresponding classifiers.

To be more specific, let us consider a feature f_i which appears in k random subspace sets $RS_{f_i} = \{rs_{i_1}, rs_{i_2}, \ldots, rs_{i_k}\}$, each subspace being used in the training set of k base classifiers $L_{f_i} = \{l_{i_1}, l_{i_2}, \ldots, l_{i_k}\}$. Intuitively, a feature is important if the corresponding classifiers using it get a high accuracy in the ensemble. Therefore, *Individual accuracy/RSE accuracy* ratio can be used for evaluating the feature *relevancy*. In the same way, $d(l_{ens}, l_{i_u}), u = 1 \ldots k$ computes the *redundancy* of a feature using an individual classifier l_{i_u} compared to the ensemble l_{ens} (d refers to any pairwise diversity measure).

Considering all of the above mentioned issues, we propose a measure called *Diversity-Accuracy (DA)* associated with each feature. The DA measure uses the ensemble components to incorporate both accuracy (to assess feature relevancy) and diversity (to assess feature redundancy) information. For a feature f_i present in the k subsets RS_{f_i} used in classifiers L_{f_i}, the DA measure is defined as follows:

$$DA(f_i) = \frac{1}{k} \sum_{u=1}^{k} d(l_{ens}, l_{i_u})^r \times \left[\frac{a(l_{i_u})}{a(l_{ens})} \right]^r, \qquad (2)$$

where $a(l)$ represents the accuracy of classifier l and r is a parameter used for adjusting the contribution of an individual classifier versus the ensemble to the feature evaluation.

It should be noted that the ensemble performance represents the contribution of the entire set of features and majority of base classifiers. A feature will get a high DA value if the classifiers using that feature have more diversity and accuracy with respect to the ensemble (majority of other features). Therefore, the DA measure considers both the ensemble accuracy and diversity as reference points for evaluating the performance of the individual classifiers for each feature.

2.3   DAFS Algorithm

The main steps of proposed DAFS method refer to randomly sampling the feature set, building the RSE, calculating the DA value for each feature appearing in at least one base classifier and ranking the features according to the DA value.

The DAFS algorithm is detailed below (Algorithm 1). The input parameters are m - the subspace size, L - the ensemble size and G - the maximum number of features that will be selected. The sample set $X_{M \times N}$ is divided in training ($\mathbb{X}_t$) and testing ($\mathbb{X}_e$) subsets (by using for example cross-validation).

The *feature evaluation and selection* phase of the algorithm refers to the application of RSE method and the evaluation of features based on the DA measure. Ensemble accuracy is computed by majority voting strategy. Based on the individual and ensemble classifier results, the DA value is calculated for each feature using formula 2. Features are then sorted in descending order of the DA value and the first γ are selected. The final number γ of selected features is determined by considering all possible feature sets (of size up to the maximum G parameter) and choosing the one that triggers the best performance on the validation set by the final classifier h. The γ features are finally used in a *classification* phase to determine the performance of a classifier using the selected features projected on the training and testing sets.

3   Experimental Results for Gene Expression Data

This section presents computational experiments for various cancer gene expression datasets. Results are analysed from multiple perspectives (e.g. subset size vs. ensemble size, DA feature evaluation, different classification algorithms) and compared to those of related methods.

Algorithm 1. DAFS Algorithm

Parameters

- m : subspace size
- L : ensemble size
- G : maximum number of selected features

Input

- $\mathbb{X}_t, \mathbb{X}_e$ represent the training and testing sets respectively
- $l_i, i = 1 \ldots L$ are the base classifiers in RSE
- h is the classifier used for the final classification

Feature evaluation and selection

- Select relevant m and L which satisfy the *feature-space coverage* P_{cov} condition (formula 1)
- Create L sets of feature subspace $rs_i, i = 1 \ldots L$ randomly and independently sampled from the entire feature space
- Build learners (subclassifiers) each using the training data projected on its selected subspace
- Apply $\mathbb{X}_v$ on the L classifiers and assign the labels by majority voting ($\mathbb{X}_v$ is a subset of $\mathbb{X}_t$ used for validation)
- Calculate the DA value for each feature using formula 2
- Sort the features in descending order of DA value
- Select the first γ features from the sorted list, $1 \leq \gamma \leq G$, which trigger the best accuracy on $\mathbb{X}_v$ by classifier h

Classification

- Train the classifier model h using $\mathbb{X}_t$ projected on the selected feature space of size γ
- Project the testing set on the selected feature space and classify them applying h

---

3.1 Dataset Description

The cancer gene expression datasets available in [13] are engaged for computational experiments. Nine out of the 11 datasets considered are multicategory while two of them are binary classification problems. Table 1 presents the different characteristics of these datasets. The column *Maximum prior* indicates the prior probability of the main diagnostic class for each problem.

The main properties of the considered datasets are as follows: *11_Tumors* contains 174 samples with 12533 genes with the task of finding 11 various human tumor types, *14_Tumors* refers to 14 various human tumor types and 12 normal tissue types, *9_Tumors* has 60 samples in 9 various human tumor types, *Brain_Tumor1* with 5 human brain tumor types, *Brain_Tumor2* contains 4 malignant glioma types, *Leukemia1* is a three-class task: Acute myelogenous leukemia (AML), acute lymphoblastic leukemia (ALL) B-cell, and ALL T-cell, *Leukemia2* with AML, ALL, and mixed-lineage leukemia (MLL) classes, *Lung_Cancer* contains 4 lung cancer types and normal tissues, *SRBCT* is for

Table 1. Cancer gene expression datasets used in computational experiments (from [13])

Dataset	Samples	Genes (features)	Classes	Features/Samples	Max. Prior
11_Tumors	174	12533	11	72	15.5 %
14_Tumors	308	15009	26	49	9.7 %
9_Tumors	60	5726	9	95	15.0 %
Brain_Tumor1	90	5920	5	66	66.7 %
Brain_Tumor2	50	10367	4	207	30.0 %
Leukemia1	72	5327	3	74	52.8 %
Leukemia2	72	11225	3	156	38.9 %
Lung_Cancer	203	12600	5	62	68.5 %
SRBCT	83	2308	4	28	34.9%
Prostate_Tumor	102	10509	2	103	51.0 %
DLBCL	77	5469	2	71	75.3 %

Small, round blue cell tumors (SRBCT) of childhood, *Prostate_Tumor* refers to a binary classification task Prostate tumor and normal tissues, and *DLBCL* has 77 samples with two possible classes: Diffuse large b-cell lymphomas (DLBCL) and follicular lymphomas.

3.2 DAFS Setup

The DAFS method needs the specification of the classifiers used in the RSE and of the diversity measure used in calculating the DA feature values.

We have used different types of classification algorithms - k-Nearest Neighbour (kNN), Support Vector Machine (SVM) and Multi-Layer Perceptron (MLP) - as base learners in order to show that the proposed method is independent of the particular base classifier. The error backpropagation algorithm is used for the training of the MLP base classifiers and the iterative estimation process is stopped when an average squared error of 0.9 over the training set is obtained, or when the maximum number of iterations is reached (adopted mainly for preventing networks from overtraining). We also varied the number of hidden neurons to experimentally find the optimal architecture of the MLPs for each problem. The other parameter values used for training are as follows: learning rate is 0.4 and momentum parameter is 0.6. In the case of SVM serving as base classifier, linear kernel is used. In the case of kNN classifier, the standard euclidean distance is used to calculate distances between samples. The value of k was varied from 3 to 9 in order to find the best neighbourhood size for each problem. All other parameters for these three algorithms have been chosen according to the standard setting of MATLAB Toolboxes.

The datasets are split into training and testing subsets by using Leave-One-Out Cross-Validation (LOOCV). Since the problems are low sample size, the evaluation of a method assessed by LOOCV provides realistic generalization for unseen data. Furthermore, 20% of the training set is used for the validation phase.

Several diversity measures are considered to compute the DA measure in the evaluation of DAFS. In [7], Kuncheva and Whitaker emphasize that there is no generally accepted formal definition for diversity and present several definitions from statistics to measure pairwise and non-pairwise diversities. Based on the analysis presented in [7], the diversity measures used in this paper are Q-statistics ($Q\text{-}sta$), Correlation coefficient ($Corr$), Disagreement measure (Dis) and Double Fault measure (DF). We briefly recall these measures below:

$$Q - sta_{la,lb} = \frac{N^{11}N^{00} - N^{01}N^{10}}{N^{11}N^{00} + N^{01}N^{10}} \tag{3}$$

$$Corr_{la,lb} = \frac{N^{11}N^{00} - N^{01}N^{10}}{\sqrt{(N^{11} + N^{10})(N^{01} + N^{00})(N^{11} + N^{01})(N^{10} + N^{00})}} \tag{4}$$

$$Dis_{la,lb} = \frac{N^{01} + N^{10}}{N^{11} + N^{10} + N^{01} + N^{00}} \tag{5}$$

$$DF_{la,lb} = \frac{N^{00}}{N^{11} + N^{10} + N^{01} + N^{00}} \tag{6}$$

where la, lb represent the two classifiers compared, N^{11} and N^{00} are the number of samples for which both classifiers la and lb concurrently make correct and incorrect decisions respectively while N^{10} and N^{01} are the number of samples for which la and lb do not agree on their labels.

3.3   Numerical Experiments

The proposed feature selection algorithm is compared to related methods based on the recognition rates of the final classifiers. The following (m, L) pairs have been tested for the values of the subspace size (m) and the ensemble size (L): $(10, 10000)$, $(50, 500)$, $(500, 100)$, $(1000, 20)$ and $(1000, 10)$. All these (m, L) value pairs are chosen in accordance with the feature coverage probability P_{cov} (formula 1). For each dataset considered, all the (m, L) pairs have been tested and the best performing one is used in the reported results. In the same way, all four diversity measures given in the previous section (equations 3 to 6) are considered for each dataset and the one triggering the best performance is used to report the results of DAFS.

The last input parameter of DAFS refers to G, the maximum number of selected features, and is set to 30. The final number γ of selected features is determined by considering all possible feature sets (of size up to G) and choosing the one that leads to the best performance on the validation set (as explained in section 2.3). The value of γ is different for each dataset considered in the experiments. Finally, the parameter r used in the DA measure (equation 2) is set to 1 for all experiments.

Table 2 presents the comparative numerical results. The accuracies of single kNN, SVM and MLP are given in the first three columns of results. The next two columns show the accuracy obtained by standard RSE and the average individual accuracy in RSE (column labelled *Avg. Ind.*). These five columns provide the baseline results of standard methods without any feature selection to comparatively assess the DAFS performance. Furthermore, two commonly used feature selection methods in machine learning i.e. Mutual information (MI) and $MRMR$ [2] have been applied for all considered cancer gene expression datasets and the results are given in the columns labelled MI and $MRMR$ respectively. Last column in Table 2 contains the accuracy reported by proposed DAFS method. Except for the single classifier results, SVM has been used as the final classifier to produce the recognition rates presented in Table 2.

The average rank of DAFS (computed based on the average results over all datasets) is the best among all considered methods compared in Table 2. The advantages of DAFS compared to the baseline methods are clear from the results obtained. Also, DAFS clearly obtains better recognition rates than mutual information. An overall better performance compared to MRMR can also be observed. For three out of the 11 considered datasets (i.e. 11_Tumors, 9_Tumors and Brain_Tumor1), DAFS and MRMR report the same accuracy while for the Lung_Cancer dataset (a five-class problem with a high maximum prior probability of the dominant class), MRMR obtains a better recognition rate of 96.5 (same accuracy with single SVM) compared to 95.4 reported by DAFS. However, the proposed DAFS outperforms MRMR for the rest of seven datasets considered.

The results presented in Table 2 show that SVM classifiers are robust against dimensionality and able to achieve a good performance even without feature selection. However, feature selection significantly improves the classification performance of non-SVM learners. The recognition rates of kNN and MLP increase dramatically (more than double) when DAFS is first used in feature selection.

Table 2. Recognition rates obtained by proposed DAFS (last column) compared to the results of some baseline methods as well as those of MI and MRMR feature selection methods

Dataset	single SVM	single kNN	single MLP	RS ensemble	Avg. Ind.	MI	MRMR	DAFS method
11_Tumors	94.6	73.2	55.8	68.1	52.8	92.2	**95.3**	**95.3**
14_Tumors	**75.6**	45.8	11.9	50.9	33.7	73.6	74.9	74.9
9_Tumors	62.7	44.1	20.5	44.9	31.6	60.0	**65.6**	**65.6**
Brain_Tumor1	**91.2**	85.3	83.9	84.4	79.1	89.9	91.0	90.6
Brain_Tumor2	77.8	65.1	62.6	63.0	54.9	77.5	**78.8**	**78.8**
Leukemia1	**97.5**	81.9	79.4	82.0	73.1	96.9	97.0	**97.5**
Leukemia2	95.2	85.9	88.6	87.7	80.3	94.8	95.0	**97.2**
Lung_Cancer	**96.5**	86.5	84.8	87.7	79.1	94.9	**96.5**	95.4
SRBCT	98.1	83.2	89.5	90.5	78.4	97.1	97.3	**100.0**
Prostate_Tumor	92.0	80.7	75.5	79.9	71.1	91.8	92.0	**92.3**
DLBCL	95.8	77.9	81.8	78.8	72.5	95.6	97.3	**98.0**
Avg. Rank	1.72	5.36	6	5	7.45	3.45	1.81	**1.36**

4   Conclusions

A simple and effective feature selection method has been proposed and evaluated for gene expression cancer diagnosis. The main strength of the introduced DAFS method is represented by an efficient combination of classification accuracy information extracted using RSE and diversity information generated by using certain measures. The first component facilitates the identification of relevant features in obtaining a good recognition rate in classification while the second component focuses on minimizing the redundancy among selected features. Experimental results for 11 microarray cancer diagnosis datasets support the conclusions that DAFS is a fast and efficient feature selection method. It has been shown that DAFS is able to reduce the computational cost of the classification by removing irrelevant features and improving the recognition rate. The best results are obtained using SVM as base classifier while DAFS performance induced by the diversity measure varies from one dataset to another.

Obviously, the DAFS functionality is independent from the particular classifier used while the recognition rate based on the selected features does depend on the final classifier. For gene expression cancer classification, experiments revealed that better accuracies are produced by SVM in comparison with kNN and MLP. Moreover, the non-SVM learners benefit more from the proposed feature selection method in the sense that classfication accuracies are significantly improved when using DAFS as a preprocessing step.

Acknowledgments. This work has been partially supported by the Italian Ministry of Education - Investment funds for basic research, under the project ITALBIONET - Italian Network of Bioinformatics.

Camelia Chira acknowledges the support of CNCS Romania through grant PN II TE 320 - Emergence, auto-organization and evolution: New computational models in the study of complex systems.

References

1. Banerjee, M., Mitra, S., Banka, H.: Evolutionary Rough Feature Selection in Gene Expression Data. IEEE Transactions on Systems, Man, and Cybernetics-Part C: Applications and Reviews 37(4), 622–632 (2007)
2. Ding, C., Peng, H.: Minimum redundancy feature selection from microarray gene expression data. J. Bioinform.Comput. Biol. 3, 185–205 (2005)
3. Golub, T.R., Slonim, D.K., Tamayo, P., Huard, C., Gaasenbeek, M., Mesirov, J.P., Coller, H., Loh, M.L., Downing, J.R., Caligiuri, M.A., Bloomfield, C.D., Lander, E.S.: Molecular classification of cancer: class discovery and class prediction by gene expression monitoring. Science 286(5439), 531–537 (1999)
4. Ho, T.: The random subspace method for constructing decision forests. IEEE Transactions on Pattern Analysis and Machine Intelligence 20(8), 832–844 (1998)
5. Huang, H.-L., Chang, F.-L.: ESVM: Evolutionary support vector machine for automatic feature selection and classification of microarray data. BioSystems 90, 516–528 (2007)

6. Kuncheva, L.I., Rodriguez, J.J., Plumpton, C.O., Linden, D.E.J., Johnston, S.J.: Random Subspace Ensembles for fMRI Classification. IEEE Transactions on Medical Imaging 29(2), 531–542 (2010)
7. Kuncheva, L.I., Whitaker, C.J.: Measures of diversity in classifier ensembles. Machine Learning 51, 181–207 (2003)
8. Larrañaga, P., Calvo, B., Santana, R., Bielza, C., Galdiano, J., Inza, I., Lozano, J.A., Armañanzas, R., Santafé, G., Pérez, A., Robles, V.: Machine learning in bioinformatics. Briefings in Bioinformatics 7(1), 86–112 (2006)
9. Lee, C.P., Leu, Y.: A novel hybrid feature selection method for microarray data analysis. Applied Soft Computing 11, 208–213 (2011)
10. Liu, H., Liu, L., Zhang, H.: Ensemble gene selection by grouping for microarray data classification. Journal of Biomedical Informatics 43, 81–87 (2010)
11. Lu, Y., Han, J.: Cancer classification using gene expression data. Information Systems 28(4), 243–268 (2003)
12. Maji, P., Paul, S.: Rough set based maximum relevance-maximum significance criterion and gene selection from microarray data. International Journal of Approximate Reasoning 52, 408–426 (2011)
13. Statnikov, A., Aliferis, C., Tsamardinos, I., Hardin, D., Levy, S.: (2004), http://www.gems-system.org
14. Statnikov, A., Aliferis, C.F., Tsamardinos, I., Hardin, D., Levy, S.: A Comprehensive Evaluation of Multicategory Classification Methods for Microarray Gene Expression Cancer Diagnosis. Bioinformatics 21(5), 631–643 (2005)

A Comparison on Score Spaces for Expression Microarray Data Classification

Alessandro Perina[1], Pietro Lovato[2], Marco Cristani[2,3], and Manuele Bicego[2]

[1] Microsoft Research, Redmond, USA
[2] University of Verona, Department of Computer Science, Verona, Italy
[3] Italian Institute of Technology, Genoa, Italy

Abstract. In this paper an empirical evaluation of different generative scores for expression microarray data classification is proposed. Score spaces represent a quite recent trend in the machine learning community, taking the best of both generative and discriminative classification paradigms. The scores are extracted from topic models, a class of highly interpretable probabilistic tools whose utility in the microarray classification context has been recently assessed. The experimental evaluation, performed on 3 literature datasets and with 7 score spaces, demonstrates the viability of the proposed scheme and, for the first time, it compares *pros* and *cons* of each space.

1 Introduction

Microarrays represent a widely employed tool in molecular biology and genetics, allowing DNA and/or RNA analysis to be carried out in microminiaturized highly parallel formats. DNA microarray applications are usually directed at gene expression analysis that usually implies to process huge amounts of data. Therefore, fast and robust methodologies are required to face diverse microarray analysis problems such as noise suppression [7], segmentation of spots/background, quantification of the spots, grid matching, clustering or classification [9, 17, 29, 31].

In this paper we focus on this last class of problems, where many approaches have been presented in the literature in the past, each one focusing on different aspects, like computational complexity, effectiveness, interpretability, optimization criterion and others – for a review see e.g. [17, 29]. Among others, in recent years some promising techniques were based on a particular class of probabilistic approaches, called topic models, showing optimal and highly interpretable results [2, 22, 24]. Such probabilistic topic models, the two most famous examples being the Probabilistic Latent Semantic Analysis (PLSA [15]) and the Latent Dirichlet Allocation (LDA [5]), have been imported from the text analysis realm as workhorses in several scientific fields [6, 8, 33]. Their wide usage is motivated by their simplicity and expressiveness in dealing with very large datasets both in samples and features number. Therefore, they appeared to be a convenient tool for the microarray data analysis problem, and especially in the context of expression microarray classification [2, 22]. Nevertheless, not all the potentialities

M. Loog et al. (Eds.): PRIB 2011, LNBI 7036, pp. 202–213, 2011.
© Springer-Verlag Berlin Heidelberg 2011

of these schemes have been exploited in such context. To overcome this problem, here we make a step forward, by applying a hybrid generative-discriminative paradigm based on the definition of *generative score spaces* [28]. Generative and discriminative classification schemes represent the two main directions for classifying data: each philosophy brings pros and cons with itself and the last research frontier aims at fusing them, following heterogeneous recipes [4,20,19]. In this paper, we adopt the "staged" strategy: the idea is that a generative framework (in this case the PLSA[1]) is instantiated and learned. Then, surrogates of the learning (in the simplest case, likelihood probabilities) are injected as features in a discriminative classifier which is eventually learned. In some cases, this is theoretically proved to rise the purely generative classification performances [16,18,20,30].

In this paper, we show how different strategies to build score spaces lead to diverse classification accuracies, considering different publicly available microarray datasets. Obtained results confirm the goodness of the classification strategies based on topic models in the expression microarray classification context.

2   Methodology

In this section the background concepts regarding topic models and generative embedding are reported. In particular, after introducing the general ideas underlying the PLSA model, we will present it by using the terminology and the notation of the document analysis context. Then we will briefly review how the framework of hybrid generative-discriminative approach can be employed alongside the PLSA, and how it is applied in the microarray classification scenario.

2.1   Probabilistic Latent Semantic Analysis

In Probabilistic Latent Semantic Analysis (PLSA – [15]) the input is a set of D documents, each one containing a set of words taken from a vocabulary of cardinality N. The documents are summarized by an occurrence matrix of size $N \times D$, where $n(w_j, d_i)$ indicates the number of occurrences of the word w_j in the document d_i. The presence of a word w_j in the document d_i is mediated by a latent *topic* variable, $z \in Z = \{z_1, ..., z_Z\}$, also called *aspect* class, i.e.,

$$p(w_j, d_i) = \sum_{k=1}^{Z} p(z_k) \cdot p(w_j | z_k) \cdot p(d_i | z_k) \tag{1}$$

In practice, each k-th topic $z_k{}^2$ is a probabilistic co-occurrence of words encoded by the distribution $\beta(w) = p(w|z_k)$, $w = \{w_1, ..., w_N\}$, and each document d_i is compactly (usually, $Z < N$) modeled as a probability distribution over the topics,

---

[1] PLSA is commonly employed as a generative model, even if it is not under a strict formal treatment. See the text for further details.

[2] Throughout the paper v_k stands for the variable v assuming the value k.

i.e., $p(z|d_i)$, $z = \{z_1,...,z_Z\}$ (note that this formulation, derived from $p(d_i|z)$, provides an immediate interpretation).

The hidden quantities of the model, $p(w|z)$, $p(d|z)$ and $p(z)$, are learnt using Expectation-Maximization (EM) [10], maximizing the model data-loglikelihood $\mathcal{L}$:

$$\mathcal{L} = \prod_{j=1}^{N} \prod_{i=1}^{D} n(w_j, d_i) \cdot \log p(w_j, d_i) \tag{2}$$

The E-step computes the posterior over the topics, $p(z|w,d)$, and the M-step updates the hidden distributions.

Once the model has been learnt one can estimates the topic proportion of an unseen document. Here, the learning algorithm is applied by fixing the previously learnt parameters $p(w|z)$ and estimating $p(d|z)$ for the document in hand. For a deeper review of PLSA, see [15].

It is important to note that d is a dummy index into the list of documents in the training set. Thus, d is a multinomial random variable with as many possible values as there are training documents and the model learns the topic mixtures $p(d|z)$ only for those documents on which it is trained. For this reason, PLSA is not a well-defined generative model of documents; there is no natural way to assign probability to a previously unseen document and the procedure just described to estimate $p(d|z)$ is an heuristic [15].

PLSA may be very useful in the expression microarray context, since it may provide powerful and interpretable descriptions of experiments [3,22,24]. In particular there is an analogy between the pairs *word-document* and *gene-sample*: actually it is reasonable to intend the samples as documents and the genes as words. In fact the expression level of a gene in a sample may be easily interpreted as the count of words in a document (the higher the number the more present/expressed the word/gene is). In our case, therefore, we can consider the expression matrix as the count matrix $<w_j, d_i>$ of topic models, after a proper normalization in order to have positive and integer values.

3   Generative Score-Spaces

Pursuing principled hybrid architectures of discriminative and generative classifiers is currently one of the most interesting, useful, and difficult challenges for Machine Learning. The underlying motivation is the proved complementarity of discriminative and generative estimations: asymptotically (in the number of labeled training examples), classification error of discriminative methods is lower than for generative ones [19]. On the other side, generative counterparts are effective with less, possibly unlabeled, data; further, they provide intuitive mappings among structure of the model and data features. Among these hybrid generative-discriminative methods, "generative score space" approaches grow in the recent years their importance in the literature [6, 16, 18, 20, 27, 28, 30].

Generative score space framework consists of two steps: first, one or a set of generative models are learned from the data; then a score (namely a vector of features) is extracted from it, to be used in a discriminative scenario. The idea is to extract fixed dimensions feature vectors from observations by subsuming the process of data generation, projecting them in highly informative spaces called score spaces. In this way, standard discriminative classifiers such as support vector machines, or logistic regressors are proved to achieve higher performances than a solely generative or discriminative approach.

Using the notation of [27, 20], such spaces can be built from data by mapping each observation x to the fixed-length score vector $\varphi^f_{\hat{F}}(x)$,

$$\varphi^f_{\hat{F}}(x) = \varphi_{\hat{F}} f(\{P_i(x|\theta_i))\}), \tag{3}$$

where $P_i(x|\theta_i)$ represents the family of generative models learnt from the data, f is the function of the set of probability densities under the different models, and $\hat{F}$ is some operator applied to it. In general, the generative score-space approaches help to distill the relationship between a model parameters θ and the particular data sample.

Generative score-space approaches are strictly linked to generative kernels family, namely kernels which compute similarity between points through a generative model – the most famous example being the Fisher Kernel [16]): Typically, a generative kernel is obtained by defining a similarity measure in the score space, e.g. the inner product.

Score spaces are also called model dependent feature extractors, since they extract features from a generative model. We can divide score spaces in two families: parameters-based and hidden variable-based. Let us review the 7 different score spaces tested in this paper.

3.1 Parameters Based Score Space

These methods derive the features on the basis of differential operations linked to the parameters of the probabilistic model.

The Fisher Score. Fisher kernel [16] was the first example of generative score space. At first, a parameter estimate $\hat{\theta}$ is obtained from training examples. Then, the tangent vector of the data log likelihood $\log p(x|\theta)$ is used as a feature vector. Referring to the notation of [27, 20], the score function is the data log likelihood, while the score argument is the gradient.

The fisher score for the PLSA model has been introduced in [14], starting from the asymmetric formulation of PLSA. In this case, the log-probability of a document d_i is defined by

$$l(d_i) = \frac{\log p(d_i, w)}{\sum_m n(d_i, w_m)} = \sum_{j=1}^{N} \hat{p}(w_j|d_i) \log \sum_{k=1}^{Z} p(w_j|z_k) \cdot p(d_i|z_k) \cdot p(z_k), \tag{4}$$

where $\hat{p}(w_j|d_i) \equiv n(d_i, w_j)/\sum_m n(d_i, w_m)$ and where $l(d_i)$ represents the probability of all the word occurrences in d_i normalized by document length.

Differentiating Eq. 4 with respect to $p(z)$ and $p(w|z)$, the PLSA model parameters, we can compute the score. In formulae:

$$\frac{\partial l(d_i)}{\partial p(w_r|z_t)} = n(d_i, w_t) \cdot \frac{p(d_i|z_t) \cdot p(z_t)}{\sum_k p(w_r|z_k) \cdot p(d_i|z_k) \cdot p(z_k)} \tag{5}$$

$$\frac{\partial l(d_i)}{\partial p(z_t)} = \sum_{j=1}^{W} n(d_i, w_j) \cdot \frac{p(d_i|z_t) \cdot p(w_j|z_t)}{\sum_k p(z_k) \cdot p(w_j|z_k) \cdot p(d_i|z_k)} \tag{6}$$

As visible from Eq. 5-6, the samples are mapped in a space of dimension $W \times Z + Z$. The fisher kernel is defined as the inner product in this space. We will refer to it as FSH.

TOP Kernel Scores. Top Kernel and the tangent vector of posterior log odds score space were introduced in [30]. One of the aim of the paper was to introduce a performance measure for score spaces. They considered the estimation error of the posterior probability by a logistic regressor and they derived the TOP kernel in order to maximize the performance.

Whereas the Fisher score is calculated from the marginal log-likelihood, TOP kernel is derived from Tangent vectors Of Posterior log-odds. Therefore the two score spaces have the same score function (i.e., the gradient) but different score argument, which, for TOP kernel $f(p(x|\theta)) = \log p(c = +1|x, \theta) - \log p(c = -1|x, \theta)$ where, c is the class label. We will refer to it as TOP.

Log Likelihood Ratio Score Space. The loglikelihood ratio score space is introduced in [28]. Its dimensions are similar to the Fisher score, except that the procedure is repeated for each class: a model per class is learnt θ_c and the gradient is applied to each class-loglikelihood $\log p(x|\theta_c)$. The dimensionality of the resulting space is C-times the dimensionality of the original Fisher score. We will refer to it as LLR.

3.2   Random Variable Based Methods

These methods, starting from considerations in [20], seek to derive feature maps on the basis of the log likelihood function of a model, focusing on the random variables rather than on the parameters in their derivation (as done in the parameter-based score spaces).

Free Energy Score Space. In the Free Energy Score Space [20], the score function is the free energy while the score argument is its unique decomposition in addends that composes it[3]. Free energy is a popular score function representing

---

[3] This is true once a family for the posterior distribution is given. See the original paper for details.

a lower bound of the negative log-likelihood of the visible variables used in the variational learning. For PLSA it is defined by the following equation:

$$\mathcal{F}(d_i) = \sum_w n(d_i, w) \cdot \sum_z p(z|d, w) \cdot \log p(z|d, w)$$

$$- \sum_w n(d_i, w) \cdot \sum_z p(z|d, w) \cdot \log p(d, w|z) \cdot p(z) \tag{7}$$

where the first term represents the entropy of the posterior distribution and the second term is the cross-entropy. For further details on the free energy and on variational learning see [12], on the PLSA's free energy see [15].

As visible in Eq. 7 both terms are composed of $Z \times N$ addends $\{f_j\}_{j=1}^{Z \times N}$, and their sum is equal to the free energy. In generative classification, a test data is assigned to the class which gives the lower free energy (i.e., higher loglikelihood). The idea of FESS is to decompose the free energy of each class in its addends, i.e., $\mathcal{F}(d_i)^c = \sum_j \{f_{j,c}\}$ and to add a discriminative layer by estimating a set of weights $\{w_{j,c}\}$ through a discriminative method.

For PLSA this results in a space of dimension equal to $C \times 2 \times Z \times W$; we will refer to this score space FESS L_3.

In [20] the authors point out that, if the dimensionality is too high, some of the sums can be carried out to reduce the dimensionality of the score vector before learning the weights. The choice of the addend to optimize is intuitive but guided by the particular application. In our case, as previously done in [18,21], we perform the sums over the word indices, optimizing the topics contribute. The resulting score space has dimension equal to $C \times 2 \times Z$; we will refer to this score space FESS L_2.

Posterior Divergence. Posterior Divergence score space is described in [18]. Like FESS it takes into account how well a sample fits the model (crossentropy terms in FESS) and how uncertain the fitting is (entropy terms in FESS, Eq. 7) but it also assesses the change in model parameters brought on by the input sample, i.e. how much a sample affects the model. These three measures are not simply stacked together, but they are derived from the incremental EM algorithm which, in the E-step only looks at one or few selected samples to update the model in each iteration. Details on posterior divergence score vector for PLSA and on its relationships with FESS case can be found in [18]. We will refer to this score space as PD.

Classifying with the Mixture of Topics of a Document. Very recently, PLSA has been used as dimensionality reduction method in several fields, like computer vision, bioinformatics and medicine [6, 2, 8]. The idea is to learn a PLSA model to capture the co-occurrence between visual words [6, 8], or gene expressions [2], which represent the (usually) high-dimensional data description; co-occurrences are captured by the topics. Subsequently, the classification is performed using the topic distribution of a document as its descriptor.

Since we are extracting features from a generative model, we are defining a score space which is the (Z-1)-dimensional simplex. In this case, the score argument f, a function of the generative model, is the topic distribution $p(z|d)$ (using Bayes' formula, one can easily derive $p(z|d)$ starting from $p(d|z)$), while the score function is the identity. We will refer to this score space as TPM, or citing [2], the first work in the context of microarray classification that used this technique.

Summary. Summarizing, here we propose to face the expression microarray classification task by first learning a PLSA model, then extracting a score space, and finally classifying the samples in this new space, using a discriminative classifier (e.g. a Support Vector Machine).

4   Experimental Evaluation

The suitability of the proposed classification schemas has been tested using three different well-known datasets, briefly summarized in Tab. 1. The whole description of each dataset may be found in the reported reference.

Table 1. Summary of the employed dataset

Dataset Name	n. of genes	n. of samples	n. of classes	citation	BIC
1. Colon cancer	2000	62	2	[1]	6
2. Ovarian cancer	1513	53	2	[11]	4
3. DLBCL	6285	77	2	[26]	4

As in many expression microarray analysis, a beneficial effect may be obtained by selecting a sub group of genes, using a prior belief that genes varying little across samples are less likely to be interesting. Hence, we decided to perform the experiments by retaining the top 500 genes ranked by decreasing variance, as done also in [24].

A crucial issue arising when learning a topic model is to decide beforehand the number of topics. Here we employed the well-known Bayesian Information Criterion [25], which penalizes the likelihood with a penalty term which depends on the number of free parameters of the model – in such way, larger models which do not lead to a substantial increase in the likelihood are discouraged. In the PLSA model, the free parameters are $(D-1)\cdot Z+(N-1)\cdot Z+(Z-1)$, where Z and N refers to the number of topics and the number of words respectively. The penalization term is then given by

$$Pen. = \frac{1}{2} \cdot ((D-1)\cdot Z + (N-1)\cdot Z + (Z-1)) \cdot \log \sum_{j=1}^{N}\sum_{i=1}^{D} n(d_i, w_j) \quad (8)$$

The best number of topic is found by searching for the maximum of the penalized likelihood, varying the number of topics from 2 to 50. The optimal number of topic for each dataset is shown in Tab. 1, column BIC.

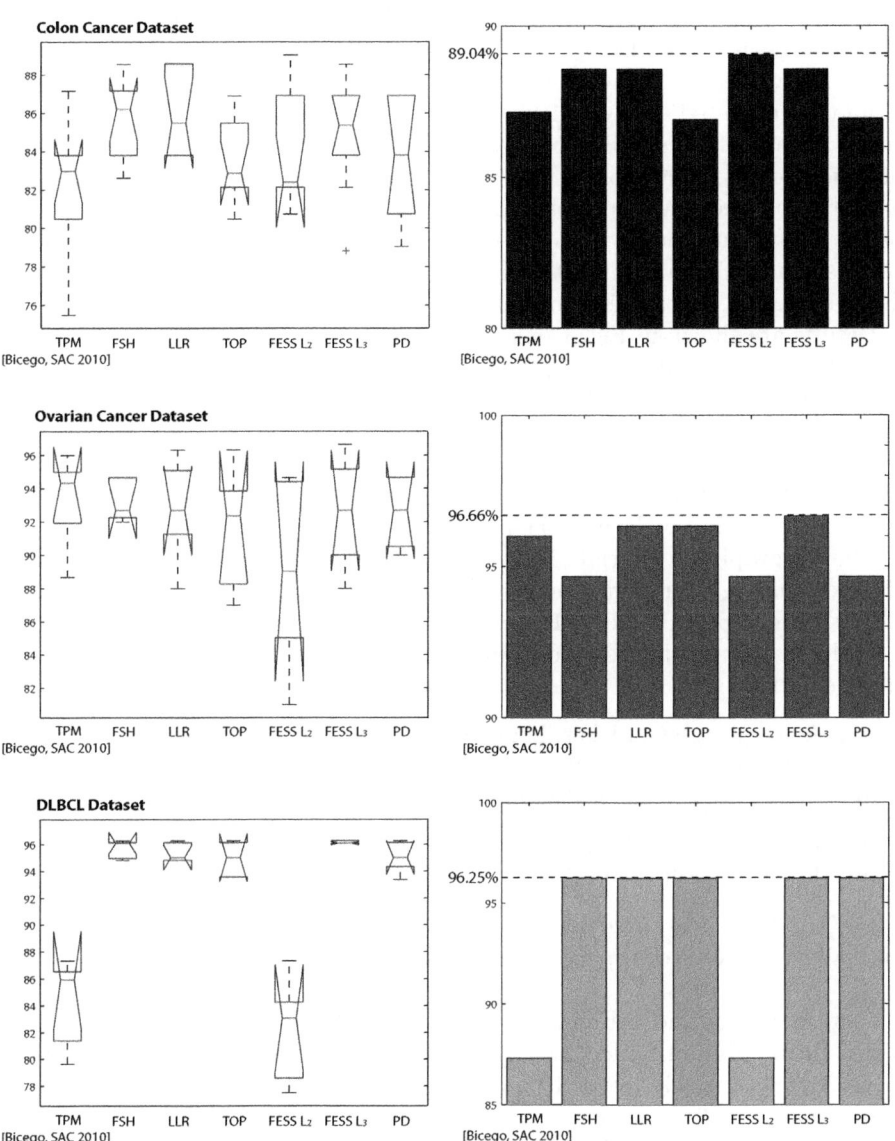

Fig. 1. Microarray classification results. TPM stands for the method present in [2], FSH is the fisher score space, LLR is the loglikelihood ration score space, TOP is the tangent of posterior log-odds score space, FESS L_2 and FESS L_3 are two complexities of the free energy score space while PD is the posterior divergence score space. See the text for details. Please print in color.

The errors have been found using a cross validation scheme: in particular the subdivision in training and testing set is carried out using 10-fold crossvalidation. Differently from [2] we learned the generative models only with the training data; this is necessary since LLR, TOP, FESS and PD require to learn a model per class, and we cannot use labels at training time. In order to have a significant results, we repeated each test 10 times and the results of this 10 repetitions have been validated through the standard anova variance test [23]. Finally, as final classifier we used a support vector machine with linear kernel; as in [16,30], the similarity between two datapoints is defined as the inner product of their scores. Before computing the kernels, the scores are normalized to have zero mean and unit variance; the constants to perform the normalization are computed with the training set and applied to each test sample.

Results are shown in Fig. 1, where each row of the figure describes a dataset. The graph on the left is a boxplot useful to assess the statistical significance of the results. The red bar is the median accuracy of the 10 repetitions, the edges of the blue box are the 25th and 75th percentiles, while the whiskers (black dotted bars) extend to the most extreme data points not considered outliers. Outliers are plotted individually with a red cross. Two medians are significantly different at the 5% significance level if their intervals (boxes) do not overlap.

The bar graphs on the right represent the accuracy obtained in correspondence of maximum loglikelihood value of the generative model. The best result among all the score spaces considered is textually reported on the left of the figure.

By looking at the figures and examining the results, the following observations may be extracted:

Colon Cancer dataset : Fisher, Loglikelihood Ratio and FESS L_3 are statistically better than the method presented in [2], while all the other methods are clearly better but without statistical significance.

Ovarian Cancer dataset : all the methods seem to be equivalent even if, once again, the best result is obtained with FESS.

DLBCL dataset : it is clear that [2] and FESS L_2 perform significantly worse than all the other score space, which in turn do not differ much.

To better understand the differences between the considered score spaces, we varied the number of topics between 2 and 20, to see how they are robust to this value. Mean accuracies (over 10 repetitions) for Fisher, FESS L_3, PD and TPM are shown in Fig. 2. Score spaces based on the parameters are clearly less sensitive to this value[4], while for TPM and FESS the results obtained for different Z's are statistically different (t-test, significance level 5%). Despite being a score space based on random variables like TPM and FESS, PD looks very robust (see also the very small variance among the repetitions) to Z. This is not surprising since PD is composed by entropy and crossentropy terms (as FESS), but also it has a set of extra terms that assess the change in model parameters brought on by the

---

[4] We have found that for this test the results of TOP are nearly identical to FSH.

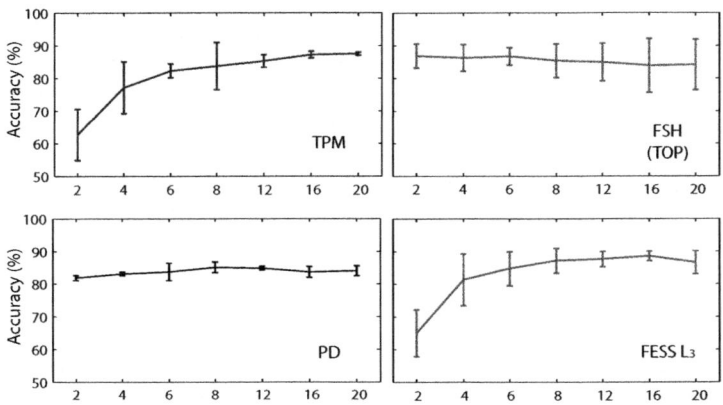

Fig. 2. Robustness of the score space to the number of topics (Z) for the Colon Cancer dataset. On the x-axis we show the number of topic, on the y-axis the accuracy. The four lines are the mean accuracy, while the vertical bars represent the variance computed across the 10 repetitions.

input sample d, which is characteristic of parameter based methods. This extra set permits to inherit the peculiar robustness to changes in Z of the parameters based method.

As a last test we tried transductive learning, namely learning a single model for all the available data, not using labels [32, 13] – on Fisher, FESS L_2, FESS L_3 and [2], to assess if this has some influence on the accuracy. For each fold and for each repetition, we learned a *single* model using all the data. Subsequently we used the training labels to train the discriminative classifier. Transductive learning has the problem that it requires to learn a model each time a test sample is available.

We performed anova test considering the three datasets together as different factors, with the following null hypothesis: *"Transductive and non-Transductive learning do not differ"*. The hypothesis is confirmed with p-values respectively of 0.8473, 0.094 and 0.8683. This means that FESS is the less robust even we cannot claim that the difference of the results is statistically significant.

5   Conclusions

In this paper different generative score spaces have been evaluated, with the aim of classifying expression microarray data. Such score spaces are built on the PLSA generative model, a probabilistic tool whose usefulness in this context has been already assessed. Experimental results confirm the viability of the proposed hybrid schemes, also in comparison with the state of the art. In particular, all the score spaces introduced here outperform the previously published frameworks on microarray classification [2, 22]. FESS reached the best classification results even if the variance across repetitions or changes in Z was sensibly higher than the other score spaces. Fisher, TOP and LLR performed similarly and they are sufficiently robust. PD presented performances slightly inferior to the other methods but it has shown the best robustness to changes in number of topics and multiple restarts. Finally, the accuracies reported here can be further improved using more complex kernels, like done in [22].

Acknowledgements. The authors want to thank Xiong Li for providing the code for the score space based on the posterior divergence.

References

1. Alon, U., Barkai, N., Notterman, D., Gish, K., Ybarra, S., Mack, D., Levine, A.: Broad patterns of gene expression revealed by clustering analysis of tumor and normal colon tissues probed by oligonucleotide arrays. Proc. Natl. Acad. Sci. 96(12), 6745–6750 (1999)
2. Bicego, M., Lovato, P., Oliboni, B., Perina, A.: Expression microarray classification using topic models. In: ACM SAC - Bioinformatics track (2010)
3. Bicego, M., Lovato, P., Ferrarini, A., Delledonne, M.: Biclustering of expression microarray data with topic models. In: Proc. of International Conference on Pattern Recognition (2010)
4. Bishop, C., Lasserre, J.: Generative or discriminative? getting the best of both worlds. Bayesian Statistics 8, 3–24 (2007)
5. Blei, D., Ng, A., Jordan, M.: Latent dirichlet allocation. Journal of Machine Learning Research 3, 993–1022 (2003)
6. Bosch, A., Zisserman, A., Muñoz, X.: Scene classification via pLSA. In: Leonardis, A., Bischof, H., Pinz, A. (eds.) ECCV 2006. LNCS, vol. 3954, pp. 517–530. Springer, Heidelberg (2006)
7. Brändle, N., Bischof, H., Lapp, H.: Robust DNA microarray image analysis. Machine Vision and Applications 15, 11–28 (2003)
8. Castellani, U., Perina, A., Murino, V., Bellani, M., Rambaldelli, G., Tansella, M., Brambilla, P.: Brain morphometry by probabilistic latent semantic analysis. In: Jiang, T., Navab, N., Pluim, J.P.W., Viergever, M.A. (eds.) MICCAI 2010. LNCS, vol. 6362, pp. 177–184. Springer, Heidelberg (2010)
9. de Souto, M., Costa, I., de Araujo, D., Ludermir, T., Schliep, A.: Clustering cancer gene expression data: A comparative study. BMC Bioinformatics 9 (2008)
10. Dempster, A., Laird, N., Rubin, D.: Maximum likelihood from incomplete data via the EM algorithm. J. Roy. Statist. Soc. B 39, 1–38 (1977)
11. Dhanasekaran, S., Barrette, T., Ghosh, D., Shah, R., Varambally, S., Kurachi, K., Pienta, K., Rubin, M., Chinnaiya, A.: Delineation of prognostic biomarkers in prostate cancer. Nature 412(6849), 822–826 (2001)
12. Frey, B.J., Jojic, N.: A comparison of algorithms for inference and learning in probabilistic graphical models. IEEE Transactions on Pattern Analysis and Machine Intelligence 27 (2005)
13. Gammerman, A., Vovk, V., Vapnik, V.: Learning by transduction. In: Proc. of Uncertainty in Artificial Intelligence (1998)
14. Hofmann, T.: Learning the similarity of documents: An information-geometric approach to document retrieval and categorization. In: Adv. in Neural Information Processing Systems (1999)

15. Hofmann, T.: Unsupervised learning by probabilistic latent semantic analysis. Mach. Learn. 42, 177–196 (2001)
16. Jaakkola, T., Haussler, D.: Exploiting generative models in discriminative classifiers. In: Adv. in Neural Information Processing Systems (1998)
17. Lee, J., Lee, J., Park, M., Song, S.: An extensive comparison of recent classification tools applied to microarray data. Computational Statistics & Data Analysis 48(4), 869–885 (2005)
18. Li, X., Lee, T.S., Liu, Y.: Hybrid generative-discriminative classification using posterior divergence. In: Proc. of Conference on Computer Vision and Pattern Recognition (2011)
19. Ng, A., Jordan, M.: On discriminative vs generative classifiers: A comparison of logistic regression and naive Bayes. In: Adv. in Neural Information Processing Systems (2002)
20. Perina, A., Cristani, M., Castellani, U., Murino, V., Jojic, N.: Free energy score space. In: Adv. in Neural Information Processing Systems (2009)
21. Perina, A., Cristani, M., Castellani, U., Murino, V., Jojic, N.: An hybrid generativediscriminative framework based on free energy terms. In: Proc. of the International Conference on Computer Vision (2009)
22. Perina, A., Lovato, P., Murino, V., Bicego, M.: Biologically-aware latent dirichlet allocation (balda) for the classification of expression microarray. Proc. of Pattern Recognition in Bioinformatics (2010)
23. Rao, C.R.: Diversity: Its Measurement, Decomposition, Apportionment and Analysis. Sankhy: The Indian Journal of Statistics, Series A 44(1), 1–22 (1982)
24. Rogers, S., Girolami, M., Campbell, C., Breitling, R.: The latent process decomposition of cdna microarray data sets. IEEE/ACM Transactions on Computational Biology and Bioinformatics 2(2), 143–156 (2005)
25. Schwarz, G.: Estimating the dimension of a model. Annals of Statistics 6, 461–464 (1978)
26. Shipp, M., Ross, K.: Diffuse large b-cell lymphoma outcome prediction by gene expression profiling and supervised machine learning. Nature Medicine 8, 68–74 (2002)
27. Smith, N., Gales, M.: Speech recognition using svms. In: Adv. in Neural Information Processing Systems (2002)
28. Smith, N.D., Gales, M.J.F.: Using SVMs to Classify Variable Length Speech Patterns. Tech. rep., Cambridge University Engineering Dept. (2002)
29. Statnikov, A., Aliferis, C., Tsamardinos, I., Hardin, D., Levy, S.: A comprehensive evaluation of multicategory classification methods for microarray gene expression cancer diagnosis. Bioinformatics 21(5), 631–643 (2005)
30. Tsuda, K., Kawanabe, M., Rotsch, G., Sonnenburg, S., Mueller, K.R.: A new discriminative kernel from probabilistic models. In: Neural Computation. MIT Press (2001)
31. Valafar, F.: Pattern recognition techniques in microarray data analysis: A survey. Annals of the New York Academy of Sciences 980, 41–64 (2002)
32. Vapnik, V.: The Nature of Statistical Learning Theory. Springer, Heidelberg (1995)
33. Xing, D., Girolami, M.: Employing latent dirichlet allocation for fraud detection in telecommunications. Pattern Recogn. Lett. 28, 1727–1734 (2007)

Flux Measurement Selection in Metabolic Networks

Wout Megchelenbrink, Martijn Huynen, and Elena Marchiori

Radboud University, Nijmegen,
The Netherlands

Abstract. Genome-scale metabolic networks can be reconstructed using a constraint-based modeling approach. The stoichiometry of the network and the physiochemical laws still enable organisms to achieve certain objectives -such as biomass composition- through many various pathways. This means that the system is underdetermined and many alternative solutions exist. A known method used to reduce the number of alternative pathways is Flux Balance Analysis (FBA), which tries to optimize a given biological objective function. FBA does not always find a correct solution and for many networks the biological objective function is simply unknown. This leaves researchers no other choice than to measure certain fluxes. In this article we propose a method that combines a sampling approach with a greedy algorithm for finding a subset of k fluxes that, if measured, are expected to reduce as much as possible the solution space towards the 'true' flux distribution. The parameter k is given by the user. Application of the proposed method to a toy example and two real-life metabolic networks indicate its effectiveness. The method achieves significantly more reduction of the solution space than when k fluxes are selected either at random or by a faster simple heuristic procedure. It can be used for guiding the biologists to perform experimental analysis of metabolic networks.

1 Introduction

An important goal for researchers in systems biology is to understand the properties of metabolic networks. These networks are built in a bottom-up fashion using various biological sources, such as genome annotations, metabolic databases and biochemical information [19]. Most metabolic networks are large and complex, having many alternative pathways for constructing certain metabolites. Therefore they are often modeled in a constraint-based fashion [24,14,13]. This approach enables researchers to perform quantitative analysis within a validated mathematical model. For instance, COBRA [21] is a computational toolbox based on this approach, that enables researchers to model and infer metabolic networks. Metabolic networks can be defined by two types of constraints. The first is the so-called mass-balance constraint in (1).

$$\frac{dx}{dt} = Sv, \tag{1}$$

M. Loog et al. (Eds.): PRIB 2011, LNBI 7036, pp. 214–224, 2011.

where dx/dt is the change in metabolite concentration over time. S is the stoichiometry matrix, and v is the flux vector. Assuming a steady-state, (1) simplifies to $Sv = 0$. The thermodynamic constraint (2 is the second type, which limits the upper and lower bounds of the flux rates.

$$v_{j,min} \leq v_j \leq v_{j,max}, \forall j \in J, \tag{2}$$

where v_j is the flux value of reaction j, and J denotes the set of reactions. Most of the networks contain more reactions than metabolites, leaving the system underdetermined and resulting in many possible solutions, that is, alternative flux distributions. In order to find the 'true' flux distribution under certain growth conditions, researchers measure certain fluxes to tighten the constraints; thereby reducing the solution space towards the 'true' flux distribution. One of the challenges we address in this paper is to detect a small set of k reactions that, when measured, will maximally reduce the solution space. Here k is a parameter selected by the user.

2   Related Work

Two early papers about discovering optimal flux measurements [18,17] suggest measuring those fluxes that are least sensitive to experimental error. The authors show that an upper bound for this sensitivity can be approximated solely on stoichiometry. When actual information is known on measurement errors, the sensitivity can be computed more accurately. However, these measurement are not optimized for reducing the solution space.

More recent methods for determining metabolic fluxes often optimize a biological objective function such a growth or ATP production. A well-known method that uses this strategy is flux balance analysis [9,12,23]. However for eukaryotic cells a biological objective function is often not easy to determine. An alternative approach often used in perturbed networks is the minimization of metabolic adjustment [22], involving the minimization of the Euclidean distance between the wild-type and perturbed network.

In the setting considered in this paper we assume that no information other than the mass-balance and thermodynamic constraints, is available. In particular, we do not use any biological objective function or external sources of knowledge such as gene expression. The goal is to find a set of k fluxes that, if measured, will reduce as much as possible the search space obtained from constraints (1) and (2). To the best of our knowledge, this is the first time such a problem is tackled in the context of metabolic networks analysis. However there are methods that try to tackle the related problem of finding the shape and size of the solution space by randomized sampling [10,20]. We will show that these type of methods are very useful also for addressing the problem we want to solve.

In [25] the authors use a sampling approach to find the size and shape of the steady-state solution space. The authors demonstrate which reactions are sensitive to the V_{max} conditions using a singular value decomposition on the human red blood cell network [8]. The authors of [15] show that the sampling approach can be used to find probability distributions for all fluxes, that it can be used to measure pairwise correlation coefficients and to compute the network wide effects of changes in flux variables. Braunstein et al. [3] show an alternative approach to approximate the volume and shape of the convex polytope using a message-passing algorithm based on belief propagation. Almaas et al. [1] have used random sampling on the E. coli network to show that there is a core set of reactions carrying a high flux, termed the 'high-flux backbone'. Finally flux variability analysis (FVA) [7,11] has been used to explore alternative optima and network redundancy.

3   Research Problem

The most reliable method to reduce the solution space is to perform experimental measurements to find the real flux values. The goal of this paper is to help the biologists to decide which k reactions to measure in order to get as close as possible to the 'true' flux distribution the organism uses, where k is a user given parameter.

The solution space of the system of equations in (1) and inequalities in (2) forms a bounded convex polytope. The number of alternative flux distributions can be expressed as the hypervolume of this polytope. Computing the exact volume of a convex polytope is a NP-hard problem and has been shown to be infeasible for dimensions bigger than 10 [2]. Since we don't need the exact volume but the smallest volume resulting when measuring a flux, we only need to estimate a relative volume. The authors of [25] have shown that the relative volume of a more constrained polytope P_c relative to the original model P_o can be approximated as follows:

$$\hat{V}_{rel} = \frac{|P_c|}{|P_o|}, \tag{3}$$

where $|P_o|$ is the number of sample points taken from P_o and $|P_c|$ is the number of these sample points that are also in the more constrained model P_c. Whenever a flux is measured, the initial bounds in the constrained model of the metabolic network become those of the measured value $\pm$ a term quantifying the uncertainty of the measurement. As a consequence, also the hypervolume of the polytope is reduced. The problem is that we don't know the value resulting from such experimental measurement. Therefore we resort to a sampling histogram of the network for a given flux: the number of samples within a certain bin reflects the expected volume of the polytope when the flux is constrained to assume its value in that particular bin. Formally, the expected volume when flux j is selected for being measured is

$$\hat{V}_j = \sum_{b=1}^{n} V_{j(b)}^2, \tag{4}$$

where n is the number of bins considered, $V_{j(b)}$ is the relative volume for bin b in flux j and $\sum_{b=1}^{n} V_{j(b)} = 1, \forall j$. This expected volume also reflects the probability that a sample is in that bin. Thus a flux having similar values across all bins will, when measured, yield the highest expected volume reduction. An example illustrates the observations above.

Example 1. Consider the histograms in Fig. 1. Figure 1a represents a flux yielding a low expected volume. Specifically, if we measure this flux, the expected volume of the polytope is 0.3. A flux yielding higher expected volume reduction when measured is shown in Fig. 1b: indeed, for that flux, $\hat{V}$ is minimal, equal to 0.25, which corresponds to the volume of any bin.

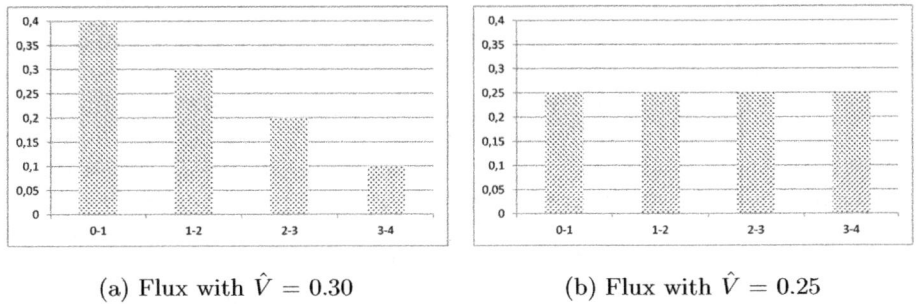

(a) Flux with $\hat{V} = 0.30$ (b) Flux with $\hat{V} = 0.25$

Fig. 1. Finding the flux with minimum expected volume

We want to find a set K^* consisting of k fluxes, that minimizes the expected volume, that is, we want to find the optimum of the following objective function:

$$\underset{K \subseteq J}{\arg\min} \, \hat{V}(K), \tag{5}$$

where J denotes the set of all fluxes in the considered metabolic network, K denotes a subset J of size k, and $\hat{V}(K)$ denotes the expected volume resulting from the measurements of the reactions in the set K. A greedy method for tackling this optimization problem is proposed in the next section.

4   Methods

For $k = 1$ one can tackle the optimization problem (5) using the formula (4) for finding the flux with minimum expected volume. However, using laboratory techniques such as isotope labeling enables researchers to measure multiple fluxes at once [16]. Therefore it is practically more useful to select $k > 1$ fluxes. An exact approach to solve optimization problem (5) amounts to search among all possible k flux combinations the one that minimizes $\hat{V}$. This exhaustive search becomes intractable on any genome-scale metabolic network even for low values of k, due to the combinatorial explosion of the number of subsets of reactions.

A fast heuristic method would be to select the k fluxes having smallest expected volume, as computed using equation (4). We call this method Reaction Minimizing Expected Volume Naive (RMEV-N in short). As shown in the computational analysis provided in the next section, this approach yields results of low quality. This is due to the fact that the shape and volume of the polytope changes whenever a flux is measured while all the computed predictions use the same initial model of the considered metabolic network. Therefore we consider a more involved, greedy search method, called Reaction Minimizing Expected Volume Greedy (RMEV-G in short) which is described in the next section.

4.1   RMEV-G: Reaction Minimizing Expected Volume Greedy

RMEV-G consists of three main phases. First, it reduces the ranges of the fluxes using a method for constraint domain reduction implemented in the COBRA toolbox. The resulting ranges will be used for computing a set of n bins $b_1 \ldots b_n$ for the range of each flux. Next, it generates a set of points from the solution space of the mass-balance (see (1)) and thermodynamic (see (2)) constraints, by means of a uniform sampling procedure. Specifically, we use an implementation of the ACHR sampler algorithm [10,21] to generate such a set of sampled solutions. This set will be used for computing the expected volume of a flux, using equation (4). Finally, RMEV-G uses the resulting ranges and sampled solutions for selecting k fluxes $j_1, \ldots, j_k$ using the following greedy search procedure.

- j_1. The reaction with lowest $\hat{V}$ is picked as the first selected reaction j_1. This can be done by computing the expected volume of each flux using equation (4) and selecting the flux with minimum expected volume.
- j_i, with $i > 1$. At iteration i, the method tries to find the reaction j_i that minimizes the expected volume $\hat{V}$ given that the reactions $j_1 \ldots j_{(i-1)}$ have been selected. Since we only know that $j_1, \ldots, j_{i-1}$ have been selected, but we do not know in which bins the selected fluxes will occur, we have to consider each bin as a possibility. Therefore we construct a search tree as follows.

Search Tree Construction. Nodes of this tree are (labeled with) fluxes and edges are (labeled with) bins. Each edge has a weight, equal to the probability of the parent node (flux) to be in the bin corresponding to that edge given the bins that have been selected in the path from the root to that edge (for instance, the probability that the flux in Fig. 1a is in bin [1,2] is 0.3).

1. The root of this tree is j_1.
2. Each node has n children, one for each bin (edge). Each of these children is the flux having minimum expected volume in the reduced solution space resulting from the selection of fluxes and bins along the path from the root to that node. The expected volume of that flux in this reduced space can be computed using the subset of the sampled solutions obtained by discarding those that are not consistent with the new restricted ranges of the fluxes occurring in that path.

An example of such a search tree in given Fig. 2, with 3 bins and $k = 3$ fluxes to be selected. The search tree is used to select j_i by means of a weighted majority vote criterion applied to the nodes occurring at depth i.

Majority Vote Selection Criterion. For each flux, the weighted sum of the occurrences of that flux at depth i in the search tree is computed, where each occurrence is weighted by the weight of the edge linking that occurrence with its parent. Then the flux with maximum weighted sum is selected (ties are broken randomly). The output of RMEV-G is an ordered list of k fluxes and a resulting expected volume after measuring each flux given the previous ones that have been selected. The computational complexity of RMEV-G is $O(mn^k)$, where m is the total number of fluxes, n the number of bins, and k the number of fluxes to be selected. Therefore the method is in general applicable for small values of k and n.

Example 2. Consider the example search tree in Fig. 2, with 3 bins and $k = 3$ fluxes to be selected. The first reaction selected is f_4, having minimum $\hat{V}$. The three bins of f_4 are split and the reaction having minimum $\hat{V}$ is multiplied by its prior probability. The reaction with highest weighted vote is used in all subsequent iterations.

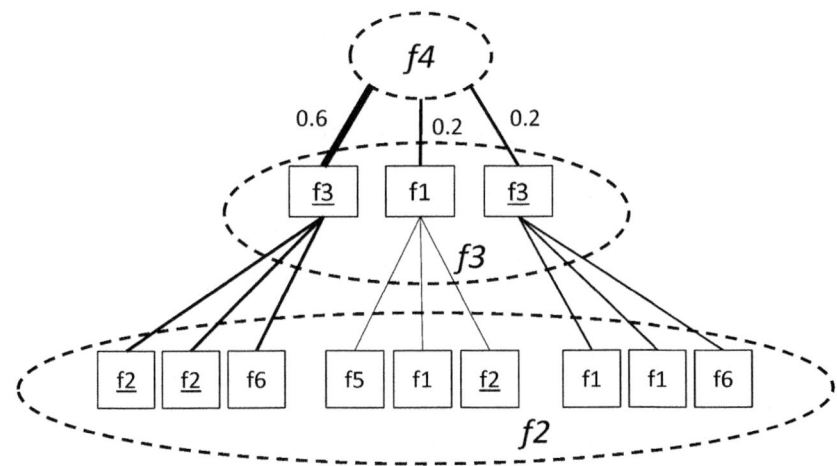

Fig. 2. Schematic representation of a search tree with three bins and three fluxes. The line width is proportional to the probability mass in each bin and thus to the weight of each vote. Underlined fluxes are those selected using weighted majority vote. In particular, in the second iteration, f_3 is chosen because it has the maximum weighted vote, equal to $0.6 + 0.2$.

In the next section we test the effectiveness of RMEV-G on artificial and real-life metabolic networks, and measure the reduced volume versus that achieved by applying faster algorithms like RMEV-N and random selection.

5   Experimental Analysis

In order to test comparatively the effectiveness of the proposed method we have conducted computational experiments on the following three networks.

1. Toy model. This is a small network explained in [25]. It contains only 8 reactions and 5 metabolites.
2. Red blood cell (RBC). This metabolic network is a constraint-based network based on the kinetic model from [8]. We have used the model that was available as supplementary material from [3].
3. E. coli central metabolism (E. coli c.m.). This is a condensed version of the genome-scale E. coli network. It contains reactions and metabolites from central metabolism and is used also in [13]. It can be publicly downloaded[1].

We considered the following three methods.

- **RMEV-G.** The proposed greedy method previously described.
- **Reaction Minimizing Expected Volume Random (RMEV-R).** This algorithm selects fluxes randomly, adding the extra constraint that fluxes can not be fully coupled [4,6]. Fully coupled reactions often occur in a linear path and their fluxes are by definition the same. Therefore it is useless to measure multiple reactions in such a set. After selecting k fluxes not containing fully coupled reactions, we used the same bins as for the greedy approach and traverse the search tree in the exact same way. At every level of the tree, from $i = 1, \ldots, i = k$ we computed the expected volume using 3.
- **Reaction Minimizing Expected Volume Naive (RMEV-N).** This algorithm amounts to performing only the first iteration of RMEV-G and then choosing the top k fluxes having smallest expected volume and not containing fully coupled reactions. Note that this method measures the expected volume for all fluxes based on the original solution space. Therefore it does not take into account the changed shape and size of the solution space after selection of a flux.

We applied these methods to each of the considered networks, and compared the expected volumes resulting from the selection of $k \leq 5$ fluxes using $n = 5$ bins. For each network, $N = 1000$ runs of RMEV-R were performed. Results of these runs were used to compute an empirical p-value, as the fraction of runs were RMEV-R achieved smaller expected volume than RMEV-G.

$$\sum_{i=1}^{N} \frac{I(\hat{V}_1 > \hat{V}_2)}{N}, \tag{6}$$

where I denotes the indicator function, 1 denote method RMEV-G and 2 method RMEV-R.

---

[1] http://www-bioeng.ucsd.edu/research/research_groups/gcrg/organisms/ecoli/ecoli_sbml.html

5.1 Results

Table 1 reports the results of our experiments. They show that RMEV-G performs better than the other methods. In particular, results of the experiments indicate that RMEV-N performs worse than RMEV-G. This is expected since the latter algorithm considers a reduced solution space after each selection of a flux.

Table 1. Results on the toy model and two real-life metabolic networks. Column "Model" contains the considered metabolic networks, "K" depicts the number of fluxes determined, "Size" the network size (metabolites, reactions), RMEV-R, RMEV-N, and RMEV-G the expected volume computed using the three methods, where for RMEV-R, the average over the considered runs is reported. Finally, "P-value" denotes the empirical p-value between RMEV-G and RMEV-R.

Model	Size	K	RMEV-R	RMEV-N	RMEV-G	P-value
Toy model	5, 8					
		1	0.2402	0.2363	0.2363	0
		3	0.0194	0.0193	0.0193	0
		5	0.0069	0.0069	0.0069	0.6450
RBC	34, 46					
		1	0.4156	0.2104	0.2104	0
		3	0.1047	0.0666	0.0229	0.0020
		5	0.0356	0.0079	0.0044	0
E. coli c.m.	62,75					
		1	0.7127	0.2001	0.2001	0
		3	0.3794	0.0702	0.0334	0
		5	0.1704	0.0325	0.0083	0.0630

The last column in the table contains the p-value computed using (6). The relative high p-value for the toy model and $k = 5$ can be justified by the small dimension of the solution space (which is 3), meaning that there is little room for more reduction of the solution space. On the two real-life metabolic networks small p-values are obtained, except for the E. coli network, where a p-value of 0.0630 is computed. This could be possibly due to the constraint that the selected fluxes should not contain fully coupled reactions. This remains to be investigated in more depth. Figure 3 shows the performance of the methods on the largest network considered, that is, the E. coli one. The theoretical lower bound for $\hat{V}$ is also plotted, computed as $lb_k = 1/n^k$. The plots clearly show that RMEV-G has a very good performance, with expected volume closer to the lower bound than the other two methods.

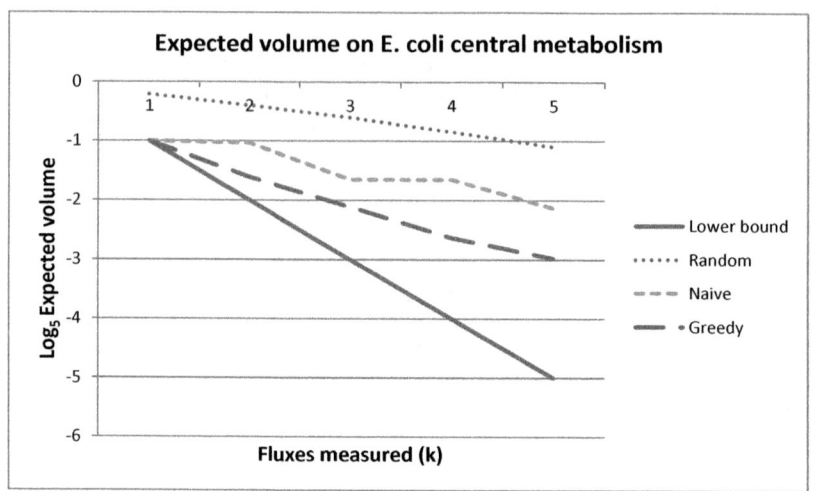

Fig. 3. Performance of the three methods and the theoretical lower bound of $\hat{V}$ on the E. coli central metabolism network. The x-axis reports the number of fluxes selected, the y-axis the $\log_5$ of the expected volume, where the base 5 is the number of bins.

6   Conclusion and Discussion

The experimental analysis shows that RMEV-G can be used to help selecting fluxes for measurement in the laboratory. In all networks examined RMEV-G performs better than the two other methods discussed, because it considers the new solution space after selecting each subsequent flux. In particular, RMEV-G reduces the solution space much more compared to random selection, even when the latter is restricted not to contain fully coupled reactions. Note that by the definition of RMEV-G it takes reaction coupling implicitly into account, because fully coupled fluxes will have sample points that coincide and therefore will give a higher expected volume than any other flux. An important point to note is that RMEV-G is based on only stoichiometry and flux bounds and is therefore highly sensitive to how the initial network is constrained. In the future, we will try to incorporate biological data, such as gene expression in order to alleviate this problem and to make predictions more accurate.

7   Future Work

Although the results of the preliminary experiments conducted in this work indicate that the proposed greedy method is effective, we are going to extend the experimental analysis by considering genome-scale metabolic networks. Moreover, we want to test the method on a measured network, to see if our predictions agree with the measured values.

Furthermore, we are currently working on an extension of the method that incorporates prior information from gene expression data. Another interesting possibility we are currently investigating is combining our algorithm with an approach that tries to minimize error amplification, like for instance the one introduced in [18,17].

Another issue we want to address is the time complexity of the method. The exponential nature (on the number of bins and number of selected fluxes) poses a problem if one wants to measure a larger number of fluxes. For instance, a possible solution could be to parallelize the algorithm by running in parallel the generation of (groups of) children nodes at each depth of the search tree prior to the integration of the decisions computed from the nodes using the majority vote criterion.

Finally, we want to investigate the relation between our method and the Bayesian experimental design approach [5].

Acknowledgments. The authors would like to thank Richard Notebaart, Sergio Rossell and Radek Szklarczyk for their valuable comments. This work has been funded by the Netherlands organization for scientific research (NWO) and the Netherlands organization for health research and innovation in healthcare (ZonMW).

References

1. Almaas, E., Kovács, B., Vicsek, T., Oltvai, Z.N., Barabási, A.-L.: Global organization of metabolic fluxes in the bacterium escherichia coli. Nature 427(6977), 839–843 (2004)
2. Beeler, B., Enge, A., Fukuda, K., Lthi, H.-J.: Exact volume computation for polytopes: a practical study. In: 12th European Workshop on Computational Geometry, Muenster, Germany (1996)
3. Braunstein, A., Mulet, R., Pagnani, A.: Estimating the size of the solution space of metabolic networks. BMC Bioinformatics 9(1), 240 (2008)
4. Burgard, A.P., Nikolaev, E.V., Schilling, C.H., Maranas, C.D.: Flux coupling analysis of genome-scale metabolic network reconstructions. Genome Research 14(2), 301–312 (2004)
5. Chaloner, K., Verdinelli, I.: Bayesian experimental design: A review. Statistical Science 10(3), 273–304 (1995)
6. David, L., Marashi, S.-A., Larhlimi, A., Mieth, B., Bockmayr, A.: FFCA: a feasibility-based method for flux coupling analysis of metabolic networks. BMC Bioinformatics 12(1), 236 (2011)
7. Gudmundsson, S., Thiele, I.: Computationally efficient flux variability analysis. BMC Bioinformatics 11, 489 (2010)
8. Jamshidi, N., Edwards, J.S., Fahland, T., Church, G.M., Palsson, B.O.: Dynamic simulation of the human red blood cell metabolic network. Bioinformatics 17(3), 286–287 (2001)
9. Kauffman, K.J., Prakash, P., Edwards, J.S.: Advances in flux balance analysis. Current Opinion in Biotechnology 14(5), 491–496 (2003)

10. Kaufman, D.E., Smith, R.L.: Direction choice for accelerated convergence in hit-and-run sampling. Operations Research 46(1), 84–95 (1998)
11. Mahadevan, R., Schilling, C.H.: The effects of alternate optimal solutions in constraint-based genome-scale metabolic models. Metabolic Engineering 5(4), 264–276 (2003)
12. Orth, J.D., Thiele, I., Palsson, B.O.: What is flux balance analysis? Nature Biotechnology 28(3), 245–248 (2010)
13. Palsson, B.O.: Systems Biology: Properties of Reconstructed Networks, 1st edn. Cambridge University Press (2006)
14. Price, N.D., Reed, J.L., Palsson, B.O.: Genome-scale models of microbial cells: evaluating the consequences of constraints. Nature Reviews Microbiology 2(11), 886–897 (2004)
15. Price, N.D., Schellenberger, J., Palsson, B.O.: Uniform sampling of Steady-State flux spaces: Means to design experiments and to interpret enzymopathies. Biophysical Journal 87(4), 2172–2186 (2004)
16. Sauer, U.: Metabolic networks in motion: 13C-based flux analysis. Molecular Systems Biology 2, 62 (2006)
17. Savinell, J.M., Palsson, B.O.: Optimal selection of metabolic fluxes for in vivo measurement. I. Development of mathematical methods. Journal of Theoretical Biology 155(2), 201–214 (1992)
18. Savinell, J.M., Palsson, B.O.: Optimal selection of metabolic fluxes for in vivo measurement. II. Application to escherichia coli and hybridoma cell metabolism. Journal of Theoretical Biology 155(2), 215–242 (1992)
19. Schellenberger, J., Lewis, N.E., Palsson, B.O.: Elimination of thermodynamically infeasible loops in steady-state metabolic models. Biophysical Journal 100(3), 544–553 (2011)
20. Schellenberger, J., Palsson, B.O.: Use of randomized sampling for analysis of metabolic networks. The Journal of Biological Chemistry 284(9), 5457–5461 (2009)
21. Schellenberger, J., et al.: Quantitative prediction of cellular metabolism with constraint-basded models: the cobra toolbox v2.0. Nature Protocols 6(9), 1290–1307 (2011)
22. Segré, D., Vitkup, D., Church, G.M.: Analysis of optimality in natural and perturbed metabolic networks. Proceedings of the National Academy of Sciences of the United States of America 99(23), 15112–15117 (2002)
23. Smallbone, K., Simeonidis, E.: Flux balance analysis: a geometric perspective. Journal of Theoretical Biology 258(2), 311–315 (2009)
24. Varma, A., Palsson, B.O.: Metabolic flux balancing: Basic concepts, scientific and practical use. Nature Biotechnology 12, 994 (1994)
25. Wiback, S.J., Famili, I., Harvey, J., Greenberg, H.J., Palsson, B.O.: Monte carlo sampling can be used to determine the size and shape of the steady-state flux space. Journal of Theoretical Biology 228(4), 437–447 (2004)

Lagrangian Relaxation Applied to Sparse Global Network Alignment

Mohammed El-Kebir[1,2,3], Jaap Heringa[2,3,4], and Gunnar W. Klau[1,3]

[1] Centrum Wiskunde & Informatica, Life Sciences Group, Science Park 123,
1098 XG Amsterdam, The Netherlands
{m.el-kebir,gunnar.klau}@cwi.nl
[2] Centre for Integrative Bioinformatics VU (IBIVU), VU University Amsterdam,
De Boelelaan 1081A, 1081 HV Amsterdam, The Netherlands
heringa@few.vu.nl
[3] Netherlands Institute for Systems Biology, Amsterdam, The Netherlands
[4] Netherlands Bioinformatics Centre, The Netherlands

Abstract. Data on molecular interactions is increasing at a tremendous pace, while the development of solid methods for analyzing this network data is lagging behind. This holds in particular for the field of comparative network analysis, where one wants to identify commonalities between biological networks. Since biological functionality primarily operates at the network level, there is a clear need for topology-aware comparison methods. In this paper we present a method for global network alignment that is fast and robust, and can flexibly deal with various scoring schemes taking both node-to-node correspondences as well as network topologies into account. It is based on an integer linear programming formulation, generalizing the well-studied quadratic assignment problem. We obtain strong upper and lower bounds for the problem by improving a Lagrangian relaxation approach and introduce the software tool NATALIE 2.0, a publicly available implementation of our method. In an extensive computational study on protein interaction networks for six different species, we find that our new method outperforms alternative state-of-the-art methods with respect to quality and running time. An extended version of this paper including proofs and pseudo code is available at http://arxiv.org/pdf/1108.4358v1.

1 Introduction

In the last decade, data on molecular interactions has increased at a tremendous pace. For instance, the STRING database [24], which contains protein protein interaction (PPI) data, grew from 261,033 proteins in 89 organisms in 2003 to 5,214,234 proteins in 1,133 organisms in May 2011, more than doubling the number of proteins in the database every two years. The same trends can be observed for other types of biological networks, including metabolic, gene-regulatory, signal transduction and metagenomic networks, where the latter can incorporate the excretion and uptake of organic compounds through, for example, a microbial community [21, 12]. In addition to the plethora of experimentally derived

M. Loog et al. (Eds.): PRIB 2011, LNBI 7036, pp. 225–236, 2011.
© Springer-Verlag Berlin Heidelberg 2011

network data for many species, also the structure and behavior of molecular networks have become intensively studied over the last few years [2], leading to the observation of many conserved features at the network level. However, the development of solid methods for analyzing network data is lagging behind, particularly in the field of comparative network analysis. Here, one wants to identify commonalities between biological networks from different strains or species, or derived form different conditions. Based on the assumption that evolutionary conservation implies functional significance, comparative approaches may help (i) improve the accuracy of data, (ii) generate, investigate, and validate hypotheses, and (iii) transfer functional annotations. Until recently, the most common way of comparing two networks has been to solely consider node-to-node correspondences, for example by finding homologous relationships between nodes (e.g. proteins in PPI networks) of either network, while the topology of the two networks has not been taken into account. Since biological functionality primarily operates at the network level, there is a clear need for topology-aware comparison methods. In this paper we present a network alignment method that is fast and robust, and can flexibly deal with various scoring schemes taking both node-to-node correspondences as well as network topologies into account.

Previous Work. Network alignment establishes node correspondences based on both node-to-node similarities and conserved topological information. Similar to sequence alignment, *local* network alignment aims at identifying one or more shared subnetworks, whereas *global* network alignment addresses the overall comparison of the complete input networks.

Over the last years a number of methods have been proposed for both global and local network alignment, for example PATHBLAST [14], NETWORKBLAST [22], MAWISH [16], GRAEMLIN [8], ISORANK [23], GRAAL [17], and SUBMAP [4]. PATHBLAST heuristically computes high-scoring similar paths in two PPI networks. Detecting protein complexes has been addressed with NETWORK-BLAST by Sharan et al. [22], where the authors introduce a probabilistic model and propose a heuristic greedy approach to search for shared complexes. Koyutürk et al. [16] use a more elaborate scoring scheme based on an evolutionary model to compute local pairwise alignments of PPI networks. The ISORANK algorithm by Singh et al. [23] approaches the global alignment problem by preferably matching nodes which have a similar neighborhood, which is elegantly solved as an eigenvalue problem. Kuchaiev et al. [17] take a similar approach. Their method GRAAL matches nodes that share a similar distribution of so-called graphlets, which are small connected non-isomorphic induced subgraphs.

In this paper we focus on pairwise global network alignment, where an alignment is scored by summing up individual scores of aligned node and interaction pairs. Among the above mentioned methods, ISORANK and GRAAL use a scoring model that can be expressed in this manner.

Contribution. We present an algorithm for global network alignment based on an integer linear programming (ILP) formulation, generalizing the well-studied quadratic assignment problem (QAP). We improve upon an existing Lagrangian

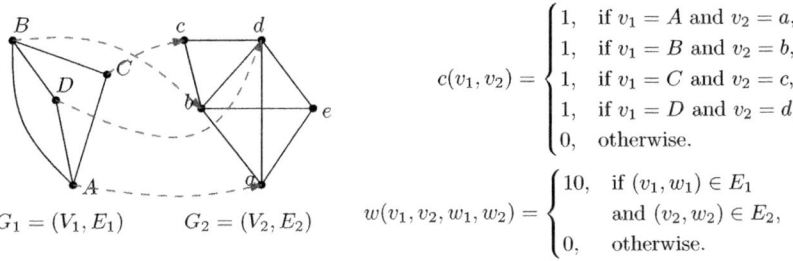

$$c(v_1, v_2) = \begin{cases} 1, & \text{if } v_1 = A \text{ and } v_2 = a, \\ 1, & \text{if } v_1 = B \text{ and } v_2 = b, \\ 1, & \text{if } v_1 = C \text{ and } v_2 = c, \\ 1, & \text{if } v_1 = D \text{ and } v_2 = d, \\ 0, & \text{otherwise.} \end{cases}$$

$$w(v_1, v_2, w_1, w_2) = \begin{cases} 10, & \text{if } (v_1, w_1) \in E_1 \\ & \text{and } (v_2, w_2) \in E_2, \\ 0, & \text{otherwise.} \end{cases}$$

Fig. 1. Example of a network alignment. With the given scoring function, the alignment has a score of $4 + 40 = 44$.

relaxation approach presented in previous work [15] to obtain strong upper and lower bounds for the problem. We exploit the closeness to QAP and generalize a dual descent method for updating the Lagrangian multipliers to the generalized problem. We have implemented the revised algorithm from scratch as the software tool NATALIE 2.0. In an extensive computational study on protein interaction networks for six different species, we compare NATALIE 2.0 to GRAAL and ISORANK, evaluating the number of conserved edges as well as functional coherence of the modules in terms of GO annotation. We find that NATALIE 2.0 outperforms the alternative methods with respect to quality and running time. Our software tool NATALIE 2.0 as well as all data sets used in this study are publicly available at http://planet-lisa.net.

2   Preliminaries

Given two simple graphs $G_1 = (V_1, E_1)$ and $G_2 = (V_2, E_2)$, an *alignment* $a : V_1 \rightarrow V_2$ is a *partial injective function* from V_1 to V_2. As such we have that an alignment relates every node in V_1 to at most one node in V_2 and that conversely every node in V_2 has at most one counterpart in V_1. An alignment is assigned a real-valued *score* using an additive scoring function s defined as follows:

$$s(a) = \sum_{v \in V_1} c(v, a(v)) + \sum_{\substack{v, w \in V_1 \\ v < w}} w(v, a(v), w, a(w)) \qquad (1)$$

where $c : V_1 \times V_2 \rightarrow \mathbb{R}$ is the score of aligning a pair of nodes in V_1 and V_2 respectively. On the other hand, $w : V_1 \times V_2 \times V_1 \times V_2 \rightarrow \mathbb{R}$ allows for scoring topological similarity. The problem of global pairwise network alignment (GNA) is to find the highest scoring alignment a^*, i.e. $a^* = \arg \max s(a)$. Figure 1 shows an example.

NP-hardness of GNA follows by a simple reduction from the decision problem CLIQUE, which asks whether there is a clique of cardinality at least k in a given simple graph $G = (V, E)$ [13]. The corresponding GNA instance concerns the alignment of the complete graph of k vertices $K_k = (V_k, E_k)$ with G using the scoring function $s(a) = |\{(v, w) \in E_k \mid (a(v), a(w)) \in E\}|$. Since an alignment is injective, there is a clique of cardinality at least k if and only if the cost of the optimal alignment is $\binom{k}{2}$. The close relationship of GNA with the quadratic assignment problem is more easily observed when formulating GNA as a mathematical program. Throughout the remainder of the text we use dummy variables $i, j \in \{1, \ldots, |V_1|\}$ and $k, l \in \{1, \ldots, |V_2|\}$ to denotes nodes in V_1 and V_2, respectively. Let C be a $|V_1| \times |V_2|$ matrix such that $c_{ik} = c(i, k)$ and let W be a $(|V_1| \times |V_2|) \times (|V_1| \times |V_2|)$ matrix whose entries w_{ikjl} correspond to interaction scores $w(i, k, j, l)$. Now we can formulate GNA as

$$\max_{x} \quad \sum_{i,k} c_{ik} x_{ik} + \sum_{\substack{i,j \\ i<j}} \sum_{\substack{k,l \\ k \neq l}} w_{ikjl} x_{ik} x_{jl} \tag{IQP}$$

$$\text{s.t.} \quad \sum_{l} x_{jl} \leq 1 \qquad \forall j \tag{2}$$

$$\sum_{j} x_{jl} \leq 1 \qquad \forall l \tag{3}$$

$$x_{ik} \in \{0, 1\} \qquad \forall i, k \tag{4}$$

where the decision variable x_{ik} indicates whether the i-th node in V_1 is aligned with the k-th node in V_2. The above formulation shares many similarities with Lawler's formulation [19] of the QAP. However, instead of finding an assignment we are interested in finding a matching, which is reflected in constraints (2) and (3) being inequalities rather than equalities. As can be seen in (1) we only consider the upper triangle of W rather than the entire matrix. An analogous way of looking at this, is to consider W to be symmetric. This is usually not the case for QAP instances. In addition, due to the fact that biological input graphs are typically sparse, we have that W is sparse as well. These differences allow us to come up with an effective method of solving the problem as we will see in the following.

3   Method

The relaxation presented here follows the same lines as the one given by Adams and Johnson for the QAP [1]. We start by linearizing (IQP) by introducing binary variables y_{ikjl} defined as $y_{ikjl} := x_{ik} x_{jl}$ and constraints $y_{ikjl} \leq x_{jl}$ and $y_{ikjl} \leq x_{ik}$ for all $i \leq j$ and $k \neq l$. If we assume that all entries in W are positive, we do not need to enforce that $y_{ikjl} \geq x_{ik} + x_{jl} - 1$. In Section 5 we will discuss this assumption. Rather than using the aforementioned constraints, we make use of a stronger set of constraints which we obtain by multiplying constraints (2) and (3) by x_{ik}:

$$\sum_{l \neq k} y_{ikjl} = \sum_{l \neq k} x_{ik} x_{jl} \leq \sum_{l} x_{ik} x_{jl} \leq x_{ik}, \quad \forall i, j, k, \ i < j \tag{5}$$

$$\sum_{j > i} y_{ikjl} = \sum_{j > i} x_{ik} x_{jl} \leq \sum_{j} x_{ik} x_{jl} \leq x_{ik}, \quad \forall i, k, l, \ k \neq l \tag{6}$$

We proceed by splitting the variable y_{ikjl} (where $i < j$ and $k \neq l$). In other words, we extend the objective function such that the counterpart of y_{ikjl} becomes y_{jlik}. This is accomplished by rewriting the dummy constraint in (6) to $j \neq i$. In addition, we split the weights: $w_{ikjl} = w_{jlik} = (w'_{ikjl}/2)$ where w'_{ikjl} denotes the original weight. Furthermore, we require that the counterparts of the split decision variables assume the same value, which amounts to

$$\max_{x,y} \quad \sum_{i,k} c_{ik}x_{ik} + \sum_{\substack{i,j \\ i<j}} \sum_{\substack{k,l \\ k \neq l}} w_{ikjl}y_{ikjl} + \sum_{\substack{i,j \\ i>j}} \sum_{\substack{k,l \\ k \neq l}} w_{ikjl}y_{ikjl} \qquad \text{(ILP)}$$

s.t. (2), (3) and (4)

$$\sum_{\substack{l \\ l \neq k}} y_{ikjl} \leq x_{ik} \qquad\qquad \forall i,j,k,\ i \neq j \qquad (7)$$

$$\sum_{\substack{j \\ j \neq i}} y_{ikjl} \leq x_{ik} \qquad\qquad \forall i,k,l,\ k \neq l \qquad (8)$$

$$y_{ikjl} = y_{jlik} \qquad\qquad \forall i,j,k,l,\ i < j, k \neq l \qquad (9)$$

$$y_{ikjl} \in \{0,1\} \qquad\qquad \forall i,j,k,l,\ i \neq j, k \neq l \qquad (10)$$

We can solve the continuous relaxation of (ILP) via its Lagrangian dual by dualizing the linking constraints (9) with multiplier λ:

$$\min_{\lambda} \quad Z_{\text{LD}}(\lambda) , \qquad\qquad\qquad\qquad \text{(LD)}$$

where $Z_{\text{LD}}(\lambda)$ equals

$$\max_{x,y} \quad \sum_{i,k} c_{ik}x_{ik} + \sum_{\substack{i,j \\ i<j}} \sum_{\substack{k,l \\ k \neq l}} (w_{ikjl} + \lambda_{ikjl})y_{ikjl} + \sum_{\substack{i,j \\ i>j}} \sum_{\substack{k,l \\ k \neq l}} (w_{ikjl} - \lambda_{jlik})y_{ikjl}$$

s.t. (2), (3), (4), (7), (8), (10)

Now that the linking constraints have been dualized, one can observe that the remaining constraints decompose the variables into $|V_1||V_2|$ disjoint groups, where variables across groups are not linked by any constraint, and where each group contains a variable x_{ik} and variables y_{ikjl} for $j \neq i$ and $l \neq k$. Hence, we have

$$Z_{\text{LD}}(\lambda) = \max_{x} \quad \sum_{i,k} [c_{ik} + v_{ik}(\lambda)]x_{ik} \quad \text{s.t.} \quad (2), (3) \text{ and } (4) \qquad (\text{LD}_\lambda)$$

which corresponds to a maximum weight bipartite matching problem on the so-called *alignment graph* $G_m = (V_1 \cup V_2, E_m)$. In the general case G_m is a complete bipartite graph, i.e. $E_m = \{(i,k) \mid i \in V_1, v_2 \in V_2\}$. However, by exploiting biological knowledge one can make G_m more sparse by excluding biologically-unlikely edges (see Section 4). For the global problem, the weight of a matching edge (i,k) is set to $c_{ik} + v_{ik}(\lambda)$, where the latter term is computed as

$$v_{ik}(\lambda) = \max_y \sum_{\substack{j \\ j>i}} \sum_{\substack{l \\ l \neq k}} (w_{ikjl} + \lambda_{ikjl}) y_{ikjl} + \sum_{\substack{j \\ j<i}} \sum_{\substack{l \\ l \neq k}} (w_{ikjl} - \lambda_{jlik}) y_{ikjl} \qquad (\text{LD}_\lambda^{ik})$$

$$\text{s.t.} \quad \sum_{\substack{l \\ l \neq k}} y_{ikjl} \leq 1 \qquad\qquad\qquad\qquad \forall j,\ j \neq i \quad (11)$$

$$\sum_{\substack{j \\ j \neq i}} y_{ikjl} \leq 1 \qquad\qquad\qquad\qquad \forall l,\ l \neq k \quad (12)$$

$$y_{ikjl} \in \{0,1\} \qquad\qquad\qquad\qquad \forall j,l. \quad (13)$$

Again, this is a maximum weight bipartite matching problem on the same alignment graph but excluding edges incident to either i or k and using different edge weights: the weight of an edge (j,l) is $w_{ikjl} + \lambda_{ikjl}$ if $j > i$, or $w_{ikjl} - \lambda_{jlik}$ if $j < i$. So in order to compute $Z_{\text{LD}}(\lambda)$, we need to solve a total number of $|V_1||V_2| + 1$ maximum weight bipartite matching problems, which, using the Hungarian algorithm [18, 20] can be done in $O(n^5)$ time, where $n = \max(|V_1|, |V_2|)$. In case the alignment graph is sparse, i.e. $O(|E_m|) = O(n)$, $Z_{\text{LD}}(\lambda)$ can be computed in $O(n^4 \log n)$ time using the successive shortest path variant of the Hungarian algorithm [7]. It is important to note that for any λ, $Z_{\text{LD}}(\lambda)$ is an upper bound on the score of an optimal alignment. This is because any alignment a is feasible to $Z_{\text{LD}}(\lambda)$ and does not violate the original linking constraints and therefore has an objective value equal to $s(a)$. In particular, the optimal alignment a^* is also feasible to $Z_{\text{LD}}(\lambda)$ and hence $a^* \leq Z_{\text{LD}}(\lambda)$. Since the two sets of problems resulting from the decomposition both have the integrality property [6], the smallest upper bound we can achieve equals the linear programming (LP) bound of the continuous relaxation of (ILP) [9]. However, computing the smallest upper bound by finding suitable multipliers is much faster than solving the corresponding LP. Given solution (x, y) to $Z_{\text{LD}}(\lambda)$, we obtain a lower bound on $s(a^*)$, denoted $Z_{\text{lb}}(\lambda)$, by considering the score of the alignment encoded in x.

3.1   Solving Strategies

In this section we will discuss strategies for identifying Lagrangian multipliers λ that yield an as small as possible gap between the upper and lower bound resulting from the solution to $Z_{\text{LD}}(\lambda)$.

Subgradient Optimization. We start by discussing subgradient optimization, which is originally due to Held and Karp [10]. The idea is to generate a sequence $\lambda^0, \lambda^1, \ldots$ of Lagrangian multiplier vectors starting from $\lambda^0 = \mathbf{0}$ as follows:

$$\lambda_{ikjl}^{t+1} = \lambda_{ikjl}^t - \frac{\alpha \cdot (Z_{\text{LD}}(\lambda) - Z_{\text{lb}}(\lambda))}{\|g(\lambda^t)\|^2} g(\lambda_{ikjl}^t) \qquad \forall i,j,k,l,\ i<j, k \neq l \quad (14)$$

where $g(\lambda_{ikjl}^t)$ corresponds to the subgradient of multiplier λ_{ikjl}^t, i.e. $g(\lambda_{ikjl}^t) = y_{ikjl} - y_{jlik}$, and α is the step size parameter. Initially α is set to 1 and it is halved if neither $Z_{\text{LD}}(\lambda)$ nor $Z_{\text{lb}}(\lambda)$ have improved for over N consecutive iterations. Conversely, α is doubled if M times in a row there was an improvement in either $Z_{\text{LD}}(\lambda)$ or $Z_{\text{lb}}(\lambda)$ [5]. In case all subgradients are zero, the optimal solution has been found and the scheme terminates. Note that this is not guaranteed to happen. Therefore we abort the scheme after exceeding a time limit or a pre-specified number of iterations. In addition, we terminate if α has dropped below machine precision.

Dual Descent. In this section we derive a dual descent method which is an extension of the one presented in [1]. The dual descent method takes as a starting point the dual of $Z_{\text{LD}}(\lambda)$:

$$Z_{\text{LD}}(\lambda) = \min_{\alpha,\beta} \quad \sum_i \alpha_i + \sum_k \beta_k \tag{15}$$

$$\text{s.t.} \quad \alpha_i + \beta_k \geq c_{ik} + v_{ik}(\lambda) \qquad \forall i,k \tag{16}$$

$$\alpha_i \geq 0 \qquad \forall i \tag{17}$$

$$\beta_k \geq 0 \qquad \forall k \tag{18}$$

where the dual of $v_{ik}(\lambda)$ is

$$v_{ik}(\lambda) = \min_{\mu,\nu} \quad \sum_{\substack{j \\ j \neq i}} \mu_j^{ik} + \sum_{\substack{l \\ l \neq k}} \nu_l^{ik} \tag{19}$$

$$\text{s.t.} \quad \mu_j^{ik} + \nu_l^{ik} \geq w_{ikjl} + \lambda_{ikjl} \qquad \forall j,l,\ j > i, l \neq k \tag{20}$$

$$\mu_j^{ik} + \nu_l^{ik} \geq w_{ikjl} - \lambda_{jlik} \qquad \forall j,l,\ j < i, l \neq k \tag{21}$$

$$\mu_j^{ik} \geq 0 \qquad \forall j \tag{22}$$

$$\nu_l^{ik} \geq 0 \qquad \forall l. \tag{23}$$

Suppose that for a given λ^t we have computed dual variables (α, β) solving (15) with objective value $Z_{\text{LD}}(\lambda^t)$, as well as dual variables (μ^{ik}, ν^{ik}) yielding values $v_{ik}(\lambda)$ to linear programs (19). The goal now is to find λ^{t+1} such that the resulting bound is better or just as good, i.e. $Z_{\text{LD}}(\lambda^{t+1}) \leq Z_{\text{LD}}(\lambda^t)$. We prevent the bound from increasing, by ensuring that the dual variables (α, β) remain feasible to (15). This we can achieve by considering the slacks: $\pi_{ik}(\lambda) = \alpha_i + \beta_k - c_{ik} - v_{ik}(\lambda)$. So for (α, β) to remain feasible, we can only allow every $v_{ik}(\lambda^t)$ to increase by as much as $\pi_{ik}(\lambda^t)$. We can achieve such an increase by considering linear programs (19) and their slacks defined as

$$\gamma_{ikjl}(\lambda) = \begin{cases} \mu_j^{ik} + \nu_l^{ik} - w_{ikjl} + \lambda_{ikjl}, & \text{if } j > i, \\ \mu_j^{ik} + \nu_l^{ik} - w_{ikjl} - \lambda_{jlik}, & \text{if } j < i, \end{cases} \quad \forall j,l,\ j \neq i, l \neq k, \tag{24}$$

and update the multipliers in the following way.

Lemma 1. *The adjustment scheme below yields solutions to linear programs* (19) *with objective values* $v_{ik}(\lambda^{t+1})$ *at most* $\pi_{ik}(\lambda^t) + v_{ik}(\lambda^t)$ *for all* i, k.

$$\lambda_{ikjl}^{t+1} = \lambda_{ikjl}^{t} + \varphi_{ikjl} \left[\gamma_{ikjl}(\lambda^t) + \tau_{ik} \left(\frac{1}{2(n_1 - 1)} + \frac{1}{2(n_2 - 1)} \right) \pi_{ik}(\lambda^t) \right]$$
$$- \varphi_{jlik} \left[\gamma_{jlik}(\lambda^t) + \tau_{jl} \left(\frac{1}{2(n_1 - 1)} + \frac{1}{2(n_2 - 1)} \right) \pi_{jl}(\lambda^t) \right] \quad (25)$$

for all $j, l, i < j, k \neq l$, where $n_1 = |V_1|$, $n_2 = |V_2|$, and $0 \leq \varphi_{ikjl}, \tau_{jl} \leq 1$ are parameters.

We use $\varphi = 0.5$, $\tau = 1$, and perform the dual descent method L successive times.

Overall Method. Our overall method combines both the subgradient optimization and dual descent method. We do this performing the subgradient method until termination and then switching over to the dual descent method. This procedure is repeated K times.

We implemented NATALIE in C++ using the LEMON graph library (http://lemon.cs.elte.hu). The successive shortest path algorithm for maximum weight bipartite matching was implemented and contributed to LEMON. Special care was taken to deal with the inherent numerical instability of floating point numbers. Our implementation supports both the GraphML and GML graph formats. Rather than using one big alignment graph, we store and use a different alignment graph for every local problem (LD_λ^{ik}). This proved to be a huge improvement in running times, especially when the global alignment graph is sparse. NATALIE is publicly available at http://planet-lisa.net.

4   Experimental Evaluation

From the STRING database v8.3 [24], we obtained PPI networks for the following six species: *C. elegans* (cel), *S. cerevisiae* (sce), *D. melanogaster* (dme), *R. norvegicus* (rno), *M. musculus* (mmu) and *H. sapiens* (hsa). We only considered interactions that were experimentally verified. Table 1 in the appendix shows the sizes of the networks. We performed, using the BLOSUM62 matrix, an all-against-all global sequence alignment on the protein sequences of all $\binom{6}{2} = 15$ pairs of networks. We used affine gap penalties with a gap-open penalty of 2 and a gap-extension penalty of 10. The first experiment in Section 4.1 compares the raw performance of IsoRank, GRAAL and NATALIE in terms of objective value. In Section 4.2 we evaluate the biological relevance of the alignments produced by the three methods. All experiments were conducted on a compute cluster with 2.26 GHz processors with 24 GB of RAM.

4.1   Edge-Correctness

The objective function used for scoring alignments in GRAAL counts the number of mapped edges. Such an objective function is easily expressible in our framework using $s(a) = |\{(v, w) \in E_1 \mid (a(v), a(w)) \in E_2\}|$ and can also be modeled using the IsoRank scoring function. In order to compare performance of the methods across instances, we normalize the scores by dividing by $\min(|E_1|, |E_2|)$. This measure is called the edge-correctness by Kuchaiev et al. [17].

Table 1. Characteristics of input networks considered in this study. The columns contain species identifier, number of nodes in the network, number of annotated nodes thereof, and number of interactions.

species	nodes	annotated	interactions
cel (c)	5,948	4,694	23,496
sce (s)	6,018	5,703	131,701
dme (d)	7,433	6,006	26,829
rno (r)	8,002	6,786	32,527
mmu (m)	9,109	8,060	38,414
hsa (h)	11,512	9,328	67,858

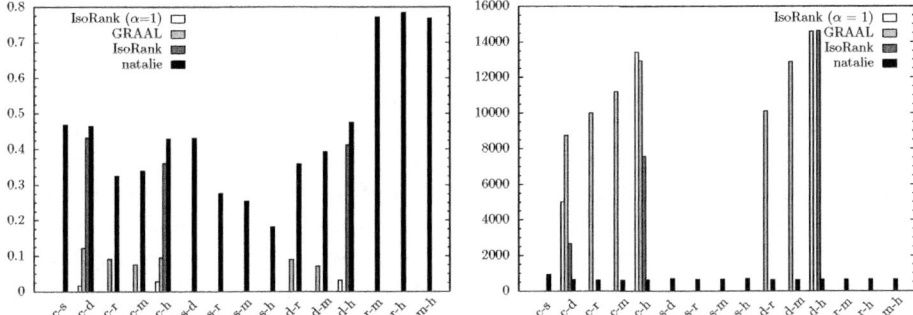

Fig. 2. Performance of the three different methods for the all-against-all species comparisons (15 alignment instances). Missing bars correspond to exceeded time/memory limits or software crashes. Left plot: Edge correctness, right plot: running times (sec.).

As mentioned in Section 3, our method benefits greatly from using a sparse alignment graph. To that end, we use the e-values obtained from the all-against-all sequence alignment to prohibit biologically unlikely matchings by only considering protein-pairs whose e-value is at most 100. Note that this only applies to NATALIE as both GRAAL and ISORANK consider the complete alignment graph. On each of the 15 instances, we ran GRAAL with 3 different random seeds and sampled the input parameter which balances the contribution of the graphlets with the node degrees uniformly within the allowed range of $[0, 1]$. As for ISO-RANK, when setting the parameter α—which controls to what extent topological similarity plays a role—to the desired value of 1, very poor results were obtained. Therefore we also sampled this parameter within its allowed range and re-evaluated the resulting alignments in terms of edge-correctness. NATALIE was run with a time limit of 10 minutes and $K = 3$, $L = 100$, $M = 10$, $N = 20$. For both GRAAL and ISORANK only the highest-scoring results were considered.

Figure 2 shows the results. ISORANK was only able to compute alignments for three out of the 15 instances. On the other instances ISORANK crashed, which may be due to the large size of the input networks. For GRAAL no alignments concerning *sce* could be computed, which is due to the large number of edges in the network on which the graphlet enumeration procedure choked: in 12 hours only for 3% of the nodes the graphlet degree vector was computed. As for the last three instances, GRAAL crashed due to exceeding the memory limit inherent to 32-bit processes. Unfortunately no 64-bit executable was available. On the instances for which GRAAL could compute alignments, the performance—both in solution quality and running time—is very poor when compared to ISORANK and NATALIE. NATALIE outperforms ISORANK in both running time and solution quality.

4.2   GO Similarity

In order to measure the biological relevance of the obtained network alignments, we make use of the Gene Ontology (GO) [3]. For every node in each of the six networks we obtained a set of GO annotations (see Table 1 for the exact numbers). Each annotation set was extended to a multiset by including all ancestral GO terms for every annotation in the original set. Subsequently we employed a similarity measure that compares a pair of aligned nodes based on their GO annotations and also takes into account the relative frequency of each annotation [11]. Since the similarity measure assigns a score between 0 and 1 to every aligned node pair, the highest similarity score one can get for any alignment is the minimum number of annotated nodes in either of the networks. Therefore we can normalize the similarity scores by this quantity. Unlike the previous experiment, this time we considered the bitscores of the pairwise global sequence alignments. Similarly to ISORANK parameter α, we introduced a parameter $\beta \in [0, 1]$ such that the sequence part of the score has weight $(1 - \beta)$ and the topology part has weight β. For both ISORANK and NATALIE we sampled the weight parameters uniformly in the range $[0, 1]$ and showed the best result in Figure 3. There we can see that both NATALIE and ISORANK identify functionally coherent alignments.

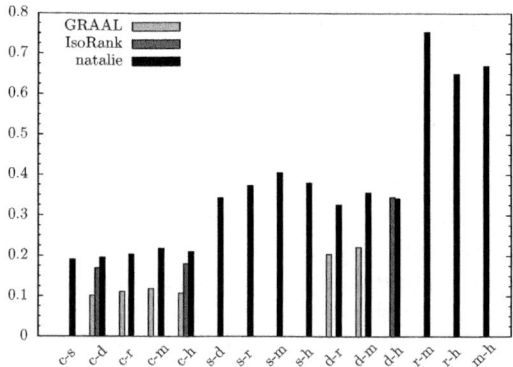

Fig. 3. Biological relevance of the alignments measured via GO similarity

5   Conclusion

Inspired by results for the closely related quadratic assignment problem (QAP), we have presented new algorithmic ideas in order to make a Lagrangian relaxation approach for global network alignment practically useful and competitive. In particular, we have generalized a dual descent method for the QAP.

We have found that combining this scheme with the traditional subgradient optimization method leads to fastest progress of upper and lower bounds.

Our implementation of the new method, NATALIE 2.0, works very well and fast when aligning biological networks, which we have shown in an extensive study on the alignment of cross-species PPI networks. We have compared NATALIE 2.0 to those state-of-the-art methods whose scoring schemes can be expressed as special cases of the scoring scheme we propose. Currently, these methods are ISORANK and GRAAL. Our experiments show that the Lagrangian relaxation approach is a very powerful method and that it currently outperforms the competitors in terms of quality of the results and running time.

Currently, all methods, including ours, approach the global network alignment problem heuristically, that is, the computed alignments are not guaranteed to be optimal solutions of the problem. While the other approaches are intrinsically heuristic—both ISORANK and GRAAL, for instance, approximate the neighborhood of a node and then match it with a similar node—the inexactness in our methods has two causes that we plan to address in future work: On the one hand, there may still be a gap between upper and lower bound of the Lagrangian relaxation approach after the last iteration. We can use these bounds, however, in a branch-and-bound approach that will compute provably optimal solutions. On the other hand, we currently do not consider the complete bipartite alignment graph and may therefore miss the optimal alignment. Here, we will investigate preprocessing strategies, in the spirit of [25], to safely sparsify the input bipartite graph without violating optimality conditions.

The independence of the local problems (LD_λ^{ik}) allows for easy parallelization, which, when exploited would lead to an even faster method. Another improvement in running times might be achieved when considering more involved heuristics for computing the lower bound, such as local search. More functionally-coherent alignments can be obtained when considering a scoring function where node-to-node correspondences are not only scored via sequence similarity but also for instance via GO similarity. In certain cases, even negative weights for topological interactions might be desired in which case one needs to reconsider the assumption of entries of matrix W being positive.

Acknowledgments. We thank SARA Computing and Networking Services (www.sara.nl) for their support in using the Lisa Compute Cluster. In addition, we are very grateful to Bernd Brandt for helping out with various bioinformatics issues and also to Samira Jaeger for providing code and advice on the GO similarity experiments.

References

1. Adams, W.P., Johnson, T.: Improved linear programming-based lower bounds for the quadratic assignment problem. DIMACS Series in Discrete Mathematics and Theoretical Computer Science (1994)
2. Alon, U.: Network motifs: theory and experimental approaches. Nat. Rev. Genet. 8(6), 450–461 (2007)

3. Ashburner, M., Ball, C.A., Blake, J.A., et al.: Gene ontology: tool for the unification of biology. Nat. Genet. 25 (2000)
4. Ay, F., Kellis, M., Kahveci, T.: SubMAP: Aligning Metabolic Pathways with Sub-network Mappings.. J. Comput. Biol. 18(3), 219–235 (2011)
5. Caprara, A., Fischetti, M., Toth, P.: A heuristic method for the set cover problem. Oper. Res. 47, 730–743 (1999)
6. Edmonds, J.: Path, trees, and flowers. Canadian J. Math. 17, 449–467 (1965)
7. Edmonds, J., Karp, R.M.: Theoretical improvements in algorithmic efficiency for network flow problems. J.ACM 19, 248–264 (1972)
8. Flannick, J., Novak, A., Srinivasan, B.S., McAdams, H.H., Batzoglou, S.: Graemlin: general and robust alignment of multiple large interaction networks.. Genome Res. 16(9), 1169–1181 (2006)
9. Guignard, M.: Lagrangean relaxation. Top 11, 151–200 (2003)
10. Held, M., Karp, R.M.: The traveling-salesman problem and minimum spanning trees: Part II. Math. Program 1, 6–25 (1971)
11. Jaeger, S., Sers, C., Leser, U.: Combining modularity, conservation, and interactions of proteins significantly increases precision and coverage of protein function prediction. BMC Genomics 11(1), 717 (2010)
12. Kanehisa, M., Goto, S., Hattori, M., et al.: From genomics to chemical genomics: new developments in KEGG. Nucleic Acids Res. 34, 354–357 (2006)
13. Karp, R.M.: Reducibility among combinatorial problems. In: Miller, R.E., Thatcher, J.W. (eds.) Complexity of Computer Computations, pp. 85–103. Plenum Press (1972)
14. Kelley, B.P., Sharan, R., Karp, R.M., et al.: Conserved pathways within bacteria and yeast as revealed by global protein network alignment. P. Natl. Acad. Sci. USA 100(20), 11394–11399 (2003)
15. Klau, G.W.: A new graph-based method for pairwise global network alignment.. BMC Bioinform. 10(suppl.1), S59 (2009)
16. Koyutürk, M., Kim, Y., Topkara, U., et al.: Pairwise alignment of protein interaction networks.. J. Comput. Biol. 13(2), 182–199 (2006)
17. Kuchaiev, O., Milenkovic, T., Memisevic, V., Hayes, W., Przulj, N.: Topological network alignment uncovers biological function and phylogeny.. J. R. Soc. Interface 7(50), 1341–1354 (2010)
18. Kuhn, H.W.: The Hungarian method for the assignment problem. Nav. Res. Logist. Q 2(1-2), 83–97 (1955)
19. Lawler, E.L.: The quadratic assignment problem. Manage. Sci. 9(4), 586–599 (1963)
20. Munkres, J.: Algorithms for the assignment and transportation problems. SIAM J. Appl. Math. 5, 32–38 (1957)
21. Sharan, R., Ideker, T.: Modeling cellular machinery through biological network comparison.. Nat. Biotechnol. 24(4), 427–433 (2006)
22. Sharan, R., Ideker, T., Kelley, B., Shamir, R., Karp, R.M.: Identification of protein complexes by comparative analysis of yeast and bacterial protein interaction data. J. Comput. Biol. 12(6), 835–846 (2005)
23. Singh, R., Xu, J., Berger, B.: Global alignment of multiple protein interaction networks with application to functional orthology detection.. P. Natl. Acad. Sci. USA 105(35), 12763–12768 (2008)
24. Szklarczyk, D., Franceschini, A., Kuhn, M., et al.: The STRING database in 2011: functional interaction networks of proteins, globally integrated and scored. Nucleic Acids Res. 39, 561–568 (2010)
25. Wohlers, I., Andonov, R., Klau, G.W.: Algorithm engineering for optimal alignment of protein structure distance matrices. Optim. Lett. (2011)

Stability of Inferring Gene Regulatory Structure with Dynamic Bayesian Networks

Jagath C. Rajapakse[1,2,3,*] and Iti Chaturvedi[1]

[1] Bioinformatics Research Center, School of Computer Engineering, Nanyang Technological University, Singapore 639798
[2] Department of Biological Engineering, Massachusetts Institute of Technology, Cambridge, MA 02142, USA
[3] Singapore-MIT Alliance, Singapore

Abstract. Though a plethora of techniques have been used to build gene regulatory networks (GRN) from time-series gene expression data, stabilities of such techniques have not been studied. This paper investigates the stability of GRN built using dynamic Bayesian networks (DBN) by synthetically generating gene expression time-series. Assuming scale-free topologies, sample datasets are drawn from DBN to evaluate the stability of estimating the structure of GRN. Our experiments indicate although high accuracy can be achieved with equal number of time points to the number of genes in the network, the presence of large numbers of false positives and false negatives deteriorate the stability of building GRN. The stability could be improved by gathering gene expression at more time points. Interestingly, large networks required less number of time points (normalized to the size of the network) than small networks to achieve the same level stability.

Keywords: Dynamic Bayesian networks, gene regulatory networks, Markov chain Monte Carlo simulation, scale-free networks, stability.

1 Introduction

Biological activities are due to interactions between genes and/or gene products (mainly proteins). These interactions are causal in the sense that the expression of one gene causes another gene to be up- or down-regulated. Transcription factors are a group of master proteins that bind to promoter regions and initiate transcription of genes. They positively or negatively mediate ensuing complex activities and enable control of several pathways, simultaneously. A set of genes that coherently interact to achieve a specific biological activity constitute to gene regulatory networks (GRN). Genetic and biochemical networks of cells are constantly subjected to random perturbations and must withstand substantial random perturbations. Therefore, the principles of efficiency and stability of such networks through feedback mechanisms are inherent in biological networks.

---

* Corresponding author.

M. Loog et al. (Eds.): PRIB 2011, LNBI 7036, pp. 237–246, 2011.
© Springer-Verlag Berlin Heidelberg 2011

GRN represents regulations among genes in a directed graph where nodes denote genes and edges regulatory connections. Microarrays are able to gather expression patterns of thousands of genes, simultaneously, and if collected over time or many experiments, underlying GRN can be constructed and inferences on gene regulations can be made. Reconstruction of genetic networks from gene expression data has many applications including inferring underlying biological mechanisms and identification of key biomarkers for drug discovery. A plethora of computational approaches have been introduced in the literature to infer GRN from microarray data, using boolean networks [1], [2], differential equations [3], concept networks, [4], and Bayesian networks [5], [6], [7], [8], [9], [10]. However, investigation of their stability of building GRN has evaded the research community.

Dynamic Bayesian networks (DBN) have recently become increasingly popular in building GRN from gene expression data because of their ability to model causal interactions [11]. Genes are represented at the nodes and regulatory interactions between two genes are represented by conditional probabilities of expressions of the two genes. Sensitivity of GRN in the Bayesian framework has been studied previously by sampling time-series from the posterior distribution [10]. However, the stability of networks of different sizes and its dependence on the number of time points in constructing GRN have not been evaluated. In this paper, we investigate the stability of building GRN with DBN by extensive simulations. Gene expression datasets of varying time points were generated by GRN, using scale-free topologies. Stability measures of estimating the structure are obtained by using bootstrapped time-series datasets. The effect of the number of time points and genes on the stability are also investigated.

2   Gene Regulatory Networks

A Bayesian network (BN) models the likelihood of a set of random variables by taking into account their conditional independences. When gene regulatory networks (GRN) are modeled by BN, genes are represented at the nodes and regulatory interactions are parameterized over the connections by conditional probabilities of gene expressions. Suppose that expressions $x = (x_i)_{i=1}^I$ of I number of genes indexed by set $\{i\}_{i=1}^I$ are given. If s denotes the structure and θ the parameters of the network, the likelihood of gene expressions is given by

$$p(x|s,\theta) = \prod_{i=1}^I p(x_i|a_i,\theta_i) \qquad (1)$$

where $p(x_i|a_i,\theta_i)$ is the conditional probability of gene expression x_i of gene i, given the set of gene expressions a_i of its parents; and $\theta = (\theta_i)_{i=1}^I$ where θ_i denotes the parameters of the conditional probability. The BN decomposes the likelihood of gene expressions into a product of conditional probabilities conditioned on the expressions of parent genes.

2.1   Dynamic Bayesian Networks

Dynamic Bayesian networks (DBN) extends the BN framework to better capture the temporal characteristics of gene expressions by assuming a first-order stationary Markov chain. Suppose that the time-course of gene expression of gene i is given by $x_i = (x_i(t))_{t=1}^{T}$ where $x_i(t)$ denotes the expression of gene i at time t and T is the total number of time points at which the expressions of genes are gathered. The time points are assumed to be equally spaced. The graph structure of DBN represents regulations from genes in the previous time point to the genes at the present point.

Suppose that gene expressions are discretized into K levels (or states). The parent gene expressions a_i will have $q_i = K^{|a_i|}$ number of states where $|a_i|$ denotes the number of parents of gene i. Let $\theta_{ijk} = p(x_i(t) = k | a_i(t-1) = j)$. By using the property of decomposability, the likelihood (1) can be written as

$$p(x|s, \theta) = \prod_{i=1}^{I} \prod_{j=1}^{J_i} \prod_{k=1}^{K} \theta_{ijk}^{N_{ijk}}. \tag{2}$$

where $N_{ijk} = \sum_{t=2}^{T} \delta(x_i(t) = k)\delta(a_i(t-1) = j)$ denotes the total number of counts of parents' state $a(i)$ being j when the gene i takes the value k in the preceding time point. Note that conditional probabilities are given by the conditional probability distribution (cpd) table: $\theta = \{\theta_{ijk}\}_{I \times J_i \times K}$.

2.2   Maximum Likelihood Estimate of the Structure

Marginalizing (2) over the parameters:

$$p(x|s) \propto \int p(x|s, \theta)p(\theta|s)d\theta \tag{3}$$

where $p(\theta|s)$ denotes the prior densities of the parameters. Assuming that the densities of conditional probabilities are independent:

$$p(\theta|s) = \prod_{i=1}^{I} \prod_{j=1}^{J_i} \prod_{k=1}^{K} p(\theta_{ijk}) \tag{4}$$

and substituting (2) and (4) in (3), we get

$$p(x|s) = \prod_{i=1}^{I} \prod_{j=1}^{J_i} \int \prod_{k=1}^{K} \theta_{ijk}^{N_{ijk}} p(\theta_{ijk})d\theta_{ijk} \tag{5}$$

Assuming Dirichlet priors for conditional densities:

$$p(\theta_{ijk}) = \frac{\Gamma(\sum_{k=1}^{K} \alpha_{ijk})}{\prod_{k=1}^{K} \Gamma(\alpha_{ijk})} \prod_{k=1}^{K} \theta_{ijk}^{(\alpha_{ijk}-1)} \tag{6}$$

where hyperparameters α_{ijk} correspond to parameters of Dirichlet prior of θ_{ijk}. Substituting (6) in (5), the likelihood is given by

$$p(x|s) = \prod_{i=1}^{n} \prod_{j=1}^{J_i} \frac{\Gamma(\alpha_{ij})}{\Gamma(\alpha_{ij} + N_{ij})} \prod_{k=1}^{K} \frac{\Gamma(\alpha_{ijk} + N_{ijk})}{\Gamma(\alpha_{ijk})} \tag{7}$$

where $N_{ij} = \sum_{k=1}^{K} N_{ijk}$ and Γ denotes the gamma function. Here we take $\alpha_{ij} = 1.0/(\sqrt{q_i})$. The Dirichlet priors enable modeling of complex interactions among the genes in regulatory networks [5].

Using Lagrange theory, the maximum likelihood (ML) estimates of the parameters can be derived from (7):

$$\hat{\theta}_{ijk} = \frac{\alpha_{ijk} + N_{ijk} - 1}{\alpha_{ij} + N_{ij} - K} \tag{8}$$

2.3   Structure Learning

The structure of GRN is obtained by the maximum a posteriori (MAP) estimation:

$$s^* = \arg \max_s p(s|x) \tag{9}$$

In order to find the optimal structure, Markov chain of structures is formed with Markov chain Monte Carlo (MCMC) simulation which converges to the optimal structure at the equilibrium. The Metropolis-Hastings method of MCMC is adopted, which associates an acceptance mechanism when new sample structures are drawn. The acceptance of new structure s^{new} is given by

$$\min \left\{ 1, \frac{p(s^{\text{new}}|x)}{p(s|x)} \cdot \frac{Q(s^{\text{new}}|s)}{Q(s|s^{\text{new}})} \right\} \tag{10}$$

where the Metropolis-Hastings acceptance ratio:

$$\frac{p(s^{\text{new}}|x)}{p(s|x)} \cdot \frac{Q(s^{\text{new}}|s)}{Q(s|s^{\text{new}})}. \tag{11}$$

Sampling new structures by using the above procedure generates a Markov chain converging in distribution to the true posterior distribution. In practice, a new network structure is proposed by applying one of the elementary operations such as deleting, reversing, or adding an edge, and then discarding those structures that violate the acyclic condition. The first term of the acceptance ratio, the ratio of likelihoods, is computed using (7). The second term or Hastings ratio is obtained by

$$\frac{Q(s^{\text{new}}|s)}{Q(s|s^{\text{new}})} = \frac{N^{\text{new}}}{N} \tag{12}$$

where N denotes the size of the neighborhood obtained by elementary operations on structure s and counting of the valid structures.

3   Stability

The stability of building GRN was evaluated by drawing time-series samples of gene expressions from a given network topology. The structure s is defined by a connectivity matrix $c = \{c_{ii'}\}_{n \times n}$ where $c_{ii'} = 0$ or $\neq 0$ depending on the presence or absence of a regulatory connection between two genes, i and i'. Stability was evaluated by sampling gene expression time-series from GRN represented by a DBN with known structure and parameters. The distance based similarity criteria were used to evaluate the stability of estimating structure and parameters of GRN.

3.1   Simulation of Gene Expression Time-Series

The likelihood of gene expressions $x(t)$ of genes at t is given by

$$p(x(t)|s, \theta) = \prod_{i=1}^{I} p(x_i(t)|x(t-1), \theta_i) \tag{13}$$

and the gene expression value of gene i is estimated by

$$\hat{x}_i(t) = \arg \max_k p(x_i(t) = k|\theta_{ijk}, x(t-1) = j) \tag{14}$$

If the gene expressions at $t = 1$ is initialized, $\{x_i(1)\}_{i=1}^{I}$, subsequent time points are generated by sampling from GRN by using (14) and adding Gaussian noise $N(0, \sigma^2)$.

Algorithm 1. Sampling gene expression data

Given: structure s, parameters θ, and Gaussian noise s.d. σ
Let $x(1) = (x_i(1) = \text{rand}[1, K])_{i=1}^{I}$
for $t = 2$ to T **do**
    for $i = 1$ to I **do**
        if $x(t-1) = j$ and $j \in J_i$ **then**
            $\hat{x}_i(t) = \arg \max_k p(x_i(t) = k|\theta_{ijk}, x(t-1) = j)$
        else
            $\hat{x}_i(t) = 0$
        end if
        $\epsilon_i \sim N(0, \sigma^2)$
        $x_i(t) = \hat{x}_i(t) + \epsilon_i$
    end for
end for
Return: dataset $(x(t))_{t=1}^{T}$

3.2   Stability of Estimation of Structure

We propose a similarity based stability criterion for evaluating GRN building algorithms, which is measured by the mean over all pairwise Hamming distances among all detected connections over different datasets.

Let $\{x^b\}_{b=1}^{B}$ be a set of B samples of gene expression datasets and s^b be the GRN derived from b-th dataset x^b. Consider two GRNs with connectivity matrices c^b and $c^{b'}$ derived from datasets x^b and $x^{b'}$, respectively; the similarity of the two networks is obtained by the average Hamming distances over all regulatory connections:

$$\rho\left(c^b, c^{b'}\right) = 1 - \frac{1}{N^b + N^{b'}} \sum_{i=1}^{I} \sum_{i'=1}^{I} d\left(c_{i,i'}^b, c_{i,i'}^{b'}\right). \tag{15}$$

where d denotes the Hamming distance between the two structures and N^b denotes the number of connections in the network s^b having the connectivity matrix c^b. The Hamming distance takes into account both the presence (1) and the absence (0) of regulatory connections between two networks. The stability is in the range of $[0, 1]$ where a lower value indicates a higher stability of GRN inference algorithm.

Using the stability of two networks in (15), the stability of GRN structure is obtained by averaging over B number of structures. The average of pairwise stability given in (15) over the bootstrapped samples is then used as the stability of the network structure:

$$\rho_{\text{structure}} = \frac{2}{B(B-1)} \sum_{b=1}^{B} \sum_{b'=b+1}^{B} \rho\left(c^b, c^{b'}\right). \tag{16}$$

The stability of an independent connection is defined as the average of bootstrap samples

$$\rho_{\text{connection}}\left(i, i'\right) = \frac{1}{B} \sum_{b=1}^{B} c_{i,i'}^b \tag{17}$$

4   Experiments and Results

4.1   Network Topology

Multiple time-series datasets of gene expressions were generated for a given network topology. In this study, various scale-free networks with varying number of genes (nodes) were generated using Barabasi-Albert model [12] as most real world networks, such as biological networks, are scale-free. Assuming that large GRNs adhere to the topology of scale-free networks, synthetic structures of GRN were built by initiating a small number of nodes. New nodes and edges were added with preferential attachment because probability of addition of new nodes to the existing node is not uniform. A node with high number of edges attracts higher number of new nodes compared to a node with no connection. This phenomena in fact leads to power-law distribution where probability $P(i)$ of preferential attachment of a new gene to existing gene i is given by

$$P(i) \sim d_i^{\gamma} + b \tag{18}$$

where d_i denotes the number of adjacent edges not initiated by gene i itself (which approximates to the in-degree of gene i). The parameters γ denotes the power of preferential attachment and b the attractiveness of the gene with no adjacent edges. New edges are added until a pre-specified number of edges are achieved. Since, GRN are causal we randomly flipped the direction of edges to create feedback loops while generating the network.

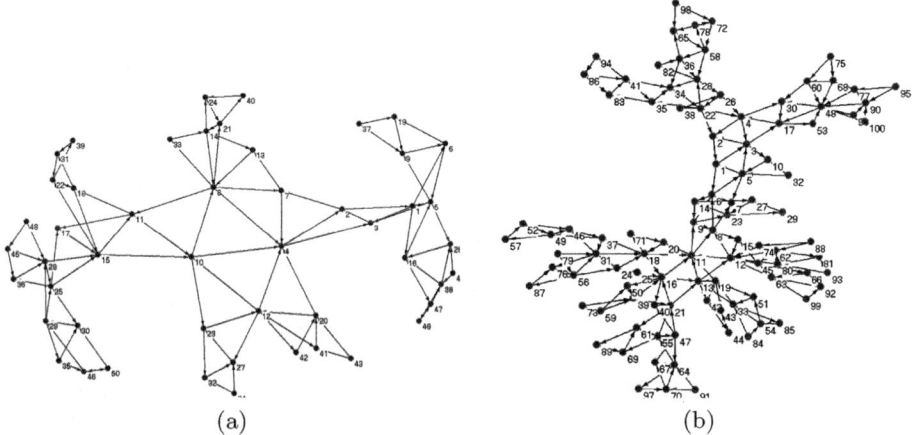

(a)   (b)

Fig. 1. Synthetic scale free networks with feedback for (a) 50 genes (b) 100 genes

4.2 Performance Evaluation

Scale-free network topologies were generated with in-degree set to a maximum of 5 genes. Figure 1 illustrates scale-free networks of 50 and 100 genes. GRNs defined by scale-free networks of 50, 100 and 250 genes were simulated and the parameters, or the cumulative probability distribution (cpd) tables $\{\theta_{ijk}\}_{i=1,j=1,k=1}^{I,J_i,K}$, were randomly initialized. The genes were assumed to take three states $\{-1, 0, +1\}$ representing down-, no-, and up-regulation of genes. Genes with no parents were set to up-regulation with corresponding CPD $= \{0.1, 0.1, 0.8\}$, for all remaining genes the state was determined by the dominant state in parent set.

Accuracy. In order to study the effects of the number of time points on the size of the network being built, gene expression values were sampled at $T \in \{10, 50, 100, 200, 250\}$ time points from GRN by using Algorithm 1. First, gene expression at the first time point were initialized randomly; second, gene expressions at other time points were sampled from DBN, using (13); and third, Gaussian noise with variance $\sigma^2 = 0.2$ were added. Structural learning was done using MCMC simulations until convergence. For a given I number of genes and T number of time points, $B = 100$ bootstrap time-series datasets were generated. Performance measures such as precision[1], recall[2], accuracy [3], and F-measure[4] were evaluated using true positives (TP), false positives (FP), true negatives (TN), and false negatives (FN), respectively.

Table 1 shows the performance of each method with varying numbers of genes and time points. As seen, the number of time points equaling the number of genes produced over 90% of accuracy of networks consisting of more than 50 genes. However, precision and recall measures were low, indicating that the performance was adversely affected by the presence of large numbers FN and FP (more by FP). Increase in the number of time points improved overall performance; note that the improvement was higher for larger networks.

Table 1. Accuracy of building DBN model with Dynamic Bayesian Networks

#Genes	#Samples	Precision	Recall	Accuracy	F-measure
	10	0.05 ± 0.01	0.05 ± 0.01	0.91 ± 0.09	0.05 ± 0.01
	50	$\mathbf{0.21 \pm 0.05}$	$\mathbf{0.41 \pm 0.07}$	$\mathbf{0.90 \pm 0.09}$	$\mathbf{0.28 \pm 0.06}$
50	100	0.59 ± 0.10	0.68 ± 0.09	0.96 ± 0.09	0.63 ± 0.09
	200	0.95 ± 0.09	0.98 ± 0.09	0.95 ± 0.09	0.98 ± 0.09
	250	0.96 ± 0.09	0.96 ± 0.09	0.98 ± 0.09	0.98 ± 0.09
	10	0.05 ± 0.01	0.05 ± 0.01	0.94 ± 0.11	0.05 ± 0.01
	50	0.23 ± 0.04	0.45 ± 0.07	0.94 ± 0.11	0.31 ± 0.05
100	**100**	$\mathbf{0.37 \pm 0.06}$	$\mathbf{0.58 \pm 0.08}$	$\mathbf{0.97 \pm 0.11}$	$\mathbf{0.46 \pm 0.06}$
	200	0.83 ± 0.10	0.83 ± 0.10	0.97 ± 0.10	0.83 ± 0.10
	250	0.89 ± 0.10	0.84 ± 0.10	0.98 ± 0.10	0.86 ± 0.10
	10	0.02 ± 0.01	0.02 ± 0.01	0.97 ± 0.09	0.02 ± 0.01
	50	0.18 ± 0.02	0.39 ± 0.05	0.97 ± 0.09	0.25 ± 0.02
250	100	0.31 ± 0.03	0.54 ± 0.05	0.97 ± 0.09	0.40 ± 0.04
	200	0.49 ± 0.06	0.63 ± 0.06	0.98 ± 0.10	0.55 ± 0.06
	250	$\mathbf{0.63 \pm 0.06}$	$\mathbf{0.71 \pm 0.06}$	$\mathbf{0.98 \pm 0.09}$	$\mathbf{0.67 \pm 0.06}$

Stability. The heatmap of figure 2 shows that the stability of learning GRN structure by using DBN decreases with the number of time points relative to the network size. Interestingly, small networks required more number of time points, relative to the number of nodes in the networks, than the large networks. Figure 3 shows stability of individual edges in networks consisting of different numbers of genes. Increase in the number of time points results in an increase in the number of edges that can be constructed stably. The stability deteriorates when

---

[1] Precision = TP/(TP+FP).
[2] Recall = TP/(TP + FN).
[3] Accuracy = TP+TN/(TP+TN+FP+FN).
[4] F-measure $= 2 \frac{Precision \times Recall}{Precision + Recall}$.

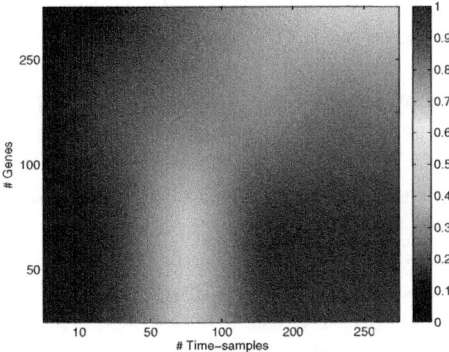

Fig. 2. Heatmaps illustrating the stability of estimating the structure of GRN by DBN. The number of discrete levels of gene expressions is three ($K = 3$).

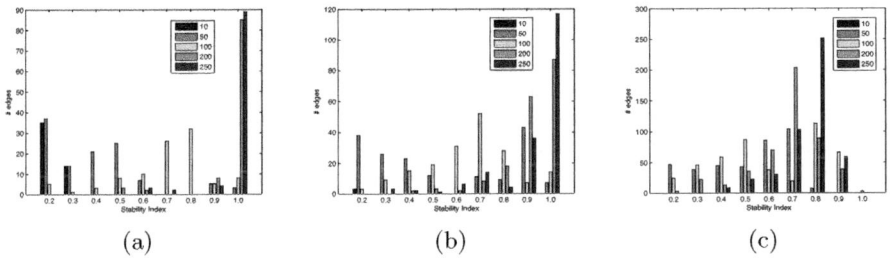

        (a)   (b)   (c)

Fig. 3. Distribution of stability of edges with the number of time points of gene regulatory networks with (a) 50 genes, (b) 100 genes, and (c) 250 genes

the number of genes relative to the networks size are increased with respective to the number time points.

5 Conclusions

Biological networks are constantly subjected to random perturbations and noise, and efficient feedback and compensatory mechanisms naturally provide stability. Gene expressions gathered from microarray experiments are confounded not only by biological noise and variations but also by measurement errors and artifacts. As seen, although the accuracy is high, GRN built from time-series of gene expressions are often unstable or non-reproducible.

In this paper, we investigated stability of building GRN in DBN framework by using simulations. By bootstrapping gene expression time-series datasets synthetically from scale-free networks (taken as topology closer to biological networks), stability of building GRN of different sizes, with varying number of time points were studied. Although a number of time points equal to the number of genes in the network provide accuracy as high as 90%, the presence of false negatives and false positives were high, especially when the number of time points are relatively less than the size of the network. This could only be reduced by increasing the number of time points at which gene expressions are gathered.

This work with synthetic data provides a foundation for future studies of stability of building GRN with real data. In practice, although gene expressions of a large number of genes are gathered simultaneously, repetition over many time points is limited due to high cost and time involved, and the nature of experiments. In such scenarios, either bootstrapping or regularization of time samples of gene expressions is needed to study the stability and reproducibility of computationally derived GRN. Investigation on real data is beyond the scope of the present manuscript but the application to real data remains an important and challenging problem in functional genomics and systems biology.

References

1. Li, P., Zhang, C., Perkins, E.J., Gong, P., Deng, Y.: Comparison of probabilistic boolean network and dynamic bayesian network approaches for inferring gene regulatorynetworks. BMC Bioinformatics 8, S13–S20 (2007)
2. Akutsu, T., Miyano, S., Kuhara, S.: Algorithms for identifying boolean networks and related biological networks based on matrix multiplication and fingerprint function. Journal of Computational Biology 7(3-4), 331–343 (2000)
3. Liu, B., Thiagarajan, P., Hsu, D.: Probabilistic approximations of signaling pathway dynamics. In: Computational Methods in Systems Biology, pp. 251–265 (2009)
4. Gebert, J., Motameny, S., Faigle, U., Forst, C.V., Schrader, R.: Identifying genes of gene regulatory networks using formal concept analysis. Journal of Computational Biology 15(2), 185–194 (2008)
5. Friedman, N., Linial, M., Nachman, I., Pe'er, D.: Using bayesian networks to analyze expression data. Journal of Computational Biology 7(3-4), 601–620 (2000)
6. Imoto, S., Kim, S., Goto, T., Miyano, S., Aburatani, S., Tashiro, K., Kuhara, S.: Bayesian network and nonparametric heteroscedastic regression for nonlinear modeling of genetic network. J. Bioinform. Comput. Biol. 1(2), 231–252 (2003)
7. Nariai, N., Kim, S., Imoto, S., Miyano, S.: Using protein-protein interactions for refining gene networks estimated from microarray data by bayesian networks. In: Pac. Symp. Biocomput., pp. 336–347 (2004)
8. Ota, K., Yamada, T., Yamanishi, Y., Goto, S., Kanehisa, M.: Comprehensive analysis of delay in transcriptional regulation using expression profiles. Genome Informatics 14, 302–303 (2003)
9. Perrin, E.D., Liva, R., Mazurie, A., Bottani, S., Mallet, J., dAlche-Buc, F.: Gene network inference using dynamic bayesian networks. Bioinformatics 12(2), 138–148 (2003)
10. Husmeier, D.: Sensitivity and specificity of inferring genetic regulatory interactions from microarray experiments with dynamic Bayesian networks. Bioinformatics 19(17), 2271–2282 (2003)
11. Friedman, N., Murphy, K., Russell, S.: Learning the structure of dynamic probabilistic networks. In: Proceedings of the 14th Annual Conference on Uncertainty in Artificial Intelligence (UAI 1998), pp. 139–140 (1998)
12. Albert-Laszlo, B., Reka, A.: Emergence of scaling in random networks. Science 286(5439), 509 (1999)

Integrating Protein Family Sequence Similarities with Gene Expression to Find Signature Gene Networks in Breast Cancer Metastasis

Sepideh Babaei[1,2], Erik van den Akker[1,3], Jeroen de Ridder[1,2,*], and Marcel Reinders[1,2,*]

[1] Delft Bioinformatics Lab, Delft University of Technology, The Netherlands
[2] Netherlands Bioinformatics Centre, The Netherlands
[3] Molecular Epidemiology, Leiden University Medical Centre, Leiden, The Netherlands
{J.deRidder,M.J.T.Reinders}@tudelft.nl

Abstract. Finding robust marker genes is one of the key challenges in breast cancer research. Significant signatures identified in independent datasets often show little to no overlap, possibly due to small sample size, noise in gene expression measurements, and heterogeneity across patients. To find more robust markers, several studies analyzed the gene expression data by grouping functionally related genes using pathways or protein interaction data. Here we pursue a protein similarity measure based on Pfam protein family information to aid the identification of robust subnetworks for prediction of metastasis. The proposed protein-to-protein similarities are derived from a protein-to-family network using family HMM profiles. The gene expression data is overlaid with the obtained protein-protein sequence similarity network on six breast cancer datasets. The results indicate that the captured protein similarities represent interesting predictive capacity that aids interpretation of the resulting signatures and improves robustness.

Keywords: protein-to-family distance matrix, protein-to-protein sequence similarity, concordant signature, breast cancer markers.

1 Introduction

Delineating gene signatures that predict cancer patient prognosis and survival is an important and challenging question in cancer research. Over the last few years, the amount of data from breast cancer patients has increased [1] and various methods for inferring prognostic gene sets from these data have been proposed [2], [3]. A major problem in this, which has been identified in several studies already, is that the prognostic signatures have relatively low concordance between different studies [4]. This is apparent from the fact that prediction performance decreases dramatically when prognostic signatures obtained from one dataset are applied to another one [5]. This reveals that a lack of a unified mechanism through which clinical outcome can be explained from gene expression profiles is still a major hurdle in clinical cancer biology.

---

[*] Corresponding author.

M. Loog et al. (Eds.): PRIB 2011, LNBI 7036, pp. 247–259, 2011.
© Springer-Verlag Berlin Heidelberg 2011

Several studies connect the lack of overlap in the gene expression to insufficient patient sample size [6], the inherent measurement noise in microarray experiments or heterogeneity in samples [5], [7]. A possible remedy is to pool breast cancer datasets in order to capture the information of as many samples as possible in the predictor [4], [8]. More recent efforts address this problem by exploiting knowledge on relations between genes and infer signatures not as individual genes but related groups of genes [9]. Of particular interest is the pioneering work of Chuang et al. [10], in which a greedy algorithm is used to detect discriminative subnetworks from protein-protein interaction (PPI) networks.

In order to overcome the drawbacks of a greedy routine to select candidate cancer networks, van den Akker et al. [11] proposed a non-greedy method that overlays the PPI network with gene-expression correlation to more accurately determine concordant breast cancer signatures across six independent expression datasets. These studies demonstrate that using additional information results in signatures with a more robust performance across dataset. Additionally, these signatures also turn out to have meaningful biological interpretations, thus providing interesting clues for the underlying molecular mechanisms of metastasis.

In this study, we propose a novel method to derive protein networks based on their functional similarities and use this to extend the van den Akker et al. method in order to further improve signature concordance and biological interpretability of breast cancer classification. To this end, we exploit protein sequence similarity since proteins with significant similar sequence are likely to have similar function [12].

Protein sequence similarity is determined using profile hidden Markov models (profile HMM) of protein families released in Pfam database [13]. More specifically, using the HMMER sequence alignment tool [14], we measure a distance (in term of an E-value) between a protein sequence and the HHM profiles constructed from the known families (Fig. 1). Proteins with common ancestors are in close proximity in the family space, sharing one or more protein domains since they are close to the same family profiles. Dimensionality reduction methods are applied to transform the high-dimensional protein-to-family matrix into a more meaningful low-dimensional representation in which only relevant dimensions are retained. The protein-to-protein similarity matrix (PPS) is derived by taking the Euclidean distance between pairs of proteins in the family space. We hypothesize that the captured protein similarity networks express predictive power that can be exploited in a cancer classification task. We evaluated these networks to identify predictive subnetworks in breast cancer metastasis.

2   Material and Methods

2.1   Data Description

In this study, six microarray datasets of breast cancer samples (Table 1) are utilized. The samples are measured on the HG U133 platform, and normalized following van den Akker et al. [11]. Briefly, all microarray data is normalized, log2 transformed and summarized per probe set. Duplicated samples are removed.

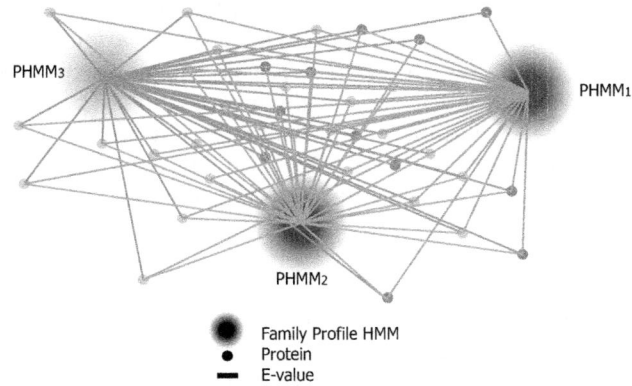

Fig. 1. The PPS is constructed based on profile HMM of protein families. These profiles are applied to scan each of the protein queries to obtain their sequence similarity in terms of an E-value.

The probe sets are mapped to protein-protein interaction networks (PPI) obtained from STRING [15]. Probe sets that do not map to known transcripts or proteins are discarded, resulting in a total of $N = 9236$ probe sets for 1107 samples. The samples are classified into a Poor or Good class according to the prognosis labels (distant metastasis and free survival events, respectively) [11].

Table 1. Summary of collected datasets

Dataset	Accession Code	Poor Sample	Good Sample
Desmedt	GSE7390	31	119
Miller	GSE3494	37	158
Loi	GSE6532	32	107
Pawitan	GSE1456	35	115
Wang	GSE2034	95	180
Schmidt	GSE11121	27	153

2.2 Protein-Protein Sequence Similarity

The protein-to-protein similarities are based on Pfam gene family information. These similarities are calculated as follows (summarized in Fig. 2):

Step 1. Profile hidden Markov models (HMM) for the families in Pfam24 dataset (2009 release, containing 11912 protein families) are constructed using the HMMER sequence alignment tool [14]. Families with less than 10 or more than 100 protein members are removed ($M = 6612$).

Step 2. These profile HMMs are used to scan N protein queries to obtain their sequence similarity to the $M = 6612$ families in terms of an E-value. The $N = 9236$ protein sequences associated with probe sets are taken from the human genes in the Ensemble database [16] (Fig. 1).

Step 3. We next estimate the intrinsic dimensionality of the protein-to-family distance matrix to capture the meaningful structure. Various dimension reduction methods are examined. In particular, classical multi dimensional scaling (MDS) and t-distributed stochastic neighbor embedding (t-SNE) methods are employed to capture a low-dimensional manifold that embeds the proteins in a low family dimensional space [17], i.e. the $M \times N$ matrix containing the distances between the N protein sequences and M protein families is mapped to a $k \times N$ ($k \leqslant M$) dimensional space.

Step 4. The $N \times N$ protein-to-protein similarity matrix (PPS) is extracted from the mapped protein-to-family distance matrix by taking the Euclidean distance between pairs of proteins in the mapped space. Depending on the employed dimension reduction approach, the PPS is referred to as $\boldsymbol{PPS}_{SNE}$ or $\boldsymbol{PPS}_{MDS}$. $\boldsymbol{PPS}_{ORG}$ refres to the protein distance matrix extracted from the non-mapped $M \times N$ matrix as well (i.e. without dimension reduction step).

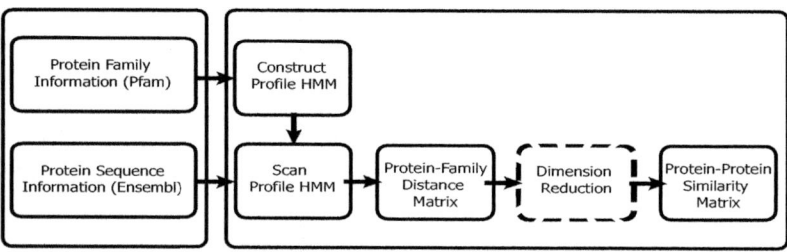

Fig. 2. Flow diagram of capturing the protein-to-protein similarity matrix ($\boldsymbol{PPS}$) framework. Three types of $\boldsymbol{PPS}$ are extracted: $\boldsymbol{PPS}_{SNE}$, $\boldsymbol{PPS}_{MDS}$ and $\boldsymbol{PPS}_{ORG}$. For the latter the dimension reduction step is skipped.

2.3   Subnetworks Construction

To construct subnetworks we mostly follow van den Akker *et al.* [11]. Briefly, we utilize different types of evidence to create the initial subnetworks including expression correlations, physical protein interactions or protein functional similarities. The $N \times N$ matrix indicating relations between genes is computed by:

$$S = C_B \circ G \circ P_B$$

where "∘" indicates Hadamard product and C_B is an $N \times N$ binary matrix in which each element in the correlation matrix C is set to zero in case it does not exceed threshold (T_{COR}). Grouping matrix G ($N \times N$) refers to the expression clustering matrix computed by hierarchically clustering genes using average linkage and a cluster cut-off value equal to $1 - T_{COR}$. Nonzero values in this matrix indicate the co-membership of a gene pair in a cluster. The binary protein association matrix P_B ($N \times N$) captures the functional relationship between proteins. This matrix is constructed based on the PPS matrix or the PPI matrix. In case of the latter, the STRING based interaction confidences, ranging from 1 to 999, are thresholded using threshold T_{PPI} such that a nonzero value indicates an interaction between the corresponding protein pair. Alternatively, $\boldsymbol{P}_B$ can be constructed from the $\boldsymbol{PPS}$ matrix, by thresholding it with threshold T_{PPS}. A nonzero value in $\boldsymbol{P}_B$ indicates the high similarity of a protein pair.

To assess the association between the gene expression data and breast cancer outcome per captured subnetworks the global test [18] is applied as a summary statistic. Only subnetworks with a global p-value less than T_S are considered as the significant networks. Thereafter, genes within union of significant subnetworks are included in the classifier and used to find out similarity in gene selection between different datasets.

2.4 Signature Concordance

In order to quantify the degree of concordance between signatures derived from different datasets two statistics, the Jaccard index [19] and odds ratio [20], are used. We specified four combinations of attributes for given subnetworks S_i and S_j as the total number of genes that are in i) both S_i and S_j (n_{11}), ii) neither S_i and S_j (n_{00}), iii) only S_i (n_{01}) and iv) only S_j (n_{10}). The Jaccard coefficient measures the degree of overlap between two networks by computing the ratio of the shared attributes between S_i and S_j: $J = \frac{n_{11}}{n_{11}+n_{01}+n_{10}}$.

The odds ratio describes the strength of association between two subnetworks. It is the ratio of the odds of an event occurring in S_i to the odds of it occurring in S_j: $OR = \frac{n_{11}n_{00}}{n_{01}n_{10}}$. Therefore, a high value of both criteria across two different datasets indicates the consistency of selected genes between them.

2.5 Classification Procedure

The predictive performances of the significant subnetworks are examined by training classifiers. More specifically, a nearest mean classifier using cosine correlation as the distance metric is trained on expression values associated with genes in the significant subnetworks. The classifiers are evaluated using the area under ROC curve, AUC metric, to capture the performance over the range of sensitivity and specificity. Two different strategies for training cross-dataset classifiers are evaluated: geneset passing and classifier passing. In geneset passing, the gene selection is based on the significant genes in dataset A while the classifier is trained and tested on the expression value associated with these selected genes in dataset B using 5-fold cross validation. The procedure is repeated 100 times and thus, the reported performance is the average of the AUC among all repeats. The classifier passing routine is carried out based on selecting genes and training the classifier on dataset A while testing on dataset B.

Finally, the datasets are also integrated to assess the classifier performance using five datasets for training while testing on the sixth. In an early integration strategy, five dataset are concatenated and then the statistically significant subnetworks are identified using the global test. Alternatively, for late integration, the gene sets are determined by intersecting the significant subnetworks per dataset.

3   Results and Discussion

We evaluate the derived protein similarity matrix to find out whether it represents biologically meaningful relationships among the proteins by means of a comparison with the Gene Ontology (GO) categorization. The protein-to-family matrix $(k \times N)$ is mapped to low-dimensional space using t-SNE method in which $k = 3$ and N is set to ten thousand randomly selected of the all human proteins in Ensembl database [16]. As shown in Fig. 3, the proteins with a common GO term are in close proximity in the PPS space. This shows our measure is indeed capable of capturing function relatedness of proteins and gives confidence that it can aid the construction of concordant and biologically interpretable signatures.

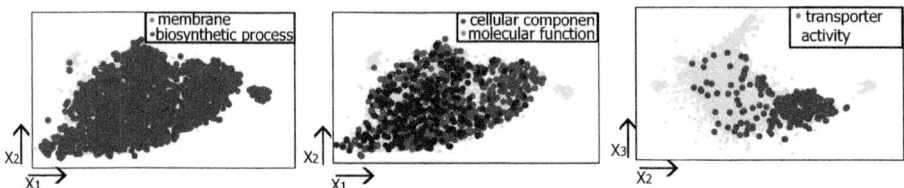

Fig. 3. Visualization of proteins colored based on their GO categorization in the mapped PPS_{SNE} space ($k = 3$). Proteins with a common GO term are closer in the mapped space. Yellow dots indicate all proteins in family space.

Following the outlined procedure, the subnetwork matrix S is created based on expression correlations (C) in the six aforementioned datasets independently. by setting $T_{COR} = 0.6$ (following van den Akker *et al.* [11]). We evaluate four different subnetwork matrices S by varying the way in which the protein association term (P_B) is calculated. More specifically, we used: PPS_{SNE}, PPS_{MDS} and PPS_{ORG} as well as PPI directly. For the latter, the threshold T_{PPI} is set to 500. In case of using the protein similarity measures, T_{PPS} is set to 0.1 quantile of all the values in the corresponding PPS matrix. The global test is

Table 2. Number of significant genes and subnetworks obtained by four different protein association terms (P_B) on the six breast cancer datasets. For each method the first column indicates number of significant genes and the percentage among all genes on the array, the second column indicates the number of selected subnetworks and their average size.

Dataset	PPI		PPS_ORG		PPS_MDS		PPS_SNE	
	#Genes(%)	#Net(mean)	#Genes(%)	#Net(mean)	#Genes(%)	#Net(mean)	#Genes(%)	#Net(mean)
Desmedt	117(1.2)	37(5.3)	44(0.4)	18(4.4)	84(0.9)	30(4.9)	85(0.9)	34(4.7)
Miller	306(3.3)	56(7.5)	130(1.4)	35(5.8)	207(2.2)	61(5.5)	192(2.1)	54(5.8)
Loi	819(8.8)	113(9.3)	556(6)	119(6.9)	776(8.4)	157(7.2)	751(8.1)	168(6.9)
Pawitan	246(2.6)	52(6.8)	109(1.2)	38(5.2)	206(2.2)	56(5.6)	190(2.1)	56(5.7)
Wang	293(3.1)	72(6.3)	122(1.3)	58(4.4)	226(2.4)	71(5.3)	227(2.4)	70(5.3)
Schmidt	209(2.2)	50(6.2)	99(1.1)	32(5.1)	186(2)	48(6)	190(2.1)	47(6.2)

performed on the obtained matrices S using $T_S = 0.05$ to identify the significant subnetworks. As a result, four different subnetwork matrices for six expression datasets are obtained. The summary of significant genes and subnetworks per dataset are reported in Table 2.

3.1   Cross Study Prediction Evaluation

We examine the classification performances of the discriminative subnetworks to compare the prediction capability of consensus genes selected by various protein association procedures. The expression values of the selected genes are employed to train and test the classifier. We evaluated two classification protocols, geneset passing and classifier passing, on the data resulting from early as well as late integration. The AUC values are given in Fig. 4. From these results we learn that in terms of classifier performance the genesets obtained with both the **PPS** matrices and the **PPI** matrix are in the same range, with a perhaps slightly better performance for the **PPS**$_{ORG}$.

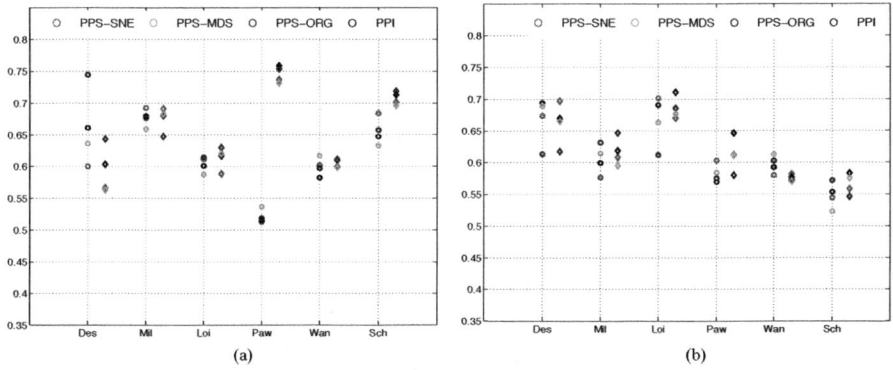

Fig. 4. Classification performance (AUC) of using four different protein association terms in determining subnetwork matrix per dataset. a) early and b) late integration approach. "circle" refers to the geneset passing and "diamond" refers to the classifier passing strategy.

3.2   Functionally Coherent Subnetworks

Table 2 shows that generally, the size of the networks obtained by **PPS** through all datasets is smaller than the obtained networks using **PPI**. This is especially apparent for the results with **PPS**$_{ORG}$. The significant subnetworks obtained with **PPS**$_{SNE}$ as well as with **PPS**$_{MDS}$ contain, on average, longer paths and nodes of lower degree than **PPS**$_{ORG}$ (Fig. 5). However, the overlap with **PPI** based subnetworks, in terms of genes, is substantially higher than **PPS**$_{ORG}$ (32% compared to 47%, Fig. 6). Noteworthy, in all approaches, genes within the significant subnetworks almost exclusively consist of genes that are either positively or negatively associated with the prognosis labels of the samples (Fig. 5).

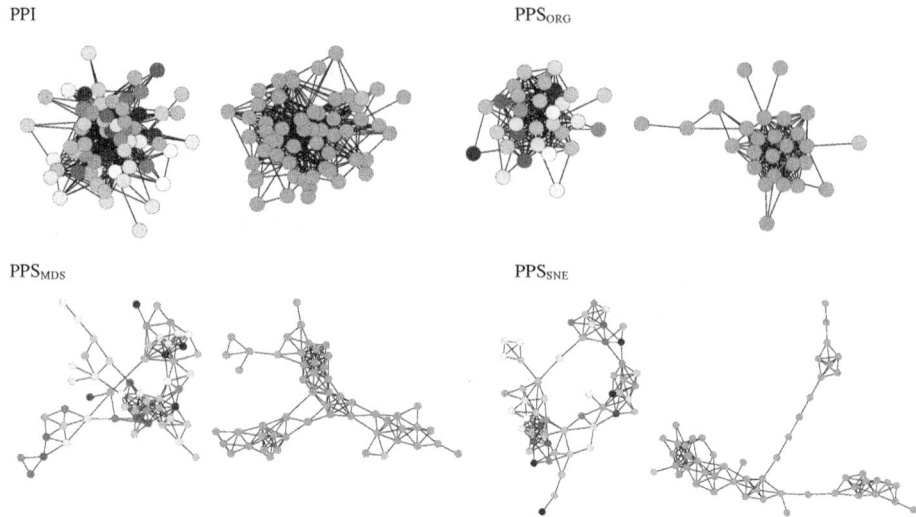

Fig. 5. The two largest significant subnetworks using the early integration approach for four different protein association matrices (P_B) in S. Genes are colored based on the p-value of the Welch t-test for their association with the Poor and Good expression outcome. Red and green indicate higher expressed in Good and Poor, respectively.

To find out whether the function of discriminative genes are enriched for cancer-related functional categories, the DAVID tool is applied [21]. This enrichment analysis reveals that for all four protein association matrices P_B enrichment for hallmark cancer categories is observed. However, this enrichment is substantially stronger for the PPS_{SNE} and PPS_{MDS} methods as indicated in Table 3. Comparing the genes in the subnetworks resulting from the PPS_{SNE}, and PPI matrices reveals that genes within the large subnetworks of PPS strongly overlap with PPI (Fig. 7). To find out if one of the two methods is better able to capture biological relevant genes (i.e. genes with a relation to cancer), we performed a functional comparison and found that, the common core of

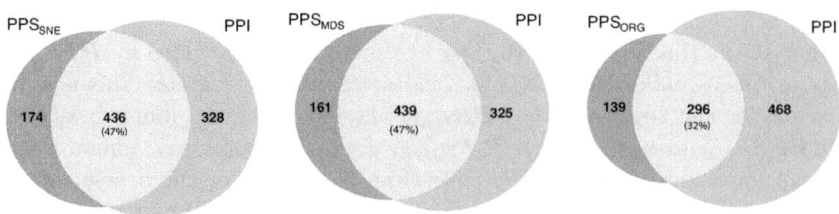

Fig. 6. Venn diagrams representing overlaps in number of the selected genes in significant subnetworks obtained by PPI and PPSs using early integration approach. The mapped PPS share more genes with PPI.

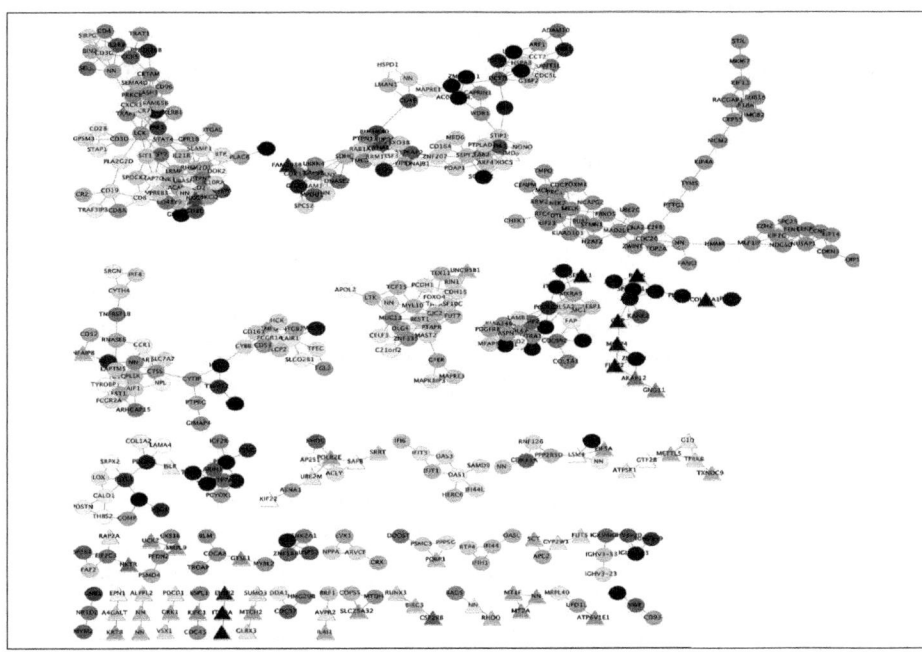

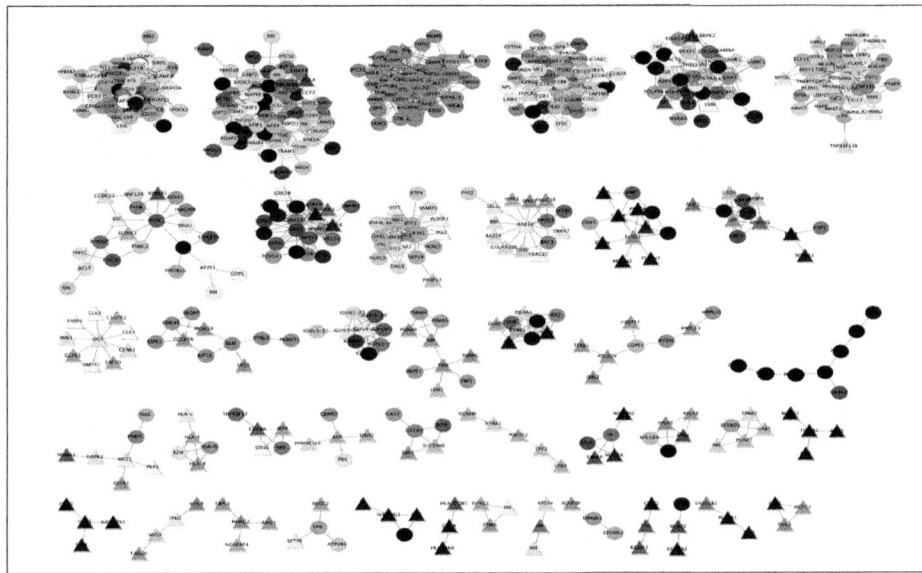

Fig. 7. Significant subnetworks ($T_S < 0.05$) with minimal size of three genes using early integration approach by a) PPS_{SNE}, b) PPI. "Circle" nodes indicate the overlapping genes between two methods (PPS_{SNE} and PPI) and "Triangular" nodes refer to gene that are exclusive for one of the methods. Genes are colored based on the p-value of the Welch t-test for their association with the Poor and Good expression outcome. Red and green indicate higher expressed in Good and Poor, respectively.

the genes in these subnetworks are significantly enriched for the hallmark process of breast cancer (Table 4) e.g., mitotic cell cycle (GO:0000278, p-value: 6.4e-18) nuclear division (GO:0000280, p-value:2.3e-16), mitosis (GO:0007067, p-value: 2.2e-16).

Most striking is the most significant GO category found for the gene set exclusive for PPS_{SNE} (GO:0070013, intracellular organelle lumen) which contains the highly relevant breast cancer associated BRCA1 and BRCA2 genes. This demonstrates the efficacy of the PPS_{SNE} to detect genes that have direct causal implications in breast cancer. In addition, the significant gene sets are analyzed using IPA (Ingenuity Systems) [22] to explore the genes strongly associated with cancer. The functional analysis identifies putative biomarkers of breast cancer process selected exclusively by PPS_{SNE} such as FAM198B (Entrez gene ID: 51313) [23], KIF22 (Entrez gene ID: 3835) [24], [25] and FLRT2 (Entrez gene ID: 23768) [26].

Table 3. Gene ontology enrichment and their approximated $\log(p - value)$ associated with the significant subnetworks in early integration method

GO Terms	PPI	PPS_{ORG}	PPS_{MDS}	PPS_{SNE}
GO:0000278 mitotic cell cycle	**20**	8	18	**20**
GO:0048285 organelle fission	13	4	16	**18**
GO:0000280 nuclear division	14	4	17	**18**
GO:0007067 mitosis	14	4	17	18
GO:0000087 M phase of mitotic cell cycle	14	4	16	**18**
GO:0000279 M phase	14	5	15	**16**
GO:0022403 cell cycle phase	13	4	14	**15**
GO:0007049 cell cycle	**19**	7	14	13
GO:0022402 cell cycle process	**18**	6	15	14
GO:0051301 cell division	9	5	11	**13**
GO:0007059 Chromosome segregation	6	-	9	**10**
GO:0005819 spindle	**10**	2	8	9

3.3   Consistent Consensus Genes across the Datasets

To investigate the consistency in gene selection across the six different expression datasets, we analyze the similarities and diversities of the significant subnetworks derived by the four alternatives of the P_B matrix in S. Pairwise similarities are computed by applying two criteria, Jaccard index and odds ratio. The Jaccard index obtained by employing mapped PPS (on average 20%) surpasses the other methods. This means that, when PPS_{SNE} or PPS_{MDS} is used to detect significant subnetworks, the signatures are more concordant across datasets (Table 5).

Table 4. Top five enriched GO FAT terms for selected genes within significant sub-networks with minimal size of three genes i) exclusively by PPS_{SNE} (dark green in Fig. 6), ii) shared by PPI and PPS_{SNE} (yellow in Fig. 6) iii)exclusively by PPI(light green in Fig. 6)

PPS_{SNE}	Common	PPI
GO:0070013	GO:0000278	GO:0010033
intracellular organelle lumen	mitotic cell cycle	response to organic substance
GO:0043233	GO:0000280	GO:0009725
organelle lumen	nuclear division	response to hormone stimulus
GO:0031974	GO:0007067	GO:0009719
membrane-enclosed	mitosis	response to endogenous stimulus
GO:0044420	GO:0000087	GO:0048545
extracellular matrix	M phase of mitotic cell cycle	response to steroid stimulus
GO:0031981	GO:0048285	GO:0032570
nuclear lumen	organelle fission	response to progesterone stimulus

Table 5. Pairwise comparison of the significant networks obtained using four protein association matrices on six datasets, the mean of the Jaccard index and odds ratio

	PPI	PPS_{ORG}	PPS_{MDS}	PPS_{SNE}
Jaccard Index	0.16	0.13	0.2	0.2
Odds Ratio	20.5	29.1	30.2	28.6

The odds ratio confirms this conclusion. The significant subnetworks selected by PPS_{MDS} (30.2%) and PPS_{SNE} (28.6%) demonstrate the ability of selecting consistent predictive genes across the different expression datasets.

4   Conclusion

We provide a novel protein similarity measure based on the protein family information contained in Pfam and use this to select discriminative markers on six breast cancer gene expression datasets. This protein similarity matrix captures functional information by focusing on shared and evolutionary conserved protein domains and other sequence similarities. We have demonstrated that the obtained significant genes exhibit more concordance across the six expression datasets. In particular, using the SNE approach to reduce dimensionality of the similarity matrix results in a promising coherence in the selected signature genes across the different datasets. The GO enrichment analysis indicates that the genes that are found for both PPI as well as PPS methods have strong links with cancer. Most importantly, however, for the genes found exclusively using the PPS information substantial evidence is available to link them to breast cancer.

The proposed method to infer protein similarities results in a promising data source to take into account while searching for marker genes and networks associated with metastasis. Since it only relies on protein sequence and Pfam information a much larger part of the protein space can be included in the marker discovery process. Therefore it is envisioned that these results will also be useful in other molecular classification problems.

Acknowledgments. This work was part of the BioRange programme of the Netherlands Bioinformatics Centre (NBIC), which is supported by the Netherlands Genomics Initiative (NGI).

References

1. Weigelt, B., et al.: Breast cancer metastasis: markers and models. Nat. Rev. Cancer 5(8), 591–602 (2005)
2. Veer, L.J., et al.: Gene expression profiling predicts clinical outcome of breast cancer. Nature 415(6871), 530–536 (2002)
3. Vijver, M.J., et al.: A gene-expression signature as a predictor of survival in breast cancer. N. Engl. J. Med. 347(25), 1999–2009 (2002)
4. van Vliet, M.H., et al.: Pooling breast cancer datasets has a synergetic effect on classification performance and improves signature stability. BMC Genomics 9, 375 (2008)
5. Ein-Dor, L., et al.: Outcome signature genes in breast cancer: is there a unique set? Bioinformatics 21(2), 171–178 (2005)
6. Hua, J., Tembe, W.D.: Performance of feature-selection methods in the classification of high-dimension data. Pattern Recog. 42(3), 409–424 (2009)
7. Symmans, W.F., et al.: Breast cancer heterogeneity: evaluation of clonality in primary and metastatic lesions. Hum. Pathol. 26(2), 210–216 (1995)
8. Shen, R., et al.: Prognostic meta-signature of breast cancer developed by two-stage mixture modeling of microarray data. BMC Genomics 5(1), 94 (2004)
9. Pujana, M.A., et al.: Network modeling links breast cancer susceptibility and centrosome dysfunction. Nat. Genet. 39(11), 1338–1349 (2007)
10. Chuang, H.Y., et al.: Network-based classification of breast cancer metastasis. Mol. Sys. Bio. 3, 140 (2007)
11. van den Akker, E., et al.: Integrating protein-protein interaction networks with gene-gene co-expression networks improves gene signatures for classifying breast cancer metastasis (submitted)
12. Rigden, D.: From protein structure to function with bioinformatics. Springer, Heidelberg (2009)
13. Finn, R.D., et al.: The Pfam protein families database. Nucleic Acids Res. 38, D211–D222 (2010)
14. Eddy, S.R.: A probabilistic model of local sequence alignment that simplifies statistical significance estimation. PLoS Comp. Bio. 4(5), e1000069 (2008)
15. von Mering, C., et al.: STRING: a database of predicted functional associations between proteins. Nucleic Acids Res. 31(1), 258–261 (2003)
16. Biomart, http://www.biomart.org/biomart/martviewrt
17. van der Maaten, L.J.P., Hinton, G.E.: Visualizing high-dimensional data using t-SNE. Journal of Machine Learning Res. 9, 2579–2605 (2008)
18. Goeman, J.J., et al.: A global test for groups of genes: testing association with a clinical outcome. Bioinformatics 20(1), 93–99 (2004)
19. Jaccard, P.: Etude comparative de la distribution florale dans une portion des Alpes et des Jura, Bulletin de la Société Vaudoise de Sciences. Naturelles 37, 547–579 (1901)
20. Edwards, A.W.F.: The measure of association in a 2×2 table. JSTOR 126(1), 1–28 (1968)

21. Huang, D.W., et al.: Systematic and integrative analysis of large gene lists using DAVID Bioinformatics Resources. Nature Protoc. 4(1), 44–57 (2009)
22. Ingenuity Pathways Analysis software, http://www.ingenuity.com
23. Deblois, G., et al.: Genome-wide identification of direct target genes implicates estrogen-related receptor alpha as a determinant of breast cancer heterogeneity. Cancer Res. 69(15), 6149–6157 (2009)
24. Yumei, F.: KNSL4 is a novel molecular marker for diagnosis and prognosis of breast cancer. American Assoc. for Cancer Res. (AACR) Meeting Abstracts, 1809 (2008)
25. Diarra-Mehrpour, M., et al.: Prion protein prevents human breast carcinoma cell line from tumor necrosis factor alpha-induced cell death. Cancer Res. 64(2), 719–727 (2004)
26. Tripathi, A., et al.: Gene expression abnormalities in histologically normal breast epithelium of breast cancer patients. Int. J. Cancer 122(7), 1557–1566 (2008)

Estimating the Class Posterior Probabilities in Protein Secondary Structure Prediction

Yann Guermeur and Fabienne Thomarat

LORIA – Equipe ABC
Campus Scientifique, BP 239
54506 Vandœuvre-lès-Nancy Cedex, France
{Yann.Guermeur,Fabienne.Thomarat}@loria.fr

Abstract. Support vector machines, let them be bi-class or multi-class, have proved efficient for protein secondary structure prediction. They can be used either as sequence-to-structure classifier, structure-to-structure classifier, or both. Compared to the classifier most commonly found in the main prediction methods, the multi-layer perceptron, they exhibit one single drawback: their outputs are not class posterior probability estimates. This paper addresses the problem of post-processing the outputs of multi-class support vector machines used as sequence-to-structure classifiers with a structure-to-structure classifier estimating the class posterior probabilities. The aim of this comparative study is to obtain improved performance with respect to both criteria: prediction accuracy and quality of the estimates.

Keywords: protein secondary structure prediction, multi-class support vector machines, class membership probabilities.

1 Introduction

With the multiplication of genome sequencing projects, the gap between the number of known protein sequences and the number of experimentally determined protein (tertiary) structures is widening rapidly. This raises a central problem, since knowledge of the structure of a protein is a key in understanding its detailed function. The prediction of protein structure from amino acid sequence, i.e., *ab initio*, has thus become a hot topic in molecular biology. Due to its intrinsic difficulty, it is ordinarily tackled through a divide and conquer approach in which a critical first step is the prediction of the secondary structure, the local, regular structure defined by hydrogen bonds. Considered from the point of view of pattern recognition, this prediction is a three-category discrimination task consisting in assigning a conformational state α-helix, β-strand or aperiodic (coil), to each residue (amino acid) of a sequence.

For almost half a century, many methods have been developed for protein secondary structure prediction. Since the pioneering work of Qian and Sejnowski [1], state-of-the-art methods are machine learning ones [2,3,4,5]. Furthermore, a majority of them shares the original architecture implemented by Qian and

M. Loog et al. (Eds.): PRIB 2011, LNBI 7036, pp. 260–271, 2011.

Sejnowski. Two classifiers are used in cascade. The first one, named sequence-to-structure, takes in input the content of a window sliding on the sequence, or the coding of a multiple alignment, to produce an initial prediction. The second one, named structure-to-structure, takes in input the content of a second window sliding on the initial prediction. Making use of the fact that the conformational states of consecutive residues are correlated, it mainly acts as a filter, increasing the biological plausibility of the prediction. Until the end of the nineties, the classifiers at the basis of most of the prediction methods implementing the cascade treatment were neural networks [6], either feed-forward, like the multi-layer perceptron (MLP) [1,2,4] or recurrent [3]. During the last decade, they were gradually replaced with bi-class support vector machines (SVMs) [7,5] and multi-class SVMs (M-SVMs) [8,9,10]. This resulted in a slight increase of the prediction accuracy. On the other and, an advantage of the neural networks over the SVMs rests in the fact that under mild hypotheses regarding the loss function and the activation function of the output units, they estimate the class posterior probabilities (see for instance [11,12]). This is a useful property in the framework of interest, for two main reasons. The first one is obvious: such estimates provide the most accurate reliability indices for the prediction (see [13,4] for indices based on them, and [7] for an index based on the outputs of bi-class SVMs). The second one is of higher importance, since it deals with the future of protein secondary structure prediction. It is commonly admitted that the main limiting factor for the prediction accuracy of any prediction method based on the standard cascade architecture is the fact that local information is not enough to specify utterly the structure. This limitation is only partly overcome by using recurrent neural networks. Several works [8,9,13] have considered a more ambitious alternative, consisting in the implementation of hybrid architectures combining discriminant models and hidden Markov models (HMMs) [14]. In short, the discriminant models are used to compute class posterior probability estimates from which the emission probabilities of the HMMs are derived, by application of Bayes' formula. This approach widens the context used for the prediction, and makes it possible to incorporate some pieces of information provided by the biologist, such as syntactic rules. It appears highly promising, although it still calls for significant developments in order to bear its fruits. It is this observation that motivated the present study. Our thesis is that in the framework of such hybrid architectures, an efficient implementation of the cascade architecture could result from using as sequence-to-structure classifiers M-SVMs endowed with a dedicated kernel, provided that the structure-to-structure classifier is chosen appropriately. In this article, we thus address the problem of identifying the optimal structure-to-structure classifier when the classifiers performing the sequence-to-structure prediction are dedicated M-SVMs, and the final outputs must be class membership probability estimates. In practice, we want to take benefit of the high recognition rate of the M-SVMs without suffering their drawback. Our contribution, the first of this kind, focuses on the use of tools from nonparametric statistics.

The organization of the paper is as follows. Section 2 proposes a general introduction to the M-SVMs. The four M-SVMs involved in the experiments are then characterized in this framework, and their implementation for sequence-to-structure classification is detailed. Section 3 is devoted to the description of the different models considered for the structure-to-structure prediction. Experimental results are gathered in Section 4. At last, we draw conclusions and outline our ongoing research in Section 5.

2   Multi-class Support Vector Machines

2.1   General Introduction

The theoretical framework of M-SVMs is the one of large margin multi-category classifiers [15]. Formally, it deals with Q-category pattern recognition problems with $3 \leqslant Q < +\infty$. Each object is represented by its description $x \in \mathcal{X}$ and the set $\mathcal{Y}$ of the categories y can be identified with the set of indices of the categories: $[\![1, Q]\!]$. The assignment of the descriptions to the categories is performed by means of a *classifier*, i.e., a function on $\mathcal{X}$ taking values in $\mathbb{R}^Q$. For such a function g, the corresponding *decision rule* f is defined as follows:

$$\forall x \in \mathcal{X}, \quad \begin{cases} \text{if } \exists k \in [\![1, Q]\!] : g_k(x) > \max_{l \neq k} g_l(x), \text{ then } f(x) = k \\ \text{else } f(x) = * \end{cases}$$

where $*$ denotes a dummy category introduced to deal with the cases of ex æquo. Like all the SVMs, all the M-SVMs published so far belong to the family of *kernel machines* [16,17], which implies that they operate on a class of functions induced by a positive type function/kernel [18]. For a given kernel, this class, hereafter denoted by $\mathcal{H}$, is the same for all the models (it only depends on the kernel and Q). In what follows, κ designates a real-valued kernel on $\mathcal{X}^2$ and $\left(\mathbf{H}_\kappa, \langle \cdot, \cdot \rangle_{\mathbf{H}_\kappa}\right)$ the corresponding reproducing kernel Hilbert space (RKHS) [18]. The RKHS of $\mathbb{R}^Q$-valued functions [19] at the basis of a Q-category M-SVM whose kernel is κ can be defined simply as a function of κ.

Definition 1 (RKHS $\bar{\mathcal{H}}$). *Let κ be a real-valued positive type function on $\mathcal{X}^2$. Then, the RKHS of $\mathbb{R}^Q$-valued functions at the basis of a Q-category M-SVM whose kernel is κ is $\bar{\mathcal{H}} = \mathbf{H}_\kappa^Q$. Furthermore, the inner product of $\bar{\mathcal{H}}$ can be expressed as a function of the inner product of $\mathbf{H}_\kappa$ as follows:*

$$\forall \left(\bar{h}, \bar{h}'\right) \in \bar{\mathcal{H}}^2, \ \bar{h} = \left(\bar{h}_k\right)_{1 \leqslant k \leqslant Q}, \ \bar{h}' = \left(\bar{h}'_k\right)_{1 \leqslant k \leqslant Q}, \ \langle \bar{h}, \bar{h}' \rangle_{\bar{\mathcal{H}}} = \sum_{k=1}^{Q} \langle \bar{h}_k, \bar{h}'_k \rangle_{\mathbf{H}_\kappa} \ .$$

Definition 2 (Class of functions $\mathcal{H}$). *Let κ be a real-valued positive type function on $\mathcal{X}^2$ and let $\bar{\mathcal{H}}$ be the RKHS of $\mathbb{R}^Q$-valued functions derived from κ according to Definition 1. Let $\{1\}$ be the one-dimensional space of real-valued constant functions on $\mathcal{X}$. The class of functions at the basis of a Q-category M-SVM whose kernel is κ is*

$$\mathcal{H} = \bar{\mathcal{H}} \oplus \{1\}^Q = \left(\mathbf{H}_\kappa \oplus \{1\}\right)^Q \ .$$

Thus, a function h in $\mathcal{H}$ can be written as

$$h\left(\cdot\right) = \bar{h}\left(\cdot\right) + b = \left(\bar{h}_k\left(\cdot\right) + b_k\right)_{1 \leqslant k \leqslant Q}$$

where the function $\bar{h} = \left(\bar{h}_k\right)_{1 \leqslant k \leqslant Q}$ is an element of $\bar{\mathcal{H}}$ and $b = (b_k)_{1 \leqslant k \leqslant Q} \in \mathbb{R}^Q$.

Let $m \in \mathbb{N}^*$. For a given sequence of examples $d_m = ((x_i, y_i))_{1 \leqslant i \leqslant m}$ in $(\mathcal{X} \times [\![1, Q]\!])^m$, we denote $\mathbb{R}^{Qm}(d_m)$ the subspace of $\mathbb{R}^{Qm}$ made up of the vectors $v = (v_t)_{1 \leqslant t \leqslant Qm}$ satisfying:

$$\left(v_{(i-1)Q+y_i}\right)_{1 \leqslant i \leqslant m} = 0_m . \tag{1}$$

Furthermore, for the sake of simplicity, the components of the vectors of $\mathbb{R}^{Qm}(d_m)$ are written with two indices, i.e., v_{ik} in place of $v_{(i-1)Q+k}$, for i in $[\![1, m]\!]$ and k in $[\![1, Q]\!]$. As a consequence, (1) simplifies into $(v_{iy_i})_{1 \leqslant i \leqslant m} = 0_m$. For n in $\mathbb{N}^*$, let $\mathcal{M}_{n,n}(\mathbb{R})$ be the algebra of $n \times n$ matrices over $\mathbb{R}$ and $\mathcal{M}_{Qm,Qm}(d_m)$ the subspace of $\mathcal{M}_{Qm,Qm}(\mathbb{R})$ made up of the matrices $M = (m_{tu})_{1 \leqslant t,u \leqslant Qm}$ satisfying:

$$\forall j \in [\![1, m]\!], \ \left(m_{t,(j-1)Q+y_j}\right)_{1 \leqslant t \leqslant Qm} = 0_{Qm} .$$

Once more for the sake of simplicity, the components of the matrices of $\mathcal{M}_{Qm,Qm}(d_m)$ are written with four indices, i.e., $m_{ik,jl}$ in place of $m_{(i-1)Q+k,(j-1)Q+l}$, for (i,j) in $[\![1, m]\!]^2$ and (k,l) in $[\![1, Q]\!]^2$. With these definitions and notations at hand, a generic model of M-SVM can be defined as follows.

Definition 3 (Generic model of M-SVM, Definition 4 in [20]). *Let $\mathcal{X}$ be a non empty set and $Q \in \mathbb{N} \setminus [\![0, 2]\!]$. Let κ be a real-valued positive type function on $\mathcal{X}^2$. Let $\bar{\mathcal{H}}$ and $\mathcal{H}$ be two classes of functions induced by κ according to Definitions 1 and 2. Let $P_{\bar{\mathcal{H}}}$ be the orthogonal projection operator from $\mathcal{H}$ onto $\bar{\mathcal{H}}$. For $m \in \mathbb{N}^*$, let $d_m = ((x_i, y_i))_{1 \leqslant i \leqslant m} \in (\mathcal{X} \times [\![1, Q]\!])^m$ and $\xi \in \mathbb{R}^{Qm}(d_m)$. A Q-category M-SVM whose kernel is κ is a large margin discriminant model obtained by solving a convex quadratic programming problem of the form*

Problem 1 (M-SVM learning problem, primal formulation).

$$\min_{h,\xi} \left\{ \|M\xi\|_p^p + \lambda \|P_{\bar{\mathcal{H}}}h\|_{\bar{\mathcal{H}}}^2 \right\}$$

$$s.t. \begin{cases} \forall i \in [\![1, m]\!], \ \forall k \in [\![1, Q]\!] \setminus \{y_i\}, \ K_1 h_{y_i}(x_i) - h_k(x_i) \geqslant K_2 - \xi_{ik} \\ \forall i \in [\![1, m]\!], \ \forall (k,l) \in ([\![1, Q]\!] \setminus \{y_i\})^2, \ K_3\left(\xi_{ik} - \xi_{il}\right) = 0 \\ \forall i \in [\![1, m]\!], \ \forall k \in [\![1, Q]\!] \setminus \{y_i\}, \ (2 - p)\xi_{ik} \geqslant 0 \\ (1 - K_1) \sum_{k=1}^{Q} h_k = 0 \end{cases}$$

where $\lambda \in \mathbb{R}_+^*$, $M \in \mathcal{M}_{Qm,Qm}(d_m)$ is a matrix of rank $(Q-1)m$, $p \in \{1, 2\}$, $(K_1, K_3) \in \{0, 1\}^2$, and $K_2 \in \mathbb{R}_+^*$. If $p = 1$, then M is a diagonal matrix.

In the chronological order of their introduction, the four M-SVMs involved in our experiments are those of Weston and Watkins [21], Crammer and Singer [22], Lee and co-authors [23], and the M-SVM^2 [24]. Let $I_{Qm}(d_m)$ and $M^{(2)}$ designate two matrices of $\mathcal{M}_{Qm,Qm}(d_m)$ whose general terms are respectively

$$m_{ik,jl} = \delta_{i,j}\delta_{k,l}(1 - \delta_{y_i,k})$$

and

$$m_{ik,jl} = (1 - \delta_{y_i,k})(1 - \delta_{y_j,l})\left(\delta_{k,l} + \frac{\sqrt{Q}-1}{Q-1}\right)\delta_{i,j},$$

where δ is the Kronecker symbol. In order to characterize the aforementioned M-SVMs as instances of the generic model, we express the primal formulation of their learning problems as a specification of Problem 1. The corresponding values of the hyperparameters are gathered in Table 1.

Table 1. Specifications of the four M-SVMs used as sequence-to-structure classifier. The first three machines are the ones of Weston and Watkins (WW), Crammer and Singer (CS), and Lee, Lin, and Wahba (LLW).

M-SVM	M	p	K_1	K_2	K_3
WW-M-SVM	$I_{Qm}(d_m)$	1	1	1	0
CS-M-SVM	$\frac{1}{Q-1}I_{Qm}(d_m)$	1	1	1	1
LLW-M-SVM	$I_{Qm}(d_m)$	1	0	$\frac{1}{Q-1}$	0
M-SVM^2	$M^{(2)}$	2	0	$\frac{1}{Q-1}$	0

2.2   Dedication to Sequence-to-Structure Prediction

In our experiments, each protein sequence is represented by a position-specific scoring matrix (PSSM) produced by PSI-BLAST [25]. To generate each PSSM, we ran three iterations of PSI-BLAST against the nr database downloaded in February 2010. The E-value inclusion threshold was set to 0.005 and the default scoring matrix (BLOSUM62) was used. The sliding window of the sequence-to-structure classifiers, i.e., the M-SVMs, is of size 13, and is centered on the residue of interest. The description (vector of predictors) x_i processed by the M-SVMs to predict the conformational state of the ith residue in the data set is thus obtained by appending rows of the PSSM associated with the sequence to which it belongs. The indices of these rows range from $i' - 6$ to $i' + 6$, where i' is the index of the residue of interest in its sequence. Since a PSSM has 20 columns, one per amino acid, this corresponds to 260 predictors. More precisely, $\mathcal{X} \subset \mathbb{Z}^{260}$. The kernel κ is an elliptic Gaussian kernel function applying a weighting on the predictors as a function of their position in the window. This weighting is learned by application of the principle of multi-class kernel target alignment [26] (the training algorithm is a stochastic steepest descent). At last, the programs implementing the different M-SVMs are those of MSVMpack [27].

3   Structure-to-Structure Classifiers

In this section, we make the hypothesis that N M-SVMs are available to perform the sequence-to-structure prediction. The function computed by the jth of these machines is denoted $h^{(j)} = \left(h_k^{(j)} \right)_{1 \leqslant k \leqslant Q}$. The structure-to-structure classifier $g = (g_k)_{1 \leqslant k \leqslant Q}$ uses a sliding window with a left context of size T_l and a right context of size T_r (for a total length of $T = T_l + 1 + T_r$). Thus, the vector of predictors available to estimate the probabilities associated with the ith residue in the data set is $z_i = \left(h_k^{(j)} (x_{i+t}) \right)_{1 \leqslant j \leqslant N, 1 \leqslant k \leqslant Q, -T_l \leqslant t \leqslant T_r} \in \mathbb{R}^{NQT}$. With slight abuse of notation, we use $g(x_i)$ in place of $g(z_i)$ to denote the outputs of the structure-to-structure classifier for this residue. The classifiers we consider to perform the task are now introduced in order of increasing complexity.

3.1   Polytomous Logistic Regression

The most natural way of deriving class posterior probability estimates from the outputs of an M-SVM consists in extending Platt's bi-class solution [28] to the multi-class case. In the framework of our study (implementation of the cascade architecture), this corresponds to applying to the predictors the parametric form of the softmax function such that

$$\forall i \in [\![1, m]\!], \ g(x_i) = \left(\frac{\exp \left(\sum_{j=1}^{N} \sum_{t=-T_l}^{T_r} a_{kjt} h_k^{(j)} (x_{i+t}) + b_k \right)}{\sum_{l=1}^{Q} \exp \left(\sum_{j=1}^{N} \sum_{t=-T_l}^{T_r} a_{ljt} h_l^{(j)} (x_{i+t}) + b_l \right)} \right)_{1 \leqslant k \leqslant Q}.$$

The values of the corresponding parameters, the coefficients a_{kjt} and the biases b_k, are obtained by maximum likelihood estimation (the training criterion used is cross-entropy), so that the model specified appears as a simplified variant of the polytomous (multinomial) logistic regression model [29]. In practice, the training algorithm that we implemented is a multi-class extension of the one exposed in [30].

3.2   Linear Ensemble Methods

In [31], we studied the class of *linear ensemble methods* (LEMs). Their use requires an additional hypothesis regarding the base classifiers, namely that they take their values in the probability simplex (which is less restrictive than assuming that their outputs are probability estimates). For the sake of simplicity, we present them without taking into account the sliding window. In practice, its introduction is equivalent to the introduction of $N(T - 1)$ additional classifiers. For all k in $[\![1, Q]\!]$, let t_k denote the *one of Q coding* of category k, i.e., $t_k = (\delta_{k,l})_{1 \leqslant l \leqslant Q}$. With this notation at hand, an LEM is defined as follows.

Definition 4 (Linear ensemble method). *Let* $\left\{ \tilde{h}^{(j)} = \left(\tilde{h}_k^{(j)} \right)_{1 \leqslant k \leqslant Q} : 1 \leqslant j \leqslant N \right\}$ *be a set of N classifiers taking their values in the probability simplex. Let $\beta =$*

$(\beta_k)_{1 \leqslant k \leqslant Q}$, with the vectors β_k belonging to $\mathbb{R}^{NQ}$. For all j in $[\![1, N]\!]$ and all (k, l) in $[\![1, Q]\!]^2$, let β_{kjl} denote the component of vector β_k of index $(j-1)Q+l$. Then, a linear ensemble method is a multivariate linear model of the form

$$\forall x \in \mathcal{X}, \ g(x) = \left(\sum_{j=1}^{N} \sum_{l=1}^{Q} \beta_{kjl} \tilde{h}_l^{(j)}(x) \right)_{1 \leqslant k \leqslant Q}$$

for which the vector β is obtained by solving a convex programming problem of the type

Problem 2 (Learning problem of an LEM)

$$\min_{\beta} \sum_{i=1}^{m} \ell_{LEM} \left(t_{y_i}, g(x_i) \right)$$

$$s.t. \begin{cases} \beta \in \mathbb{R}_+^{NQ^2} \\ \forall j \in [\![1, N]\!], \ \forall l \in [\![1, Q-1]\!], \ \sum_{k=1}^{Q} (\beta_{kjl} - \beta_{kjQ}) = 0 \\ 1_{NQ^2}^T \beta = Q \end{cases}$$

where the loss function ℓ_{LEM} is convex.

Note that a special case of LEM is the standard convex combination obtained by adding the following constraint:

$$\forall (j, k, l) \in [\![1, N]\!] \times [\![1, Q]\!] \times [\![1, Q]\!], \ k \neq l \Longrightarrow \beta_{kjl} = 0 \ .$$

The constraints of Problem 2 are sufficient to ensure that g takes its values in the probability simplex. Furthermore, if ℓ_{LEM} is the quadratic loss (Brier score) or the cross-entropy loss, then the outputs are actually estimates of the class posterior probabilities (see [31] for the proof). In order to make comparison with the polytomous logistic regression straightforward, we selected the second option for our experiments. Since the M-SVMs do not take their values in the probability simplex, their outputs must be post-processed prior to being combined by an LEM. This can be done by means of the polytomous logistic regression model. In that case, the flowchart of the cascade architecture is the one depicted in Figure 1 (right).

3.3   Multi-layer Perceptron

As pointed out in the introduction, the MLP is the standard structure-to-structure classifier. In our experiments, we implemented it with the softmax activation function for the output units (a sigmoid for the hidden units) and the cross-entropy loss function. In that way, it could be seen as a complexified extension of the polytomous logistic regression model described in Section 3.1, making once more the comparison of performance straightforward.

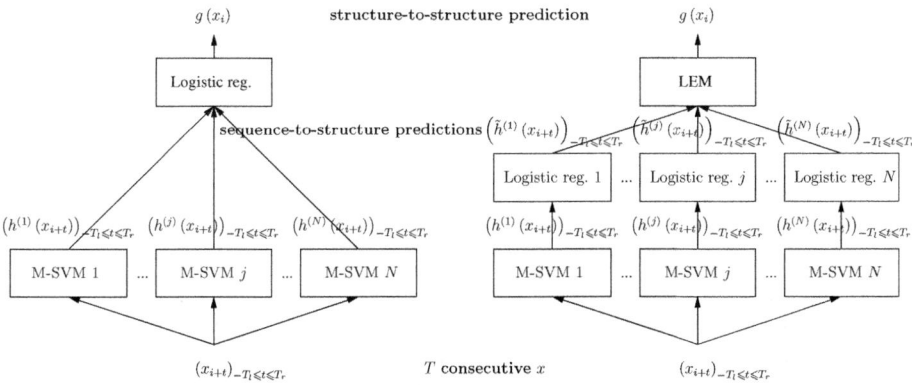

Fig. 1. Flowchart of the computation of the class posterior probabilities using a polytomous logistic regression model (left) and an LEM (right) as structure-to-structure classifier

4   Experimental Results

To assess the different structure-to-structure classifiers described in Section 3, we used the CB513 data set [32]. The 513 sequences of this set are made up of 84119 residues. The derivation of the descriptions x_i of these residues (inputs of the sequence-to-structure classifiers) has been detailed in Section 2.2. As for their labels y_i, the initial secondary structure assignment was performed by the DSSP program [33], with the reduction from 8 to 3 conformational states following the CASP method, i.e., H+G $\rightarrow$ H (α-helix), E+B $\rightarrow$ E (β-strand), and all the other states in C (coil). In all cases, the two sliding windows used were centered on the residue of interest, and their respective sizes were 13 (sequence-to-structure) and 15 (structure-to-structure, $T_l = T_r = 7$). The performance of reference was provided by a cascade architecture involving two MLPs. The respective sizes of their hidden layers are 16 (sequence-to-structure) and 6 (structure-to-structure). The structure-to-structure network is precisely the one described in Section 3.3.

A secondary structure prediction method must fulfill different requirements in order to be useful for the biologist. Thus, several standard measures giving complementary indications must be used to assess the prediction accuracy [34]. We computed the three most popular ones: the recognition rate Q_3, Pearson-Matthews correlation coefficients $C_{\alpha/\beta/\text{coil}}$, and the segment overlap measure (Sov) in its most recent version (Sov'99). The quality of the probability estimates was measured by means of the (averaged) cross-entropy (CE). To train the two levels of the cascade and assess performance, a seven-fold cross-validation procedure was implemented. At each step, two thirds of the training set were used to train the sequence-to-structure classifiers, and one third to train the structure-to-structure classifier. The experimental results obtained are gathered in Table 2.

Table 2. Prediction accuracy and quality of the probability estimates for the different implementations of the cascade architecture considered

Cascade architecture	Q_3 (%)	C_α	C_β	C_{coil}	Sov'99 (%)	CE
MLP + MLP	74.6	0.69	0.59	0.54	71.1	0.615
M-SVMs + logistic reg.	76.5	0.71	0.62	0.57	73.0	0.576
M-SVMs + logistic reg. + LEM	76.5	0.71	0.63	0.57	**73.1**	0.576
M-SVMs + MLP	**76.7**	0.72	0.63	0.57	71.9	**0.572**

Using the two sample proportion test (the one for large samples), the gain in prediction accuracy resulting from using dedicated M-SVMs in place of an MLP for the sequence-to-structure prediction of the cascade architecture appears always (i.e., irrespective of the choice of the structure-to-structure classifier) statistically significant with confidence exceeding 0.95. The value of the cross-entropy confirms this superiority. If we restrict to the architecture we advocate, then the recognition rates of the three structure-to-structure classifiers are almost identical. The value of the cross-entropy does not really help to break the tie. However, this goal can be achieved by resorting to the Sov. In that case, the logistic regression and the LEM appear almost equivalent, and significantly superior to the MLP. The reason for this asymmetry is still to be highlighted.

Thus, the main conclusion that can be drawn regarding the choice of the structure-to-structure classifier is that the prediction accuracy does not benefit significantly from an increase in the complexity. From this point of view, the small size of the hidden layer of the MLP used for this task is telling. This phenomenon can be explained by a well-known fact in secondary structure prediction: the main limiting factor when applying a cascade architecture is overfitting (see for instance [35]). On the one hand, the sequence-to-structure classifiers must be complex enough to cope with the complexity of the task, but on the other hand, the classifier at the second level must be of far lower capacity, otherwise its recognition rate in test will be disconnected with its recognition rate on the training set. With this restriction in mind, the behavior of the LEM appears promising, since its accuracy with respect to both major criteria (Q_3 and cross-entropy) is similar to the one of the other combiners although it requires an additional level of training, resulting from the need for a post-processing of the outputs of the M-SVMs prior to their combination (see Figure 1). Given the experimental protocol described above, we used the same training set to train both the M-SVMs and their post-processing, a strategy which is prone to overfitting. As a consequence, we conjecture that the prediction accuracy of the LEM is the one which should benefit most from the availability of additional training data.

To sum up, these experiments back our thesis that M-SVMs endowed with a dedicated kernel should be used as sequence-to-structure classifiers in a cascade architecture, even in the case when the final outputs must be class posterior probability estimates. In that case, several options are available for the structure-to-structure classifier. So far, it appears that the difference between them is not significant. The levelling of performance primarily highlights the problem of overfitting.

5   Conclusions and Ongoing Research

This article has addressed the problem of optimizing the cascade architecture commonly used for protein secondary structure prediction with respect to two criteria: prediction accuracy and quality of the class posterior probability estimates. The main result is that using dedicated M-SVMs as sequence-to-structure classifiers, it is possible to outperform the standard solution, consisting in using two MLPs in sequence, according to both criteria. Making the best of the new approach should require a precise choice for the structure-to-structure classifier. From that point of view, capacity control appears to play a major part, since overfitting is a strong limiting factor. Obviously, a touchstone for the architecture we advocate is a comparison with solutions based on *pairwise coupling* (Bradley-Terry model) [36]. To that end, we are currently conducting additional comparative experiments with architectures differing from ours at the sequence-to-structure level: the M-SVMs are replaced with decomposition schemes involving bi-class machines.

This contribution paves the way for the specification of new hybrid prediction methods combining discriminant models and HMMs. By the way, the principle of these methods could be extended to many other fields of bioinformatics, including alternative splicing prediction.

Acknowledgements. This work was supported by the MBI project of the PRST MISN. The authors would like to thank the anonymous reviewers for their comments.

References

1. Qian, N., Sejnowski, T.J.: Predicting the secondary structure of globular proteins using neural network models. Journal of Molecular Biology 202, 865–884 (1988)
2. Jones, D.T.: Protein secondary structure prediction based on position-specific scoring matrices. Journal of Molecular Biology 292, 195–202 (1999)
3. Pollastri, G., Przybylski, D., Rost, B., Baldi, P.: Improving the prediction of protein secondary structure in three and eight classes using recurrent neural networks and profiles. Proteins 47, 228–235 (2002)
4. Cole, C., Barber, J.D., Barton, G.J.: The Jpred 3 secondary structure prediction server. Nucleic Acids Research 36, W197–W201 (2008)
5. Kountouris, P., Hirst, J.D.: Prediction of backbone dihedral angles and protein secondary structure using support vector machines. BMC Bioinformatics 10, 437 (2009)
6. Anthony, M., Bartlett, P.L.: Neural Network Learning: Theoretical Foundations. Cambridge University Press, Cambridge (1999)
7. Hua, S., Sun, Z.: A novel method of protein secondary structure prediction with high segment overlap measure: support vector machine approach. Journal of Molecular Biology 308, 397–407 (2001)

8. Guermeur, Y.: Combining discriminant models with new multi-class SVMs. Pattern Analysis and Applications 5, 168–179 (2002)
9. Guermeur, Y., Pollastri, G., Elisseeff, A., Zelus, D., Paugam-Moisy, H., Baldi, P.: Combining protein secondary structure prediction models with ensemble methods of optimal complexity. Neurocomputing 56, 305–327 (2004)
10. Nguyen, M.N., Rajapakse, J.C.: Two-stage multi-class support vector machines to protein secondary structure prediction. In: 10th Pacific Symposium on Biocomputing, pp. 346–357 (2005)
11. Richard, M.D., Lippmann, R.P.: Neural network classifiers estimate Bayesian *a posteriori* probabilities. Neural Computation 3, 461–483 (1991)
12. Rojas, R.: A short proof of the posterior probability property of classifier neural networks. Neural Computation 8, 41–43 (1996)
13. Lin, K., Simossis, V.A., Taylor, W.R., Heringa, J.: A simple and fast secondary structure prediction method using hidden neural networks. Bioinformatics 21, 152–159 (2005)
14. Rabiner, L.R.: A tutorial on hidden Markov models and selected applications in speech recognition. Proceedings of the IEEE 77, 257–286 (1989)
15. Guermeur, Y.: VC theory of large margin multi-category classifiers. Journal of Machine Learning Research 8, 2551–2594 (2007)
16. Schölkopf, B., Smola, A.J.: Learning with Kernels: Support Vector Machines, Regularization, Optimization, and Beyond. The MIT Press, Cambridge (2002)
17. Shawe-Taylor, J., Cristianini, N.: Kernel Methods for Pattern Analysis. Cambridge University Press, Cambridge (2004)
18. Berlinet, A., Thomas-Agnan, C.: Reproducing Kernel Hilbert Spaces in Probability and Statistics. Kluwer Academic Publishers, Boston (2004)
19. Wahba, G.: Multivariate function and operator estimation, based on smoothing splines and reproducing kernels. In: Casdagli, M., Eubank, S. (eds.) Nonlinear Modeling and Forecasting, SFI Studies in the Sciences of Complexity, vol. XII, pp. 95–112. Addison-Wesley (1992)
20. Guermeur, Y.: A generic model of multi-class support vector machine. International Journal of Intelligent Information and Database Systems (accepted)
21. Weston, J., Watkins, C.: Multi-class support vector machines. Technical Report CSD-TR-98-04, Royal Holloway, University of London, Department of Computer Science (1998)
22. Crammer, K., Singer, Y.: On the algorithmic implementation of multiclass kernel-based vector machines. Journal of Machine Learning Research 2, 265–292 (2001)
23. Lee, Y., Lin, Y., Wahba, G.: Multicategory support vector machines: Theory and application to the classification of microarray data and satellite radiance data. Journal of the American Statistical Association 99, 67–81 (2004)
24. Guermeur, Y., Monfrini, E.: A quadratic loss multi-class SVM for which a radius-margin bound applies. Informatica 22, 73–96 (2011)
25. Altschul, S.F., Madden, T.L., Schäffer, A.A., Zhang, J., Zhang, Z., Miller, W., Lipman, D.J.: Gapped BLAST and PSI-BLAST: a new generation of protein database search programs. Nucleic Acids Research 25, 3389–3402 (1997)
26. Guermeur, Y., Lifchitz, A., Vert, R.: A kernel for protein secondary structure prediction. In: Schölkopf, B., Tsuda, K., Vert, J.-P. (eds.) Kernel Methods in Computational Biology, pp. 193–206. The MIT Press, Cambridge (2004)
27. Lauer, F., Guermeur, Y.: MSVMpack: a multi-class support vector machine package. Journal of Machine Learning Research 12, 2293–2296 (2011)

28. Platt, J.C.: Probabilities for SV machines. In: Smola, A.J., Bartlett, P.L., Schölkopf, B., Schuurmans, D. (eds.) Advances in Large Margin Classifiers, pp. 61–73. The MIT Press, Cambridge (2000)
29. Hosmer, D.W., Lemeshow, S.: Applied Logistic Regression. Wiley, London (1989)
30. Lin, H.-T., Lin, C.-J., Weng, R.C.: A note on Platt's probabilistic outputs for support vector machines. Machine Learning 68, 267–276 (2007)
31. Guermeur, Y.: Combining multi-class SVMs with linear ensemble methods that estimate the class posterior probabilities. Communications in Statistics (submitted)
32. Cuff, J.A., Barton, G.J.: Evaluation and improvement of multiple sequence methods for protein secondary structure prediction. Proteins 34, 508–519 (1999)
33. Kabsch, W., Sander, C.: Dictionary of protein secondary structure: pattern recognition of hydrogen-bonded and geometrical features. Biopolymers 22, 2577–2637 (1983)
34. Baldi, P., Brunak, S., Chauvin, Y., Andersen, C.A.F., Nielsen, H.: Assessing the accuracy of prediction algorithms for classification: an overview. Bioinformatics 16, 412–424 (2000)
35. Riis, S.K., Krogh, A.: Improving prediction of protein secondary structure using structured neural networks and multiple sequence alignments. Journal of Computational Biology 3, 163–183 (1996)
36. Hastie, T., Tibshirani, R.: Classification by pairwise coupling. The Annals of Statistics 26, 451–471 (1998)

Shape Matching by Localized Calculations of Quasi-Isometric Subsets, with Applications to the Comparison of Protein Binding Patches

Frédéric Cazals[1] and Noël Malod-Dognin[1,*]

INRIA Sophia Antipolis - Méditerranée, France
{Frederic.Cazals,Noel.Malod-Dognin}@inria.fr

Abstract. Given a protein complex involving two partners, the receptor and the ligand, this paper addresses the problem of comparing their binding patches, i.e. the sets of atoms accounting for their interaction. This problem has been classically addressed by searching quasi-isometric subsets of atoms within the patches, a task equivalent to a maximum clique problem, a NP-hard problem, so that practical binding patches involving up to 300 atoms cannot be handled.

We extend previous work in two directions. First, we present a generic encoding of shapes represented as cell complexes. We partition a shape into concentric shells, based on the *shelling order* of the cells of the complex. The shelling order yields a *shelling tree* encoding the geometry and the topology of the shape. Second, for the particular case of cell complexes representing protein binding patches, we present three novel shape comparison algorithms. These algorithms combine a Tree Edit Distance calculation (TED) on shelling trees, together with *Edit* operations respectively favoring a topological or a geometric comparison of the patches. We show in particular that the geometric TED calculation strikes a balance, in terms of accuracy and running time, between purely geometric and topological comparisons, and we briefly comment on the biological findings reported in a companion paper.

Keywords: Protein, 3D-structure, comparison, maximum clique, tree-edit-distance.

1 Introduction

1.1 Comparing Proteins and Protein Complexes

The general question of shape matching encompasses a large variety of problems whose specifications require defining the objects handled, the types of features favored, and the level of noise to be accommodated. In the realm of structural bioinformatics, the shapes handled are macro-molecules or complexes of such. Comparing shapes aims at fostering our understanding of the structure-to-function relationship, which is key since the 3D structure underpins all biological

---

* Corresponding author.

M. Loog et al. (Eds.): PRIB 2011, LNBI 7036, pp. 272–283, 2011.

functions. The structural comparison of proteins is a classical endeavor which conveys complementary information with respect to sequence alignments, in particular in the context of medium to low sequence similarity. Most of the proposed methods are either based on optimal rigid-body superimposition (like VAST [1] or STRUCTAL [2]), whose computation is based on the least Root Mean Square Deviation of residue coordinates as first defined by Kabsch [3], or on the comparison of the internal distances between the residues (like DALI [4], CMO [5] or DAST [6]). The comparison of protein complexes and their interfaces, on the other hand, is a slightly different problem since (i) the focus is on the interface atoms rater than the whole molecules, and (ii) such atoms cannot be ordered as opposed to the amino-acids along the backbone. For these reasons, the comparison of protein interfaces usually boils down to seeking *quasi-isometric* subsets of atoms, that is two subsets of atoms, one on each structure, which can roughly be super-imposed through a rigid motion. Such analysis are commonplace while investigating correlations between structural and bio-physical properties in protein complexes [7], but also in docking and drug design [8]. The goal of this paper is to improve such tools and to present new ones.

1.2 Comparing Protein Binding Patches

Consider a binary protein complex, and prosaically define the *binding patch* of a partner in the complex as the set of atoms which account for the interaction. So far, the comparison of binding patches has been addressed in two ways. On the one hand, direct methods aim at identifying two subsets of atoms, one on each partner, which are *quasi-isometric*, in the sense that there exists a rigid motion sending one onto the other, with prescribed accuracy. Alas, this problem is known to be equivalent to a maximum clique calculation [9]. On the other hand, a number of indirect methods compare fingerprints of the patches, using geometric hashing [10], spherical harmonics [11] or cross-correlation coefficient analysis via FFT calculations [12]. While these approaches are successful, in particular for docking, one cannot directly use them to report quasi-isometric subsets of atoms.

The contributions in this paper fall in the first category, as we develop a generic strategy which aims at improving the complexity of geometric comparisons, and also at performing comparisons of topological nature. More precisely, our strategy consists of computing a topological encoding of a binding patch, so as to reduce the comparison of patches to the comparison of ordered trees. This latter problem can be solved in polynomial time using dynamic programming to implement the so-called Tree Edit Distance [13]. As we shall see, tuning the semantics of the *Edit* operation actually yields algorithms favoring a topological or a geometric comparison of the patches.

The paper is organized as follows. Sections 1, 2, 3 respectively present the encoding algorithm, the comparison algorithms, and the results.

2   Topological Encoding of Cells Complexes

We first present a general topological encoding of so-called cell complexes, and detail its implementation for the particular case of the binding patches of a protein complex.

2.1   Shelling a Cell Complex

Rationale. In the continuous setting, a natural way to encode the geometry and the topology of a shape consists of considering the level sets of the distance function to its boundary. In particular, the meeting points of these level sets encode the so-called medial axis of the shape, and the shape can be reconstructed by the union of an infinite number of balls centered on the medial axis [14]. To mimic this process in the discrete setting, consider the partition of a complex object into cells making up a d-dimensional cell complex K. That is, the collection of all cells satisfy the following two conditions (i) every $(d-1)$-dimensional face of a d-dimensional cell in K is also in K, and (ii) the intersection of two d-cells is either empty or a common $(d-1)$-face of both. For example, 3D triangulations provide a broad class of cells complexes to represent 3D shapes. For the particular case of molecular shapes, we shall introduce a two-dimensional cell complex in the sequel of the paper.

Shelling graph. Consider a cell complex K consisting of d-dimensional cells incident across $(d-1)$-dimensional faces. These cells and their face-incidence define a dual graph G whose nodes are the cells, and edges correspond to face-incidences between cells. Term a cell a *boundary cell* if at least one of its faces does not have any incident cell in K, and assign a value of zero to such a cell. The *Shelling Order* (SO) of a cell c in K is the smallest number of edges of G traversed to reach c from a boundary cell. To integrate this information at the level of the complex K, define a *shell* as a maximal connected component (c.c.) of cells having an identical SO and term two shells of *incident* if they contain two incident cells whose SO differ of one unit. We arrange shells into a *shelling graph*, whose roots correspond to shells with unit SO, and whose parent-child relationship encodes the incidence between shells. In particular, a branching point corresponds to a split of a shell into sub-shells while increasing the SO.

Shelling graphs and shelling forests. Define the *size* of a node of the shelling graph as the number of cells of the corresponding shell, and the size $| T |$ of any sub-graph T as the sum of the size of its nodes. If the cells whose SO is zero make up several connected components, the shelling graph G_S is a directed acyclic graph. However, one of these c.c. dominates the other ones provided that G_S contains a (dominant) tree T such that $| T | / | G_S | > 1 - \varepsilon$ for a small value of ε. In that case, the shelling graph can be edited into an *ordered shelling tree*, a transformation motivated by the fact that comparing ordered trees has polynomial time complexity resorting to dynamic programming [13]. The steps taken are as follows: *Graph-to-forest.* Each c.c. of the shelling graph is edited into trees: if the c.c. has several roots, the one with largest size remains as root, and the other ones are disconnected —each becoming a single-node tree.

Forest-to-tree. If the forest features one dominant tree, that tree only is retained.

Tree-to-ordered-tree. The descendants of a given node are sorted by increasing size, so that the resulting tree is called an *ordered* tree.

A comment w.r.t medial axis based shape matching. Having outlined our strategy, one comment is in order w.r.t. the shape matching method of [15], which uses the encoding of a smooth shape through its Medial Axis Transform (the infinite union of balls centered on the medial axis) to identify *isomorphic* subsets of the medial axis of two shapes. First, the strategy of [15] is concerned with smooth shapes, while we work in a discrete setting—the medial axis is not even defined. Second, that method aims at identifying isomorphic sub-structures of the medial axis and does not directly address the question of noise, which can clearly be hindrance given the inherent instabilities of the medial axis [16]. We instead directly target topological and geometric noise. Third, the granularity of a local match in [15] is that of a maximal ball—centered on the portions of the medial axis matched; we instead match cells rather than maximal balls, thus providing a finer matching granularity.

2.2 Shelling Binding Patches of Protein Complexes: Outline

Solvent Accessible Models. Recall that the Van der Waals model of a molecular complex consists of a collection of balls, one per atom. The corresponding Solvent Accessible Model consists of growing all atoms by the radius of a water probe $r_w = 1.4 \text{Å}$. The corresponding Solvent Accessible Surface (SAS) is defined as the boundary of the volume occupied by the balls. Note that the SAS is a cell complex made of k-cells for $k = 0, 1, 2$. Its 2-cells are spherical polygons also called caps; if two such caps intersect, they share a circle arc supported by the intersection circle of their defining spheres. Its 1-cells are circle arcs; it two such arcs intersect, they share one vertex defined by the intersection of (generically) three spheres. We encode this complex with a half-edge data structure [17] or HDS, containing in particular two so-called half-edges for each circle arc.

Binding patches and their shelling. Consider a protein complex, and assume that its interface atoms have been identified. This process consists of spotting the atoms of each partner facing one-another, and a reference method to do so consists of resorting to the α-complex of the protein complex, which is a simplicial complex contained in the dual of the Voronoi (power) diagram of the atoms in the SAS model [18]. Having identified the interface atoms, the computation of shelling forests/trees requires three main steps for each partner, namely (i) computing the cell complex representing the SAS of the partner, (ii) computing the SO of interface atoms, and (iii) building the shells, the shelling graph, and the shelling forest/trees. For the initialization of the second step, notice that a spherical cap has SO of zero iff it belongs to an interface atom and has a neighboring cap belonging to a non interface atom. See Fig. 1 (available in color in the following research report: http://hal.inria.fr/inria-00603375).

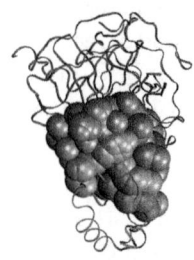 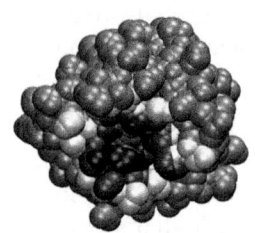 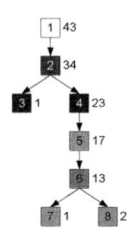

Fig. 1. Illustration of the binding patch shelling. Left: side view of protein complex 1acb, with interface atoms displayed in red for partner A (chain I) and in blue for
partner B (chain E). Grey atoms correspond to the water molecules. Middle: rotated
view of the binding patch of partner B only, with colors depending on the shelling
order (in grey-blue: non interface atoms; in white: atoms having a shelling order of
zero; etc...). Right: the corresponding atom shelling tree contains 134 atoms divided
into 8 shells.

Difficulties. The previous discussion actually eludes two difficulties. The first
one is that of binding patches involving several connected components. This
typically happens in the presence of packing defects: in that case, the interface
atoms, in addition to the main binding patch facing the partner, also define tiny
regions on the inner side of the molecule. But packing defects also punch holes
into a given connected component (a c.c. of the patch is a topological annulus).
These two difficulties actually motivate the *Graph-to-forest* and *Forest-to-tree*
operations mentioned in section 2.1, and we now present the detailed algorithm
handling them.

2.3   Shelling Protein Binding Patches: Detailed Algorithm

Step 1: Computing the HDS. The half-edge data structure encodes the
boundary of the union of balls, as computed in [19]. A certified embedding in
3D is obtained thanks to the robust geometric operations described in [20].

Step 2: Computing the Connected Component of the Boundary (CCB).
The CCB are the cycles which bound a given binding patch. Given the HDS,
finding all CCB of a patch requires running a Union-Find algorithm [21], which
has (almost) linear complexity.

Step 3: Computing the Connected Components of Half-edges (CC).
To identify the CCs of a binding patch, we run a Union-Find algorithm on all
the half-edges of the patch. Each CC will yield a shelling graph.

Step 4: Initializing the Shelling Order. From steps 2-3, the largest CCB of
each connected component is selected, and the corresponding faces are assigned
a SO of zero. This step settles the case of connected component with several
rims, and corresponds to the *Graph-to-forest* step mentioned in section 2.1.

Step 5: Computing the Shelling Order. Using the connectivity of faces
encoded in the HDS, a priority queue is used to assign the SO to all the faces.
The queue is initialized with the boundary faces identified at step 4.

Step 6: Computing the Shells. A shell being a connected component of faces having the same SO, a Union-Find algorithm is also called to create the shells.

Step 7: Computing the Face Shelling Graph. A parent-child relationship between two shells is witnessed by a half-edge incident on two faces having a SO which differs by one unit. Collecting all such pairs requires a linear pass over all half-edges. Constructing the Face Shelling Graph from the parent-child list is straightforward.

Step 8: Selecting the Face Shelling Tree. So far, one tree has been computed for each connected component of the binding patch. We select the tree corresponding to the largest component in the Face Shelling Graph. This settles the case of patch with several connected components, and corresponds to the *Forest-to-tree* step discussed in section 2.1.

Step 9: Computing the Atom Shelling Tree. In order to base the comparison of binding patches on atoms rather than faces, the selected face shelling tree is edited into an *atom shelling tree*. The process consists of substituting atoms to faces, with the following special cases: if atom is present several times in the same shell, it is counted once; if an atom belongs to several shells in a branch of the face shelling tree, it is assigned to the shell closest to the root of the tree.

Step 10: Ordering Atom Shelling Tree. This step requires sorting the sons of a node by increasing size.

Complexity-wise, the first step is the most demanding one since it requires computing the Delaunay triangulation of the balls, which has $O(n \log n + k)$ complexity, with n the number of atoms and k the number of simplices in the output. Practically, we use the *Delaunay_triangulation_3* and *Alpha_shape_3* packages of the Computational Geometry Algorithms Library [22]. The remaining steps are (almost) linear in the size of the HDS, since a Union-Find process has (almost) linear complexity—if one omits the cost of the inverse of Ackermann's function.

3   Comparison Algorithms

Based on the previous encoding, we present three methods to compare binding patches: the *topological TED* relies on a straight Tree Edit Distance calculation on our topological encoding; the *Clique* method implements a clique calculation at the binding patch level, yet incorporating topological constraints; the *geometrical TED* mixes the previous two methods.

Given two binding patches represented by their atom shelling trees T_1 and T_2, we wish to find the largest subset of atoms that is both quasi-isometric and isotopologic—see definition below. Notation-wise, for both the intersection $\cap^x$, the symmetric difference Δ^x and the similarity score SIM_x, x stands for the used methodology (t: topological TED; c: Clique; g: geometrical TED). Abusing terminology, we shall indistinctly speak of a shell or its node representation in the shelling tree.

3.1   Constraints

Quasi-isometry. Given two shelling trees T_1 and T_2, two atoms i and j from T_1 are quasi-isometric to the atoms k and l from T_2 if the euclidean distances $d_{i,j}$ (between i and j) and $d_{k,l}$ (between k and l) are such that $|d_{i,j} - d_{k,l}| \leq \varepsilon$, with ε a distance threshold circa 2Å. Between T_1 and T_2, a subset of atoms $V_1 \subset T_1$ is quasi-isometric to a subset $V_2 \subset T_2$ if there exists a one-to-one matching between the atoms of V_1 and the ones of V_2 such that any two atoms in V_1 are quasi-isometric to their counterparts in V_2. Such matchings have the property that the corresponding Root Mean Squared Deviation of internal distances $(RMSD_d)^1$ is smaller than ε.

Isotopology. A shell v_1 of a shelling tree is an ancestor of a shell v_2 if there is an oriented path from v_1 to v_2 in the shelling tree. Two shells $(v_1, w_1) \in T_1$ are isotopologic to the shells $(v_2, w_2) \in T_2$ if either: (i) $v_1 = w_1$ and $v_2 = w_2$, or (ii) v_1 is an ancestor of w_1 and v_2 is an ancestor of w_2, or (iii) or v_1 is to the left of w_1 iff v_2 is to the left of w_2 (Recall that trees are ordered). Two atoms of T_1 and two atoms of T_2 are isotopologic iff the shells containing them are isotopologic. Between T_1 and T_2, a subset of atoms $V_1 \subset T_1$ is isotopologic to a subset $V_2 \subset T_2$ if there exists a one-to-one mapping between the atoms of V_1 and the ones of V_2 such that any two atoms in V_1 are isotopologic to their counterparts in V_2.

3.2   Topological Comparison: TED$_t$

To compare two binding patches from a topological viewpoint, we are interested in finding the number of isotopologic atoms between T_1 and T_2. This problem reduces to an ordered Tree Edit Distance problem (TED) having the following edition costs. Adding or deleting a shell s has a cost of $|s|$, since all the atoms of s are added/removed. Morphing a shell s_1 into a shell s_2 has cost equal to the size of their symmetric difference: $|s_1 \Delta^t s_2| = |s_1| + |s_2| - 2|s_1 \cap^t s_2|$. Since we are matching pairs of atoms coming from the same shells, condition (1) of isotopology is always satisfied, and thus $|s_1 \cap^t s_2| = \min(|s_1|, |s_2|)$. TED returns both the sequence of edit operations having minimum sum of costs for editing T_1 into T_2 and the corresponding sum of costs, which is denoted by $\text{TED}_t(T_1, T_2)$. By definition, the ordered tree edit distance also preserves the isotopologic conditions (2) and (3), and thus $\text{TED}_t(T_1, T_2)$ is the size of the symmetric difference $T_1 \Delta^t T_2 = |T_1| + |T_2| - 2|T_1 \cap^t T_2|$, where $|T_1 \cap^t T_2|$ is the number of isotopologic atoms between T_1 and T_2. The similarity between two trees is then the number of their isotopologic atoms normalized by the size of the two trees to be in [0,1]:

$$\text{SIM}_t(T_1, T_2) = 2|T_1 \cap^t T_2| / (|T_1| + |T_2|) \tag{1}$$

---

[1] $RMSD_d(V_1, V_2) = \sqrt{\sum_{i<j}(|d_{i,j} - d_{k,l}|^2)/\binom{n}{2}}$, where $n = |V_1| = |V_2|$, and where $k \in V_2$ and $l \in V_2$ are the counterparts of $i \in V_1$ and $j \in V_1$.

3.3 Geometric Comparison: Clique

To favor geometric comparisons, we are interested in finding the largest subset of atoms between T_1 and T_2 that is both quasi-isometric and isotopologic. This problem is rephrased as a maximum clique problem as follows. Let $G_{T_1,T_2} = (V, E)$ be a graph whose vertex set V is depicted by a grid in which each row represents an atom of T_1 and each column represents an atom of T_2. Matching the atoms $i \in T_1$ and $k \in T_2$ is represented by the vertex $i.k \in V$ (on row i, column k). $\forall i.k \in V$ and $\forall j.l \in V$ such that $i \neq j$ and $k \neq l$, if i and j are quasi-isometric and isotopologic to k and l then the edge $(i.k, j.l)$ is in E. The largest subset of atoms between T_1 and T_2 that is both quasi-isometric and isotopologic, denoted by $T_1 \cap^c T_2$, corresponds to a maximum clique in G_{T_1,T_2}. Mimicking equation (1), the similarity between T_1 and T_2 is:

$$\text{SIM}_c(T_1, T_2) = 2|T_1 \cap^c T_2|/(|T_1| + |T_2|) \tag{2}$$

The maximum clique problem is NP-Hard [23], and is solved by using Östergård's algorithm [24]. Note that the isotopology constraints reduce the number of edges in G_{T_1,T_2}, thus easing the maximum clique solving process.

3.4 Hybrid Approach: TED$_g$

To strike a balance between geometric and topological criteria, we are interested in finding the subsets of atoms that are isotopologic and where the quasi-isometric constraints are satisfied between the matched shells. Meeting these criteria is amenable to a TED calculation, using the following costs. The costs for inserting/deleting a shell s is $|s|$. The cost for morphing a shell s_1 into a shell s_2 is equal to the size of their symmetric difference $s_1 \Delta^c s_2 = |s_1| + |s_2| - 2|s_1 \cap^c s_2|$, where $s_1 \cap^c s_2$ is the subset of isotopologic and quasi-isometric atoms, as found by applying the Clique method between the two shells s_1 and s_2. In this case, $\text{TED}_g(T_1, T_2)$ is the size of the symmetric difference $|T_1 \Delta^g T_2| = |T_1| + |T_2| - 2|T_1 \cap^g T_2|$, where $|T_1 \cap^g T_2|$ is number of atoms between T_1 and T_2 that are both isopologic and partially isometric. The similarity between two trees is then the size of $|T_1 \cap^g T_2|$ normalized by the size of the two trees:

$$\text{SIM}_g(T_1, T_2) = 2|T_1 \cap^g T_2|/(|T_1| + |T_2|) \tag{3}$$

3.5 Relation between the Approaches

All the methods respect the isotopologic constraints, but only Clique respects all the isometric constraints. TED$_g$ verifies the isometric constraints $|d_{i.j} - d_{k.l}| \leq \varepsilon$ only when i and j (and thus k and l) come from the same shell. Finally, in TED$_t$, the isometric constraints are not verified at all. This implies that the size of the atomic subsets returned by Clique are smaller than the size of the subsets returned by TED$_g$ which are smaller than the values found by TED$_t$. Thus, $\text{SIM}_c(T_1, T_2) \leq \text{SIM}_g(T_1, T_2) \leq \text{SIM}_t(T_1, T_2)$ holds. Also, because of the possibly broken isometric constraints, TED$_g$ may return matchings having $RMSD_d$ values larger than ε.

4   Results: Performances and Scores

Material and method. We selected 92 high resolution (≤ 2Å) protein complexes[2] from which we extracted 184 binding patches—two per complex. The smallest binding patch contains 26 atoms divided into 3 shells, and the largest one contains 271 atoms divided into 14 shells. Comparing all the binding patches requires solving $N_t = 17020$ comparison instances (including the self comparisons). The computations were done on a cluster of Intel Xeon processors at 2.66Ghz, with a time limit of 600 seconds per comparison instance, and when relevant, with a distance threshold ε of 2Å. Note that since the maximum clique problem is NP-Hard, using a larger (but still reasonable) time limit does not guaranty that all comparison instances will be optimally solved.

Solved instances and running times. Table 1 shows the number of instances solved by the three methods. As expected, TED_t solves more instances than TED_g, which solves more instances than Clique; we denote N_s the number of instances solved by the three methods. Overall, one sees that TED_t is about 31 time faster than TED_g and 3695 time faster than Clique, and that TED_g is about 118 time faster than Clique. The running time comparison between TED_g and Clique is better illustrated in Figure 4 which plots, for all the commonly solved instances, the running time of TED_g against the one of Clique.

Table 1. Solved instance comparison. Column 2 presents the number of solved instances by each method when computations are limited to 600 seconds per instance.

Method	Solved instances
TED_t	17020 (100%)
TED_g	17018 ($> 99.9\%$)
Clique	12166 ($\simeq 71.5\%$)

Table 2. Matching comparison. Over the 12165 instances that are solved by both TED_g and Clique, columns 2 to 4 (resp. 5 to 7) present the minimum, median and maximum observed $RMSD_d$ (resp. coverage, in percent of atomic length) of the matchings.

Method	$RMSD_d$ (in Å)			Coverage (%)		
	min	med	max	min	med	max
TED_g	0.38	5.87	32.3	15%	42%	100%
Clique	0.27	0.91	1.20	9%	23%	100%

Figure 2 shows, for each method, the running times as a function of the size of the two binding patches. When fixing the size of one of the binding patches, the running time of TED_t appears to be linear in the size of the second binding patches, see Fig. 2(top-left), while for TED_g and for Clique it is clearly exponential, see Fig. 2(top-right) and Fig. 2(bottom), respectively.

$RMSD_d$ *values.* Table 2 shows the $RMSD_d$ values and the coverage of the mappings returned by TED_g and by Clique over the 12165 instances that are

---

[2] More precisely: 77 antibody/antigen complexes extracted from the IMGT_3D database (http://www.imgt.org/3Dstructure-DB/); 15 protease/inhibitor complexes coming from [25].

optimally solved by the two methods. The observed $RMSD_d$ for Clique are always smaller than the prescribed distance threshold (2Å). With a median value of 5.87Å, those of TED_g are larger—recall that the quasi-isometry constraint is only guaranteed in-between matched shells. The smaller $RMSD_d$ values of Clique are obtained by atomic matchings that are about twice shorter (median coverage of 42% for TED_g versus 23% for Clique).

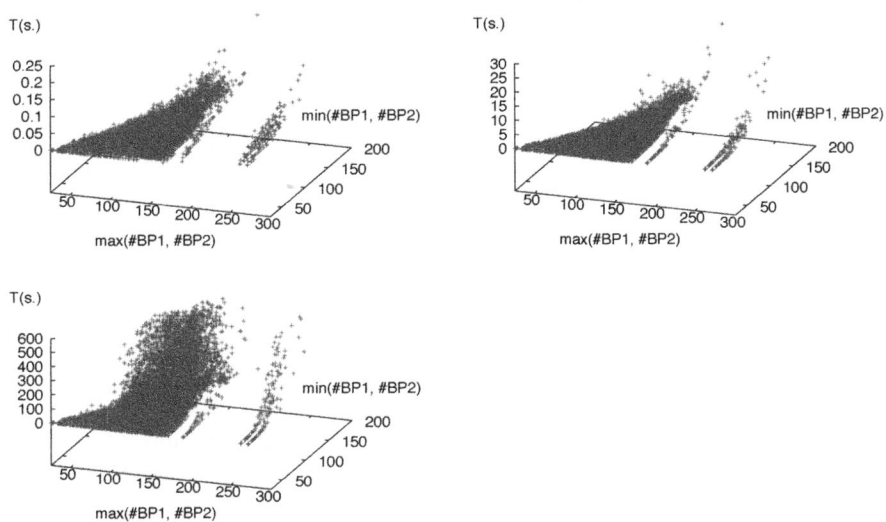

Fig. 2. Running time vs binding patches size. For all the solved instances, the running time of each method is plotted against the size (in atoms) of the largest of the two binding patches (x-axis) and the size of the smallest one (y-axis). Top-left: for TED_t. Top-right: for TED_g. Bottom: for Clique.

5   Discussion and Outlook

Topological versus Geometric Encoding. The TED_t, Clique and TED_g approaches are complementary: TED_t favors the topological comparison of shapes while Clique favors a metric criterion; TED_g strikes a balance. The nestedness of the solutions also allows using TED_t to quickly discard non-similar shapes, with subsequent finer comparisons provided by TED_g and Clique.

Biological Findings. A detailed account of our structural studies is beyond the scope of this paper. Instead, we briefly comment on the questions addressed in a companion paper:

Morphology of Binding Patches. The core-rim model has been central in dissecting protein interfaces [26]. Our shelling tree refines this binary model, and allows one to exhibit typical morphologies of binding patches.

Topology versus Geometry. In a nearby vein, our model allows the identification a patches with similar topology, yet different geometries. Such pieces

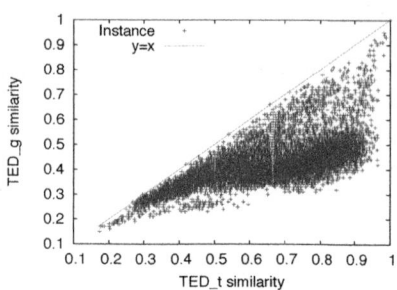

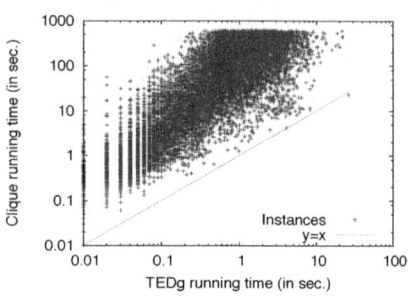

Fig. 4. Running time comparison of TED$_g$ and Clique. The running time (in log scale) of TED$_g$ is plotted against the one of Clique. All instances except one are above the $y = x$ line; TED$_g$ is about about 118 time faster than Clique.

Fig. 3. Illustration of SIM$_g \leq$ SIM$_t$. The similarity of TED$_t$ is plotted against the one of TED$_g$ for the N_s instances solved by the three algorithms.

of informations should prove instrumental to mine putative patches on orphan proteins [27].

Symmetry of Partners in a Complex. Comparing the binding patches of a co-crystallized complex allows us to comment on the symmetry of the interaction in a novel fashion [28].

Protein Families. Finally, our similarity scores can be used to cluster protein families and compare binding patches across such families, a topic barely touched upon at the atomic level [28, 29].

References

1. Gibrat, J.F., Madej, T., Bryant, S.: Surprising similarities in structure comparison. Current Opinion in Structural Biology 6, 377–385 (1996)
2. Gerstein, M., Levitt, M.: Using iterative dynamic programming to obtain accurate pair-wise and multiple alignments of protein structures. In: ISMB 1996 Proceedings, pp. 59–67 (1996)
3. Kabsch, W.: A discussion of the solution for the best rotation to relate two sets of vectors. Acta Crystallographica Section A 34(5), 827–828 (1978)
4. Holm, L., Sander, C.: Protein structure comparison by alignment of distance matrices. Journal of Molecular Biology 223, 123–138 (1993)
5. Godzik, A., Skolnick, J.: Flexible algorithm for direct multiple alignment of protein structures and seequences. CABIOS 10, 587–596 (1994)
6. Malod-Dognin, N., Andonov, R., Yanev, N.: Maximum clique in protein structure comparison. In: International Symposium on Experimental Algorithms, pp. 106–117 (2010)
7. Janin, J., Bahadur, R.P., Chakrabarti, P.: Protein-protein interaction and quaternary structure. Quarterly reviews of biophysics 41(2), 133–180 (2008)
8. Douguet, D.: Ligand-based approaches in virtual screening. Current Computer-Aided Drug Design 4, 180–190 (2008)

9. Brint, A., Willett, P.: Algorithms for the identification of three-dimensional maximal common substructures. J. of Chemical Information and Computer Sciences 27(4) (1987)
10. Norel, R., et al.: Shape Complementarity at Protein-Protein Interfaces. Biopolymers 34 (1994)
11. Ritchie, D., Kemp, G.: Protein docking using spherical polar Fourier correlations. Proteins 39(2) (2000)
12. Katchalski-Katzir, E., et al.: Molecular Surface Recognition: Determination of Geometric Fit Between Proteins and Their Ligands by Correlation Techniques. PNAS 89 (1992)
13. Bille, P.: A survey on tree edit distance and related problems. TCS 337(1-3) (2005)
14. Cazals, F., Pouget, M.: Differential topology and geometry of smooth embedded surfaces: selected topics. Int. J. of Computational Geometry and Applications 15(5) (2005)
15. Siddiqi, K., Shokoufanded, A., Dickinson, S., Zucker, S.: Shock graphs and shape matching. International Journal of Computer Vision 35(1), 13–32 (1999)
16. Chazal, F., Lieutier, A.: The λ-medial axis. Graphical Models 67(4), 304–331 (2005)
17. de Berg, M., van Kreveld, M., Overmars, M., Schwarzkopf, O.: Computational Geometry: Algorithms and Applications. Springer, Berlin (1997)
18. Loriot, S., Cazals, F.: Modeling Macro-Molecular Interfaces with Intervor. Bioinformatics 26(7) (2010)
19. Akkiraju, N., Edelsbrunner, H.: Triangulating the surface of a molecule. Discrete Applied Mathematics 71(1), 5–22 (1996)
20. Castro, P.M.M.D., Cazals, F., Loriot, S., Teillaud, M.: Design of the cgal spherical kernel and application to arrangements of circles on a sphere. Computational Geometry: Theory and Applications 42(6-7), 536–550 (2009)
21. Tarjan, R.E.: Data Structures and Network Algorithms. CBMS-NSF Regional Conference Series in Applied Mathematics, vol. 44. Society for Industrial and Applied Mathematics, Philadelphia (1983)
22. CGAL, Computational Geometry Algorithms Library, http://www.cgal.org
23. Karp, R.: Reducibility among combinatorial problems. Complexity of Computer Computations 6 (1972)
24. Östergård, P.R.J.: A fast algorithm for the maximum clique problem. Discrete Applied Mathematics 120(1-3) (2002)
25. Chen, R., et al.: A protein-protein docking benchmark. Proteins 52 (2003)
26. Chakrabarti, P., Janin, J.: Dissecting protein-protein recognition sites. Proteins 47(3), 334–343 (2002)
27. Jones, S., Thornton, J.: Principles of protein-protein interactions. PNAS 93(1), 13–20 (1996)
28. Keskin, O., Nussinov, R.: Similar binding sites and different partners: Implications to shared proteins in cellular pathways. Structure 15(3), 341–354 (2007)
29. Konc, J., Janezic, D.: ProBiS algorithm for detection of structurally similar protein binding sites by local structural alignment. Bioinformatics 26(9) (2010)

Using Kendall-τ Meta-Bagging to Improve Protein-Protein Docking Predictions

Jérôme Azé[1], Thomas Bourquard[2], Sylvie Hamel[3],
Anne Poupon[4], and David W. Ritchie[2]

[1] Université de Paris-Sud, Laboratoire de Recherche en Informatique, Bâtiment 650
Équipe Bioinformatique – INRIA AMIB group
F-91405 Orsay Cedex, France
aze@lri.fr
[2] INRIA Nancy-Grand Est, LORIA
615 Rue du Jardin Botanique, 54506 Vandoeuvre-lès-Nancy, France
[3] Université de Montréal
C.P. 6128, Succ. Centre-Ville
Montréal, Québec, Canada, H3C 3J7
[4] BIOS group, INRA, UMR85
Unité Physiologie de la Reproduction et des Comportements
F-37380 Nouzilly, France
CNRS, UMR6175, F-37380 Nouzilly, France
Université François Rabelais, 37041 Tours, France

Abstract. Predicting the three-dimensional (3D) structures of macromolecular protein-protein complexes from the structures of individual partners (docking), is a major challenge for computational biology. Most docking algorithms use two largely independent stages. First, a fast sampling stage generates a large number (millions or even billions) of candidate conformations, then a scoring stage evaluates these conformations and extracts a small ensemble amongst which a good solution is assumed to exist. Several strategies have been proposed for this stage. However, correctly distinguishing and discarding false positives from the native biological interfaces remains a difficult task. Here, we introduce a new scoring algorithm based on learnt bootstrap aggregation ("bagging") models of protein shape complementarity. 3D Voronoi diagrams are used to describe and encode the surface shapes and physico-chemical properties of proteins. A bagging method based on Kendall-τ distances is then used to minimise the pairwise disagreements between the ranks of the elements obtained from several different bagging approaches. We apply this method to the protein docking problem using 51 protein complexes from the standard Protein Docking Benchmark. Overall, our approach improves in the ranks of near-native conformation and results in more biologically relevant predictions.

1 Introduction

Many biological processes and structures depend on proteins and their ability to form complexes with other proteins or macromolecules such as DNA, and

M. Loog et al. (Eds.): PRIB 2011, LNBI 7036, pp. 284–295, 2011.
© Springer-Verlag Berlin Heidelberg 2011

RNA, for example. Solving the 3D structures of such macromolecular assemblies thus represents a key step in understanding biological machinery and how malfunctions in this machinery can lead to pathologies.

Although several structural genomic projects have led to advances in solving the structures of protein-protein complexes, this still cannot be done on a high-throughput scale. Therefore, because most proteins can often have several interaction partners [11], *in silico* techniques play a crucial role in studying molecular mechanisms. Here, we focus on the so-called protein docking problem, which aims to predict computationally how two proteins might combine to form a binary protein complex.

Several docking algorithms have been described in the literature. These use various techniques to generate and score candidate conformations [26,10,21,12,23]. Since 2001, the CAPRI experiment (Critical Assessment of PRedicted Interactions) [17] has provided an important forum in which to assess different docking algorithms in an unbiased manner, by allowing blind docking predictions to be compared with unpublished NMR and X-ray structures. Although many docking algorithms are now able to generate high quality candidate orientations, current scoring functions are still not yet sensitive enough to extract the few true positives [18] from amongst the hundreds of thousands of generated incorrect conformations.

In the present work, we introduce a novel method to score protein-protein candidate conformations based on bootstrap aggregation ("bagging") of Voronoi fingerprint representations of candidate docking orientations. We also introduce a "meta-bagging" scoring function which uses an approach based on Kendall-τ distance to minimise the pairwise disagreement between the solutions ranked by several machine learning bagging approaches. This approach is tested on the Protein Docking Benchmark set of Hwang *et al.* [16], consisting of 51 target protein complexes.

2   Methods and Data

In many protein docking algorithms, the initial stage of generating candidate conformations is often one of the most CPU-intensive steps. However, thanks to the tremendous computational power of modern graphics processor units (GPU), the Hex spherical polar Fourier correlation algorithm can generate and score billions of candidates conformations to produce a list of few hundred high quality candidate conformations in just a few seconds [27]. Hence, we chose to use Hex to generate the initial candidates for re-ranking. Additionally, we applied two search constraints in Hex in order to focus candidate docking orientations around the most biologically relevant positions. This ensures that at least one "acceptable" solution according to the CAPRI criteria (i.e. a solution with a root mean squared deviation from the correct X-ray structure of less than 10Å) appears within the top 10 best ranked conformations for 18 out of 51 targets, and within the 50 best ranked solutions for 31 targets. We take 50 solutions to be the maximum practical number to consider for re-ranking because experimental biologists would generally not be prepared to analyse any more than this number.

For the scoring stage, it has been shown previously that by using a coarse-grained Voronoi tesselation representation and evolutionary algorithm, a scoring function may be optimized with satisfying performance [4,9]. In the present work, we use the same formalism for the representation of protein structure and parameter value computations.

2.1   Training Dataset

The Protein Data Bank (PDB) [3] currently contains some 65,000 3D protein structures[1]. However, only a relatively small fraction of these represent different structural fold families. Hence, the structures of some $1,400$ binary hetero-complexes were extracted from the 3D-Complex database [22] for training.

Following [8], we applied some additional constraints to test protein docking methods:

- The 3D structures of the individual partners have to be known in the absence of the other partner.
- Among the heterodimer structures available in PDB, only structures with a chain length greater than 30 amino acids and an X-ray resolution lower than $3.5\mathring{A}$ are retained.
- Non-biological complexes are identified and removed [24].
- The SCOP protein structure classification [25] is used to identify and remove duplicate complexes.
- The 51 complexes of the benchmark were excluded.

This gave a final set of 178 complexes which were treated as positive examples for the learning procedures. Negative examples were generated using the Hex docking algorithm. They correspond to the lowest energy incorrect conformations found by Hex, each having a root mean squared deviation (RMSD) from the native X-ray structure of at least $10\mathring{A}$, and a surface contact area between the two partners of at least $400\mathring{A}^2$. Ten decoys for each native structure were selected randomly and added to the learning set.

2.2   Voronoi Fingerprints and Learning Parameters

Voronoi fingerprint representations of both the positive and negative examples are calculated as follows. For each conformation, we first build a Delaunay triangulation (i.e. the dual of Voronoi tesselation) of the assembly using the CGAL library [7] and we assign one Voronoi node for each amino acid of the constituent proteins. Following [8], the 96 attributes used in learning procedures are: number of residues at the interface, area of the interface, frequencies of the different amino acids at the interface, average volume of the Voronoi cells of the different types of residues at the interface, frequencies of pairs of interacting residues and distances between interacting residues. For the last two parameter types, residues are binned in six categories according to their physico-chemical

---

[1] http://www.rcsb.org/pdb/

properties (large, small, hydrophobic, aromatic, polar, positively and negatively charged). All attributes are computed only for the residues belonging to the core of the interface (i.e. residues of one partner in contact with at least one residue of the other partner and not in contact with the surrounding solvent).

Since a native protein-protein interface typically contains about 30 to 35 amino acid residues with around three or four pair-wise contacts per residue, not all possible pair-wise interactions might appear in a given interface. Hence missing values can pose a problem. We have tested several different approaches to handle missing values (details not shown), and we find that the best approach is to replace any missing values with the maximal value observed in the data set. Once all of the attributes have been assigned values, they are normalised to the range $[0, 1]$ by setting $x_i = \frac{x_i - min_i}{max_i - min_i}$, where min_i and max_i represent the minimum and maximum values observed for attribute i, respectively. Using this representation of the complexes, a set of learning models was built using several state-of-the art machine learning approaches. These are briefly described in Section 2.3.

As the overall goal is to find the best ordered list of candidate docking conformations, the performance of the different approaches were compared using the area under the curve (AUC) of receiver-operator-characteristic (ROC) plots.

2.3 Learning Algorithms

We used five state-of-the-art machine learning approaches, as implemented in Weka [15], to learn models in order to rank the candidate conformations: Naive Bayes, Decision Trees (with Weka command line options: J48, -C 0.25 -M 2), Random Forest (-I 10 -K 0 -S 1), PART Rules (-M 2 -C 0.25 -Q 1), and Ripper Rules (-F 3 -N 2.0 -O 2 -S 1 -P). Because these approaches were designed to learn models that can predict the class associated with each example but not to rank them, we used the probability distribution output by Weka to sort the examples by decreasing probability of being predicted as a positive docking conformation. In cases of ties (equal probabilities for different examples), the examples are ranked from negative to positive.

Bagged models of these five approaches were also learnt. The distribution probabilities of the bagged models, observed on our protein-protein docking dataset, is larger and the associate ranking contains less ties. We also tested a family of ROC-based evolutionary learning functions, using an in-house evolutionary algorithm optimising the area under the ROC curve. The various functions are learned by K independent runs initialized according to different random seeds. It has been shown previously that non-linear functions can out-perform linear functions in several different domains (medical data mining, text mining, etc.), as well as protein docking [4,2]. We therefore focus only on non-linear scoring functions of the form:

$$S_{w,c}(x) = \sum_{i=1}^{d} w_i |x_i - c_i|$$

where $x \in \mathbb{R}^d$ is the current example, $w \in \mathbb{R}^d$ is a weight vector, and c is a center point in $\mathbb{R}^d$. $S_{w,c}$ defines an order on the training set, and the fitness $\mathcal{F}(S_{w,c})$ is defined as the area under the ROC curve associated to the scoring function $S_{w,c}$.

2.4   Bagging

Given m scoring functions, each of which gives different rankings of the same conformations, the "consensus" score for a given example is calculated by medRank or sumRank aggregation of the m lists of scores as:

$$medRank(x) = \underset{S_{w,c} \in \mathcal{S}}{median}\, rank(S_{w,c}(x)), \; sumRank(x) = \sum\nolimits_{S_{w,c} \in \mathcal{S}} rank(S_{w,c}(x))$$

2.5   Finding a Median under the Kendall-τ Distance

Using the above ranking functions, m different ranks can be associated with each example. A good distance measure for combining or comparing the ranked lists from two bagging approaches, which are considered in the following section as permutations, is the Kendall-τ distance [19]. This measure counts the number of pairwise disagreements between the positions of elements in permutations.

More formally, we define the Kendall-τ distance, denoted d_{KT}, as

$$d_{KT}(\pi, \sigma) = \# \{(i,j) : i < j \; and \; ((\pi[i] < \pi[j] \; and \; \sigma[i] > \sigma[j]) \\ or\,(\pi[i] > \pi[j] \; and \; \sigma[i] < \sigma[j]))\} \tag{1}$$

where $\pi[i]$ is the position of integer i in permutation π and $\#\mathcal{A}$ the cardinality of set $\mathcal{A}$.

One way to generate a consensus permutation for a given set of permutations is to find a *median permutation* for the set – i.e. to find a permutation that minimizes the sum of Kendall-τ distances between it and all other permutations in the given set. More formally, let $\mathcal{S}_n$ denoted, as usual, the set of all permutation of $\{1, 2, 3, \ldots, n\}$. Given any set of permutations $A \subseteq \mathcal{S}_n$ and a permutation π, we have $d_{KT}(\pi, A) = \sum_{\sigma \in A} d_{KT}(\pi, \sigma)$. Then, finding a consensus permutation for a given set under the Kendall-τ distance can be stated formally as follow: Given $A \subseteq \mathcal{S}_n$, we want to find a permutation π^* of $\{1, 2, 3, \ldots, n\}$ such that $d_{KT}(\pi^*, A) \leq d_{KT}(\pi, A)$, for all $\pi \in \mathcal{S}_n$

The problem of finding the median of a set of m permutations of $\{1, 2, 3, \ldots, n\}$ under the Kendall-τ distance is a NP-hard problem when $m \geq 4$. However, this problem has been well-studied in recent years, and several good heuristics are known to exist [1,5,6,13,20,28].

In this work, we choose to apply the heuristic presented in [6] to find our medians: Let $A = \{\pi^1, \ldots \pi^m\}$ be a set of m permutations for which we want to find a median. We apply the heuristic of [6] taking either the identity permutation or each $\pi^\ell \in A$, $1 \leq \ell \leq m$ as a starting point, and we take the "median" to be the best result obtained from these $m + 1$ runs.

The idea of this heuristic is to apply a series of *"good"* movements on some starting permutations in order to bring them closer to a median. A movement is considered good if when applying it to a permutation, the Kendall-τ distance of the obtained permutation to the considered set is less than the previous distance. See [6] for more details.

3   Results

Our approach has been evaluated using three-fold cross-validation (and each experiment is repeated 10 times) and has been validated on the learning set. We first compared the performance of the different approaches using the ROC AUC criterion.

Table 1. Total number of complexes with at least one near-native solution in the top N solutions, and number of near-native solutions in the top N solutions (between brackets, total number of near-native solutions is 109) for different values of N. For example, when N=15, for 23 out of 31 complexes, Kendall-τ ranks an acceptable solution 15 or better, and over all the targets, 48 near-native solutions are ranked 15 or better.

top N	Hex	Kendall-τ	sumRank	medRank	b-RF	b-PART	b-JRip	b-J48	b-NaiveBayes
5	13(15)	13(20)	13(18)	9(16)	11(20)	12(19)	11(14)	10(12)	12(14)
10	18(30)	19(37)	20(35)	18(36)	19(36)	17(31)	20(29)	18(27)	13(20)
15	22(39)	23(48)	21(46)	23(51)	23(50)	20(39)	22(41)	23(40)	20(37)
25	26(58)	27(71)	25(67)	26(72)	26(69)	27(60)	27(62)	25(61)	25(64)
35	28(77)	29(90)	28(86)	28(85)	28(86)	29(85)	30(82)	28(78)	28(87)
45	30(96)	31(105)	30(105)	30(105)	31(106)	30(102)	30(102)	29(99)	30(106)

As expected, the bagging versions of the machine learning approaches are always better than the non-bagging versions. Consequently, in the subsequent experiments, only bagging versions will be used. We also found that sumRank (AUC 0.79) and medRank (AUC 0.782) give slightly better results than the Weka bagging approaches.

3.1   Ranking Hex Docking Solutions

For each of the 31 complexes of the benchmark (those complexes for which Hex doesn't place a good solution in the 50 top-ranking solutions were excluded), the 50 top-ranking conformations given by Hex were re-ranked using the Weka algorithms, *sumRank, medRank*. We also used Kendall-τ to bag all the ranks proposed by the seven bagging approaches. When two conformations have the same rank for a given method, these two conformations are ordered by their Hex scores. For each method, we counted the total number of solutions within $10\mathring{A}$ of the native structure in the top N solutions, and the number of complexes for which at least one such solution is found in the top N, with N varying from 1 to 50 (see Table 1).

Evolutionary-based scoring functions increase the number of complexes for which a near-native solution is found in the top N solutions, together with the overall number of near-native solutions ranked N or better. Results show that Kendall-τ meta-bagging method performs as well or better than any other method for every value of N, which makes it the best performing method on the overall test.

The targets of the benchmark have been classified in *Difficult*, *Medium* and *Easy* by its authors depending mostly on the RMSD between the partners composing the complex in their free and bound forms. We have slightly revised this classification by considering as *Difficult* the complexes having more than 2 partners (since our method was trained strictly on heterodimers), and complexes at which interfaces are found large ligands such as FAD or NAD (since such molecules are ignored by our method). Table 2 lists the rank of the first acceptable solution for each method. Results show that on average this rank is better for Kendall-τ than for any other method. Looking only at results obtained for *easy* targets with our method, for all targets an acceptable solution in found in the top 10, and the average rank is 3.7 (against 5.9 for Hex). Finally, for 4 out of these 14 targets, an acceptable solution is found at rank 1, and for 11 out these 14 targets, an acceptable solution is found in the top 5.

3.2   Further Insight on Difficult Targets

We then gave further attention to the 12 targets for which we were unable to rank an acceptable solution in the top 10. Removing the structural redundancy (conformations closer than 5 Å RMSD) allows to place a near-native solution in the top 10 for 3 of these targets (1RV6, 2A9K, 1CLV).

Amongst the remaining 9 complexes, 3 include large non-peptidic molecules at the interface (1LFD, 1R6Q, 2O3B), 1 has more than two subunits (3CPH), and 5 exhibit important fold modifications upon complexation (1JK9, 2A5T, 2I9B, 2IDO, 2Z0E). Although some complexes presenting more than two subunits (1RV6, 2ABZ), or large molecules at the interface (1F6M, 1ZHH, 2A9K, 2G77, 2J7P) can be successfully predicted, our method cannot be reliably used in these cases. Fold modifications are more problematic since they cannot be easily predicted from the structures of the partners alone.

From a biologist's point of view, a docking prediction is useful if it allows the residues lying at the interface and the interactions between those residues to be identified. Inspection of the top 10 conformations of the 9 unsuccessful targets show that for 8 of them, a solution with the two correct epitopes is present in the top 10. Figure 1 shows some biologically interesting conformations of three different complexes which are ranked within the first 10 conformations by the Kendall-τ method. For these complexes, none of the top 10 conformations is near-native. However, inspection of Figure 1 shows that the interacting regions

Table 2. Rank of the first acceptable solution found for each docking target according to each approach

Target	Hex	Kendall-τ	sumRank	medRank	b-RF	b-PART	b-JRip	b-NaiveBayes	b-J48
Difficult and medium targets									
1CLV	2	12	17	12	14	6	12	6	13
1F6M	31	5	8	9	7	2	10	4	20
1JK9	17	17	9	17	7	20	1	36	10
1LFD	33	29	33	30	21	20	20	26	31
1R6Q	1	16	27	23	14	15	17	3	7
1RV6	24	12	4	7	8	21	48	26	50
1ZHH	6	10	3	3	5	31	21	1	29
2A5T	17	45	47	46	43	24	22	46	12
2A9K	1	11	16	12	34	44	10	11	1
2ABZ	8	8	11	13	5	14	4	13	9
2G77	4	2	3	7	1	1	1	3	1
2I9B	11	13	22	12	13	19	8	11	47
2IDO	49	29	38	37	17	16	35	28	12
2J7P	16	7	3	5	8	9	10	13	10
2O3B	12	39	36	37	36	12	7	37	32
2Z0E	37	22	20	14	26	16	27	15	14
3CPH	38	24	28	28	39	26	7	25	3
Easy targets									
1FFW	8	1	1	1	4	2	4	2	9
1FLE	4	1	1	2	10	1	5	5	5
1GL1	3	5	6	8	2	3	4	1	3
1H9D	14	1	1	1	2	1	8	22	3
1JTG	2	4	9	6	1	10	25	3	3
1OC0	9	1	1	2	1	10	1	4	37
1OYV	2	4	3	7	1	1	10	3	10
1SYX	3	4	7	3	17	5	8	20	4
2AYO	11	2	4	6	9	1	11	11	4
2J0T	7	10	7	18	13	46	34	17	7
2OUL	1	2	2	3	7	5	2	25	11
3D5S	2	7	5	6	5	7	1	12	22
3SGQ	3	2	6	6	1	3	2	2	1
4CPA	5	8	4	4	6	1	1	2	8
av.	12,29	11,39	12,32	12,42	12,16	12,65	12,13	13,97	13,81

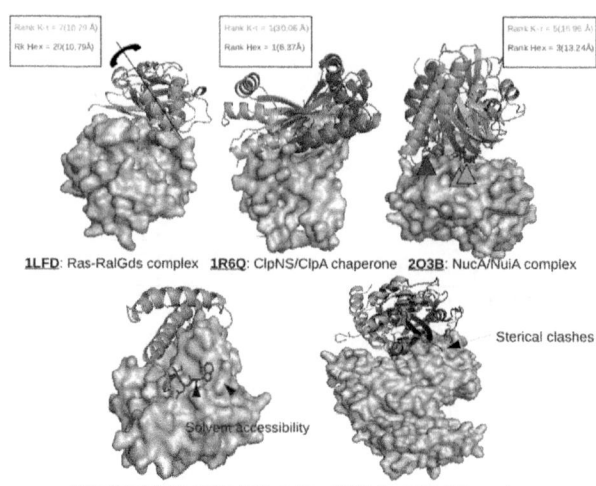

Fig. 1. Three examples for which meta-bagging of Voronoi fingerprints correctly identifies the location of the native interface, although with a relatively high RMSD (wheat) of solutions obtained by the meta-bagging scoring algorithm (wheat) and Hex (blue), compared to the correct native structures (green). The catalytic site residues are indicated in red and are located by colored arrows in order to highlight the position of the biological interface. The ranks of the represented conformations using Kendall-τ and Hex are given, and the corresponding RMSD is shown between brackets. Highest ranking near-native solution found by Hex are shown in blue, and native solution in green.

on both monomers have been correctly identified, and even some of the catalytic residues involved in the interaction have been correctly positioned. The complex between Polymerase III epsilon and the Hot protein (PDB code 2IDO) is also a special case in which we were not able to find a near-native solution. In this complex, important residues of the native interface remain accessible to the solvent instead of being occluded by the docking partner.

As shown in Figure 2, Hex is able to rank in the top 20 a near-native conformation of the NR1/NR2A complex, which plays an crucial role into mammalian central nervous system [14], whereas this conformation is ranked 45 by the Kendall-τ meta-bagging method. While the superposition of this solution with the native structure is generally satisfying, closer inspection of the interface moderates this conclusion. More specifically, the ion channel activity of this complex is regulated by Van der Waals contacts between residues Y535 and P527. In the conformation proposed by Hex, this interaction is lost. However, this crucial interaction is correctly predicted in another conformation, ranked 9 by the Kendall-τ method, even though the global RMSD is rather high.

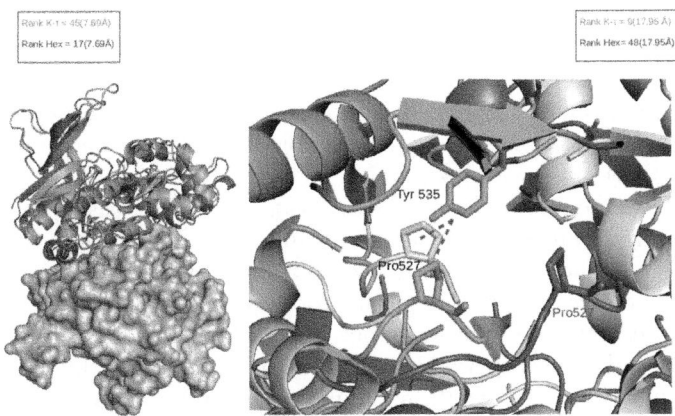

2A5T: The NR1/NR2A ligand-binding cores complex

Fig. 2. NR1/NR2A complex (PDB code 2A5T). Left: superposition of native structure (green) and near-native structure given by Hex (ranked 17 by Hex and 45 by Kendall-τ meta-bagging method). Right: zoom on the core of the interface, including a conformation ranked 9 by the Kendall-τ meta-bagging algorithm (wheat) and 48 by Hex. The key interaction between tyrosine (TYR535) and Proline (PRO527) is represented. TYR 535 belongs to the A chain of the complex/conformations and have thus been represented only once, since these chains are identical and superposed for all conformations.

4   Conclusion

Despite the fact that the Weka machine learning approaches used in this work were designed to separate examples into positive and negative instances rather than to rank them, for the current learning set, we find that using probabilities for ranking gives as good results as those achieved by the evolutionary ROC-based learning approaches. The different machine learning approaches let us find at least one acceptable solution within the top 10 for all easy targets.

The improvement of the rank of the first quasi-native conformation can seem modest. However, it should be borne in mind that this method is intended to be used by experimental biologists as a guide for experimentation. Laboratory-based experiments can be costly and time-consuming. Therefore, reducing the number of experiments is a crucial objective, and will give the method an effective practical interest. Our method not only guarantees that a near-native conformation is present in the top 10 solutions, it even places a near native solution in the top 5 for 11 out of 14 easy targets. For the experimenter, this means that the number of experiments can be cut by half with a risk of "losing" predictions for only 3 targets. Moreover, we show that the conformations selected by our method, even when their structural similarity with the native conformation is low, have a real biological relevance. Our method also ranks a near-native conformation in the top 10 for 8 of the 17 difficult targets, and a conformation with the correct epitopes for 8 of the remaining 9 targets. For the experimenter, this means that considering the top 10 conformations guarantees having the correct epitopes (which is sufficient for directed mutagenesis for example) for 30 of the 31 targets. However, the heterogeneity observed in some of the individual predictions means it is not obvious how to reach the best overall ranking from the different approaches.

The problem we aimed to address in this paper is how best to aggregate such lists of ranks from an ensemble of ranking functions. We propose that the notion of Kendall-τ distance provides one way to achieve this goal. The driving heuristic in the Kendall-τ approach is to find good cyclic movements of ranks towards the median permutation in order to obtain a consensus ranking which minimises the Kendall-τ distance between the individual rankings. Thus, the Kendall-τ approach quickly dominates all of the individual approaches tested here by aggregating the best predictions from each of them. The present study focused on the 50 best conformations found by Hex because the Kendall-τ approach is currently limited to work with permutations containing the same elements. Lifting this limitation would allow the Kendall-τ approach to focus on the best predictions from each learning algorithm, either directly by comparing Kendall-τ distances or, as was done here, by finding the optimal combination of predictions from the independent bagging algorithms.

References

1. Ailon, N., Charikar, M., Newman, A.: Aggregating inconsistent information: Ranking and clustering. J. ACM 55(5) (2008)
2. Azé, J., Roche, M., Sebag, M.: Bagging evolutionary roc-based hypotheses application to terminology extraction. In: Proceedings of ROCML (ROC Analysis in Machine Learning), Bonn, Germany (2005)
3. Berman, H., Westbrook, J., Feng, Z., Gilliland, G., Bhat, T., Weissing, H., Shindyalov, I., Bourne, P.: The protein data bank. Nucleic Acids Research 28, 235–242 (2000)
4. Bernauer, J., Azé, J., Janin, J., Poupon, A.: A new protein-protein docking scoring function based on interface residue properties. Bioinformatics 5(23), 555–562 (2007), http://bioinformatics.oxfordjournals.org/cgi/content/full/23/5/555
5. Betzler, N., Fellows, M.R., Guo, J., Niedermeier, R., Rosamond, F.A.: Fixed-parameter algorithms for kemeny scores. In: Fleischer, R., Xu, J. (eds.) AAIM 2008. LNCS, vol. 5034, pp. 60–71. Springer, Heidelberg (2008)
6. Blin, G., Crochemore, M., Hamel, S., Vialette, S.: Median of an odd number of permutations. Pure Mathematics and Applications 21(2), 161–175 (2011)
7. Boissonnat, J.D., Devillers, O., Pion, S., Teillaud, M., Yvinec, M.: Triangulations in CGAL. Comput. Geom. Theory Appl. 22, 5–19 (2002)
8. Bourquard, T., Bernauer, J., Azé, J., Poupon, A.: Comparing Voronoi and Laguerre tessellations in the protein-protein docking context. In: Sixth International Symposium on Voronoi Diagrams (ISVD), pp. 225–232 (2009)
9. Bourquard, T., Bernauer, J., Azé, J., Poupon, A.: A Collaborative Filtering Approach for Protein-Protein Docking Scoring Functions. PLoS ONE 6(4), e18541 (2011), doi:10.1371/journal.pone.0018541

10. Camacho, C.: Modeling side-chains using molecular dynamics improve recognition of binding region in capri targets. Proteins 60(2), 245–251 (2005)
11. Devos, D., Russell, R.B.: A more complete, complexed and structured interactome. Current Opinion in Structural Biology 17, 370–377 (2007)
12. Dominguez, C., Boelens, R., Bonvin, A.: HADDOCK: a protein-protein docking approach based on biochemical or biophysical information. J. Am. Chem. Soc. 125(7), 1731–1737 (2003)
13. Dwork, C., Kumar, R., Naor, M., Sivakumar, D.: Rank aggregation methods for the Web. In: World Wide Web, pp. 613–622 (2001), http://citeseerx.ist.psu.edu/viewdoc/summary?doi=10.1.1.28.8702
14. Furukawa, H., Singh, S., Mancusso, R., Gouaux, E.: Subunit arrangement and function in nmda receptors. Nature 438(7065), 185–192 (2005)
15. Hall, M., Frank, E., Holmes, G., Pfahringer, B., Reutemann, P., Witten, I.H.: The weka data mining software: An update. SIGKDD Explorations 11(1), 10–18 (2009)
16. Hwang, H., Vreven, T., Janin, J., Weng, Z.: Protein-protein docking benchmark version 4. 0. Proteins 78(15), 3111–3114 (2010)
17. Janin, J., Henrick, K., Moult, J., Eyck, L., Sternberg, M., Vajda, S., Vakser, I., Wodak, S.: CAPRI: a Critical Assessment of PRedicted Interactions. Proteins 52(1), 2–9 (2003)
18. Janin, J., Wodak, S.: The Third CAPRI Assessment meeting. Structure 15, 755–759 (2007)
19. Kendall, M.: A New Measure of Rank Correlation. Biometrika 30(1/2), 81–93 (1938), http://www.jstor.org/stable/2332226
20. Kenyon-Mathieu, C., Schudy, W.: How to rank with few errors. In: Johnson, D., Feige, U. (eds.) STOC, pp. 95–103. ACM (2007)
21. Komatsu, K., Kurihara, Y., Iwadate, M., Takeda-Shikata, M.: Evaluation of the third solvent clusters fitting procedure for the prediction of protein-protein interactions based on the results at the capri blind docking study. Proteins 52(1), 15–18 (2003)
22. Levy, E., Pereira-Leal, J., Chothia, C., Teichmann, S.: 3d complex: a structural classification of protein complexes. PLoS Comput. Biol. 2(11) (2006)
23. Mihalek, J., Res, I., Lichtarge, O.: A structure and evolution-guided monte carlo sequence selection strategy for multiple alignement-based analysis of proteins. Bioinformatics 22(2), 149–156 (2006)
24. Mintseris, J., Weng, Z.: Atomic contact vectors in protein-protein recognition. Proteins 53(3), 214–216 (2003)
25. Murzin, A., Brenner, S., Hubbard, T., Chothia, C.: Scop: a structural classification of proteins database for the investigation of sequences and structures. J. Mol. Biol. 247, 536–540 (1995)
26. Ritchie, D.: Evaluation of protein docking predictions using hex 3.1 in capri rounds 1 and 2. Proteins 52(1), 98–106 (2003)
27. Ritchie, D., Venkatraman, V.: Ultra-fast FFT protein docking on graphics processors. Bioinformatics 26(19), 2398–2405 (2010)
28. van Zuylen, A., Williamson, D.P.: Deterministic algorithms for rank aggregation and other ranking and clustering problems. In: Kaklamanis, C., Skutella, M. (eds.) WAOA 2007. LNCS, vol. 4927, pp. 260–273. Springer, Heidelberg (2008)

PAAA: A Progressive Iterative Alignment Algorithm Based on Anchors

Ahmed Mokaddem and Mourad Elloumi

Research Unit of Technologies of Information and Communication,
Higher School of Sciences and Technologies of Tunis, Tunisia
moka.ahmed@yahoo.fr,
Mourad.Elloumi@fsegt.rnu.tn

Abstract. In this paper, we present a new iterative progressive algorithm for *Multiple Sequence Alignment* (MSA), called *Progressive Iterative Alignment Algorithm Based on Anchors* (PAAA). Our algorithm adopts a new distance, called *anchor distance* to compute the distance between two sequences, and a variant of the UPGMA method to construct a guide tree.

PAAA is of complexity $O(N^4 + N * L^2)$ in computing time, where N is the number of the sequences and L is the length of the longest sequence.

We benchmarked PAAA using different benchmarks, e.g., BALIBASE, HOMSTRAD, OXBENCH and BRALIBASE, and we compared the obtained results to those obtained with other alignment algorithms, e.g., CLUSTALW, MUSCLE, MAFFT and PROBCONS, using us criteria the *Column Score* (CS) and the *Sum of Pairs Score* (SPS). We obtained good results for protein sequences.

Keywords: Multiple sequence alignment, progressive alignment, algorithms, complexities, distances.

1 Introduction

Multiple Sequences Alignment (MSA) is a good way to compare biological sequences. It consists in optimizing the number of matches between the residues occurring in the same order in each sequence. MSA is an NP-complete problem [22]. There are several approaches to solve this problem. *Progressive approach* is the most used and the most effective one, it operates in three steps:

1. Pairwise comparisons: during this step, we compute a distance between every pair of sequences. We can compute this distance directly, i.e., without constructing pairwise alignments, or indirectly, i.e., after the pairwise alignments. The goal of this step is to estimate the similarity between pairs of sequences in order to distinguish the sequences that are the first to be aligned. The distances computed between all pairs of sequences are stored in a symmetric diagonal matrix, called *distance matrix*. Among used distances, we mention *percent identity*, *percent of similarity* [19], *Kimura distance* [9], *k-mer distances* [2] and *normalized score* [23].

M. Loog et al. (Eds.): PRIB 2011, LNBI 7036, pp. 296–305, 2011.

2. Sequences clustering: during this step, we define the branching order of the sequences by construction a *guide tree*. Clustering method uses the distance matrix constructed in the previous step. The most used methods are UP-GMA [16] and *Neighbor-Joining* [15].

3. Sequences integration: during this step, we construct the alignment of sequences using the branching order. We mention *aligning alignment approach* [23] and *profile-profile alignment approach* [6]. The last one is the most used method.

Among progressive alignment algorithms, we mention CLUSTALW [18], TCOF-FEE [12], GRAMALIGN [14] and KALIGN [11]. A Progressive approach can apply a refinement step. Refinement step is used in order to enhance the multiple alignment scores. In fact, a set of operations can be applied iteratively on the alignment under convergence, i.e., the score of the alignment obtained at iteration N is less than the score of the alignment obtained at iteration N-1. Different strategies of refinement have been developed [21]. These algorithms are called *hybrid algorithms* or *iterative progressive alignment algorithms* and are almost the most efficient. Among hybrid alignment algorithms, we mention MUSCLE [2], PROBCONS [1], MAFFT [8], PRRP [7] and MSAID [24].

In this paper, we present a new iterative progressive alignment algorithm, called PAAA. The main differences between our algorithm and other progressive ones consist in:

– First, the use of a new distance, called *anchor distance*, in the pairwise comparisons step.
– Then, the use of the *maximal distance* instead of the *minimal distance*, in the construction of the *guide tree*, in the sequences clustering step.

PAAA implements also a refinement step. We benchmarked PAAA using different benchmarks, e.g., BALIBASE, HOMSTRAD, OXBENCH and BRAL-IBASE, and we compared the obtained results to those obtained with other alignment algorithms, e.g., CLUSTALW, MUSCLE, MAFFT and PROBCONS, using us criteria the *Column Score* (CS) and the *Sum of Pairs Score* (SPS). We obtained good results for protein sequences.

The rest of the paper is organized us follows. In section 2, we present some preliminaries. In section 3, we present PAAA and in section 4, we compute its time complexity. In section 5, we present an illustrative example. In section 6, we present the experimental results obtained with PAAA using different benchmarks and we compare the obtained results with those of other alignment algorithms. Finally, in section 7, we present our conclusion and give some perspectives.

2   Preliminaries

Let A be a finite alphabet, a *sequence* is an element of A^*, it is a concatenation of elements of A. The length of a sequence w, denoted by $|w|$, is the number of the characters that constitute this sequence. A portion of w beginning at the position i and ending at the position j, $0 \leq i \leq j \leq |w|$, is called *subsequence*.

Let $f = \{w_1, w_2, \ldots, w_N\}$ be a family of sequences, we say that a subsequence x is an *anchor* to the sequences of f, if and only if, x is a subsequence of every sequence of f.

Let f be a set of sequences, a list of anchors that appears in the same order, without overlapping in all the sequences of f will be denoted by *AC(f)*.

Let f be a set of sequences, *aligning* the sequences of f consists in optimizing the number of matches between the characters occurring in the same order in each sequence. When $|w| \succ 2$, aligning the sequences of f is called *Multiple Sequence Alignment* (MSA).

A *profile* is a sequence that represents an alignment. It is constructed by selecting for each column of the alignment the residue that has the maximum occurrences in this column.

A *guide tree* is a binary tree where a leaf represents a sequence, a node represents a group of sequences, and therefore the corresponding alignment. The nodes define the branching order and the root defines the final alignment.

The *anchor distance* between two sequences w and w' is calculated using the following formula:

$$D(w, w') = 1 - \frac{\sum_{a \in AC(\{w, w'\})} |a|}{min(|w|, |w'|)} \tag{1}$$

Selected anchors are fundamental in the computation of the anchor distance. In our case, we select the longest common anchors. Hence, the more two sequences have longer anchors, the more these sequence are similar.

3   Algorithm

Let f be a set of sequences, by adopting PAAA, we operate as follows:

- During the first step, we compute the anchor distance between each pair of sequences of f and store the computed distances in a distance matrix M. The computation of the anchor distance between a pair of sequences can be done by adopting our algorithm described in [3].
- During the second step, we use the distance matrix M to build the guide tree T. This is done by adopting a variant of UPGMA [16] method. We made some modifications to the UPGMA method as follows : First, we select the maximum value from the distance matrix M, instead of the minimum one, and we align the corresponding sequences. Then, we align the aligned sequences with another one following the branching order of the guide tree. We reiterate this process until we obtain a MSA of all the sequences. We adopt two dynamic programming algorithms, the first one [5] for aligning two sequences, and the second one [6] for aligning two profiles. The profile of two alignments is constructed using the profile of each alignment.
- Finally, during the last step, we make a refinement to the resulting alignment in order to correct the mistakes that could have been produced in the progressive approach. The method adopted is the one of algorithm MUSCLE [2]. We operate as follow: First, we use the MSA obtained to construct

a new guide tree using a new distance matrix. Then, we remove an edge from the new guide tree to obtain two new guide subtrees. Next, we align the sequences in each guide subtree. Finally, we align the two alignments associated with the two guide subtrees to obtain a global MSA. We repeat this step until convergence, i.e., no improvement is observed on the MSA obtained during the last iteration, compared with the MSA obtained during the previous one.

4   Time Complexity

In this section, we study time complexity of our algorithm PAAA. Let N be the number of the sequences and L be the length of longest sequence.

- Time complexity of the first step is $O(N^2 * L^2 * log(L))$. In fact, we compute the anchor distance between every pair of sequences by adopting our algorithm, described in [3], that is of complexity $O(L^2 * log(L))$ in computing time. We have $O(N^2)$ anchor distances to compute then time complexity of the first step is $O(N^2 * L^2 * log(L))$.
- Time complexity of the second step is $O(N^2)$. Indeed, the construction of the guide tree requires a running time of $O(N^2)$ [16].
- Finally the refinement step requires a time complexity of $O(N^4 + N * L^2)$ [2]. Hence our algorithm PAAA is of complexity $O(N^4 + N * L^2)$ in computing time.

5   An Illustrative Example

Let f be a set of sequences $f = \{w_1, w_2, w_3, w_4, w_5\}$
   w_1: seqvecccce; w_2: stqesmcedd; w_3: tescd; w_4: teqvenced; w_5: tescmcccctes
We compute anchor distance between every pair of sequences:
$D(w_1, w_2) = 0{,}5$; $D(w_2, w_3) = 0{,}6$; $D(w_3, w_4) = 0{,}2$; $D(w_4, w_5) = 0{,}56$
$D(w_1, w_3) = 0{,}6$; $D(w_2, w_4) = 0.34$; $D(w_3, w_5) = 0{,}2$;
$D(w_1, w_4) = 0.34$; $D(w_2, w_5) = 0{,}4$;
$D(w_1, w_5) = 0{,}4$;.
We store these distances in the distance matrix M

Table 1. Distance matrix

	w_1	w_2	w_3	w_4	w_5
w_1					
w_2	50				
w_3	60	0			
w_4	34	34	20		
w_5	40	40	20	56	

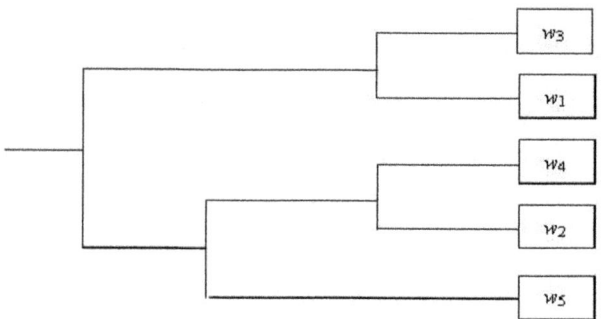

Fig. 1. Guide Tree

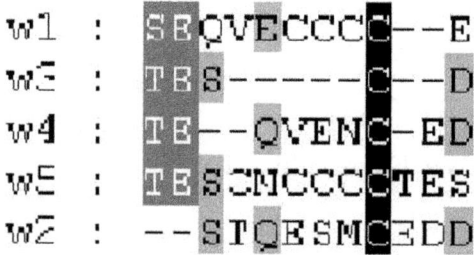

Fig. 2. MSA

Finally, we align the sequences following the branching order. Hence we obtain the MSA in Fig.2..

Figure 2 was obtained thanks to GeneDoc [25].

6   Experimental Study

In order to asses PAAA performances; we used different benchmark for protein and DNA sequences. Each benchmark contains multiple alignments that have been constructed in a manual, or automatic, way with the help of biologists and have been grouped in different categories, according to the nature of the set of the sequences.

These alignments are considered as *reference alignments*. We used the following benchmarks: BALIBASE [20], OXBENCH [13], HOMSTRAD [17] and BRALIBASE II [4] for DNA sequences.

We compared the results of PAAA with other multiple alignment programs that are the most widely used and the most efficient namely, CLUSTALW2 [18], MUSCLE [2], MAFFT [8] and PROBCONS [1]. In order to compare alignments, we use two criteria:

- CS: The *Column Score* [19] :

$$CS = \sum \frac{C_i}{L} :$$ (2)

Where, $C_i=1$ if all the residues in the test aligned are aligned in the reference alignment otherwise $C_i=0$ and L represents the total number of column in the *core blocks*.

- SPS: The *Sum of Pairs Score* [19] :

$$SPS = \frac{\sum_1^M S_i}{\sum_1^{M_r} S_r}$$ (3)

Where S_i is the pairs of residues correctly aligned in the column i, M is the number of columns in test alignment, S_r is the residue in the reference alignment and M_r is the number of columns in the reference alignment.

We ran PAAA on datasets from the BALIBASE benchmark. In order to compute the SPS and the CS, we used the *bali_score* program, a program provided with the BALIBASE benchmark, that generates these scores based on predetermined reference alignments. For OXBENCH and HOMSTRAD benchmarks, we computed the Q-scores and TC-scores, using the *Q-score* program. TC score and Q-score correspond respectively to the CS and SPS of the BALIBASE benchmark. The results of CS and SPS obtained with a dataset from BALIBASE are respectively represented in Table 2 and Table 3.

Table 2. Results obtained with BALIBASE using CS

Dataset	CLUSTALW2	MUSCLE	MAFFT	PROBCONS	PAAA
1DOX(REF1)	0,919	0,860	0.860	0.907	0,950
1PYSA(REF1)	0.922	0.920	0.910	0.930	0.940
1ZIN(REF1)	0.920	0,970	0,920	0,940	0,980
1UBI(REF2)	0.482	0.120	0.260	0.380	0,210
1CSY(REF2)	0,110	0,240	0,110	0,110	0,290
1ABOA(REF2)	0.650	0.230	0.330	0.330	0.420
1VLN(REF4)	0.879	0.710	0.700	0.800	0.770
KINASE1(REF4)	0.000	0.650	0.710	0.750	0.490
1DYNA(REF4)	0.000	0.090	0.130	0.090	0.180
1PYSA(REF4	0.000	0.350	0.360	0.360	0.330
1UKY (REF3)	0,190	0,180	0,170	0,190	0,210

Table 3. Results obtained with BALIBASE using SPS

Dataset	CLUSTALW2	MUSCLE	MAFFT	PROBCONS	PAAA
1DOX(REF1)	0.914	0.924	0.920	0.907	0.969
1PYSA(REF1)	0.957	0.930	0.931	0.952	0.952
1ZIN(REF1)	0,960	0,985	0,960	0,965	0,987
1UBI(REF2)	0.822	0.805	0.791	0.843	0.754
1CSY(REF2)	0,826	0,835	0,770	0,809	0,828
1VLN(REF4)	0.925	0.907	0.901	0.963	0.947
KINASE1(REF4)	0.869	0.862	0.875	0.908	0.802
1DYNA(REF4)	0.226	0.396	0.517	0.578	0.461
1PYSA(REF4)	0.555	0.667	0.675	0.673	0.646
1ABOA(REF2)	0.825	0.814	0.804	0.804	0.787
1UKY(REF3)	0,529	0,529	0,581	0,566	0,571

We obtained more good results for CS than SPS thanks to the anchors selection step. In fact, we selected exactly repeated subsequences and hence promoted the CS.

We also benchmarked PAAA on a dataset from OXBENCH and we compared obtained TC scores and Q-scores with those of typical softwares CLUSTALW, MUSCLE and MAFFT. OXBENCH uses TC score and Q-score. We obtain the following results that represent the average of TC scores and Q-scores of a set of 100 datasets from OXBENCH.

This table shows that PAAA returns the best TC score and Q score for the dataset from OXBENCH. We assessed also PAAA in on set of families extracted from HOMSTRAD database and we obtained the following results.

Table 4. Results obtained with OXBENCH using TC score and Q-score

PROGRAM/SCORES	Nbr of Dataset	TC	Q
MUSCLE	100	0,693	0,828
MAFFT	100	0,687	0,818
CLUSTALW	100	0,682	0,819
PAAA	100	0,706	0,832

Table 5. Results obtained with HOMSTRAD database using TC score and Q score

FAMILY/ALGORITHM	PAAA		MAFFT		CLUSTALW		MUSCLE	
SCORES	Q	TC	Q	TC	Q	TC	Q	TC
ornithine carbamoyltransferase	0,939	0,872	0,937	0,861	0,931	0,849	0,927	0,84
cryst	0,876	0,613	0,495	0,0166	0,874	0,657	0,652	0,021
C-typelectin	0,867	0,662	0,83	0,628	0,838	0,697	0,862	0,634
HLH	0,937	0,869	0,824	0,753	0,785	0,684	0,931	0,869
Nucleotidekinase	0,897	0,699	0,849	0,541	0,849	0,533	0,869	0,618
LUXS	0,911	0,848	0,892	0,845	0,913	0,869	0,881	0,797
Peroxidase	0,899	0,736	0,866	0,67	0,885	0,713	0,896	0,734

Table 5 shows that for this set of families extracted from HOMSTRAD benchmark our algorithm have the best scores compared to other typical ones.

We also ran PAAA on reference families from BRALIBASE II benchmark of DNA sequences. For DNA sequences, we obtained less good results than for protein. As we can see in Table 6 and Table 7, these tables represent respectively the average of CS and SPS.

These tables show that PAAA results are better than those of DIALIGN, POA and T-COFFEE. In addition, for TRNA and G2intron families PAAA is better than CLUSTAL W2, but gives less good scores than MUSCLE, MAFFT and PROBCONS.

Table 6. Results obtained with SPS on BRALIBASE II (DNA)

Method(DNA)	G2In	Rrna	SRP	tRNA	U5
PAAA	0,747	0,917	0,854	0,855	0,774
DIALIGN2.2	0,717	0,899	0,815	0,786	0,762
CLUSTALW2	0,727	0,933	0,874	0,870	0,796
T-COFFEE5.56	0,738	0,909	0,839	0,817	0,791
POAV2	0,672	0,889	0,855	0,769	0,773
MAFFT6.240L-INSi	0,789	0,939	0,875	0,918	0,828
MAFFT6.240E-INSi	0,774	0,938	0,872	0,906	0,805
MUSCLE3.7	0,764	0,940	0,871	0,873	0,797
PROBCONSRNA1.10	0,801	0,945	0,881	0,926	0,848

Table 7. Results obtained with CS on BRALIBASE II (DNA)

Method(DNA)	G2In	Rrna	SRP	tRNA	U5
PAAA	0,607	0,848	0,746	0,769	0,614
DIALIGN2.2	0,609	0,811	0,685	0,676	0,601
CLUSTALW2	0,612	0,867	0,766	0,762	0,651
T-COFFEE5.56	0,602	0,826	0,716	0,692	0,629
POAV2	0,552	0,804	0,738	0,660	0,616
MAFFT6.240L-INSi	0,652	0,875	0,768	0,846	0,685
MAFFT6.240E-INSi	0,638	0,873	0,766	0,833	0,657
MUSCLE3.7	0,632	0,880	0,766	0,780	0,643
PROBCONSRNA1.10	0,687	0,886	0,776	0,855	0,717

The obtained alignments for families of proteins and families of DNA are good, compared to those of other programs. Let us notice that the results obtained for families of proteins are better than those obtained for families of DNA thanks to the selected anchors step. In fact, the selected anchors have great influence on the distances computation. Indeed, the probability of having repeated subsequences in a DNA sequence is higher than the one of having repeated subsequences in proteins sequence. On the other hand, the more there are occurrences representing the same repeated subsequence in a sequence the more PAAA is likely to be induced in error, because of the choice of the occurrence associated with the repeated subsequence. This is why we obtain better results with proteins compared to DNA.

7   Conclusions and Perspectives

In this paper, we have presented a new iterative progressive alignment algorithm called PAAA. This algorithm uses a new distance called *anchor distance*, in the sequence comparison step, and a new variant of UPGMA method, in the guide tree construction step. This variant uses the maximum distance from the distance matrix. PAAA makes also a refinement step. We assessed PAAA on different benchmarks and we compared its performances with other typical multiple alignment programs, using the SPS and the TC score. We obtained good results. Indeed, in several cases, for protein sequences, we obtained better scores than those of typical multiple alignment programs, e.g., MUSCLE, MAFFT, CLUSTALW and PROBCONS.

As a future work, we plan to improve PAAA by developing new methods for sequences classification and for profile-profile alignments. Finally, we intend to improve the iterative step.

References

1. Do, C.B., Mahabhashyam, M.S., Brudno, M., Batzoglou, S.: PROBCONS: Probabilistic Consistency-Based Multiple Sequence Alignment. Genome Res. 15, 330–340 (2005)
2. Edgar, R.C.: MUSCLE: Multiple Sequence Alignment with high Accuracy high Throughput. Nucleic Acids Research 32, 1792–1797 (2004)
3. Elloumi, M., Mokaddem, A.: A Heuristic Algorithm for the N-LCS Problem. Journal of the Applied Mathematics, Statistics and Informatics (JAMSI) 4, 17–27 (2008)
4. Gardner, P.P., Wilm, A., Washiet, S.: A benchmark of multiple sequence alignment programs upon structural RNAs. Nucleic Acids Research 33, 2433–2439 (2005)
5. Gotoh, O.: An Improved Algorithm for Matching Biological Sequences. J. Mol. Biol. 162, 705–708 (1982)
6. Gotoh, O.: Further Improvement in Methods of Group-to-Group Sequence Alignment with Generalized Profile Operations. CABIOS 1, 379–387 (1994)
7. Gotoh, O.: Significant improvement in accuracy of multiple protein sequence alignments by iterative refinement as assessed by reference to structural alignments. J. Mol. Biol. 264, 823–838 (1996)
8. Katoh, K., Kuma, K., Toh, H., Miyata, T.: MAFFT version 5: Improvement in Accuracy of Multiple Sequence Alignment. Nucleic Acids Res. 33, 511–518 (2005)
9. Kimura, M.: The Neutral Theory of Molecular Evolution. Cambridge University Press, Cambridge (1983)
10. Lassman, T., Frings, O., Sonnhammer, L.L.: KALIGN2: High-Performance Multiple Alignment of Protein and Nucleotide Sequences Allowing External Features. Nucleic Acids Research 37, 858–865 (2009)
11. Lassman, T., Sonnhammer, L.L.: KALIGN: An Accurate and Fast Multiple Sequence Alignment Algorithm. BMC Bioinformatics 6 (2005)
12. Notredame, C., Higgins, D., Heringa, J.: T-COFFEE: A novel mMthod for Multiple Sequence Alignments. J. Mol. Biol. 302, 205–217 (2000)
13. Raghava, G.P., Searle, S.M., Audley, P.C., Barber, J.D., Barton, G.J.: OXBENCH: a Benchmark for Evaluation of Protein Multiple Sequence Alignment Accuracy. BMC Bioinformatics 4 (2003)

14. Russell, D.J., Out, H.H., Sayood, K.: Grammar-Based Distance in Progressive Multiple Sequence Alignment. BMC Bioinformatics 9 (2008)
15. Saitou, N., Nei, M.: The Neighbor-Joining Method: A New Method for Reconstructing Phylogenetic Trees. Mol. Biol. E 4, 406–425 (1987)
16. Sneath, P., Sokal, R.: Numerical Taxonomy, pp. 230–234. Freeman, San Francisco (1973)
17. Stebbings, L.A., Mizuguchi, K.: HOMSTRAD: Recent Developments of the Homologous Protein Structure Alignment Database. Nucleic Acids Research 32, 203–207 (2004)
18. Thompson, J.D., Higgins, T.J., Gibson, D.G.: CLUSTALW: Improving the Sensitivity of Progressive Multiple Sequence Alignment through Sequence Weighting, Position Specific Gap Penalties and Weight Matrix Choice. Nucleid Acids Research 22, 4673–4680 (1994)
19. Thompson, J.D., Plewniak, F., Poch, O.: A Comprehensive Comparison of Multiple Sequence Alignment Programs. Nucleic. Acids. Res. 27, 2682–2690 (1999)
20. Thompson, J.D., Plewniak, F., Poch, O.: BAliBASE: a Benchmark Alignment Database for the Evaluation of Multiple Alignment Programs. Bioinformatics 15, 87–88 (1999)
21. Wallace, I.M., O'Sullivan, O., Higgins, D.G.: Evaluation of Iterative Alignment Algorithms for Multiple Alignments. Bioinformatics 21, 1408–1414 (2005)
22. Wang, L., Jiang, T.: On the complexity of multiple sequence alignment. J. Comput. Biol. 1, 337–348 (1994)
23. Wheeler, T.J., Kececioglu, J.D.: Multiple alignment by aligning alignments. Bioinformatics 23, 559–568 (2007)
24. Min, Z., Weiwu, F., Junhua, Z., Zhongxian, C.: MSAID: Multiple Sequence Alignment Based on a Measure of Information Discrepancy. Computational Biology and Chemistry 29, 175–181 (2005)
25. Nicholas, K.B., Nicholas, K.B., Deerfield, D.W.: GeneDoc: Analysis and Visualization of Genetic Variation. Embnew News 4 (1997)

Heat Diffusion Based Dissimilarity Analysis for Schizophrenia Classification

Aydın Ulaş[1,*], Umberto Castellani[1], Vittorio Murino[1,2],
Marcella Bellani[3], Michele Tansella[3], and Paolo Brambilla[4]

[1] University of Verona, Department of Computer Science, Verona, Italy
[2] Istituto Italiano di Tecnologia (IIT), Genova, Italy
[3] Department of Public Health and Community Medicine, Verona, Italy
[4] IRCCS "E. Medea" Scientific Institute, Udine, Italy

Abstract. We apply shape analysis by means of heat diffusion and we show that dissimilarity space constructed using the features extracted from heat diffusion present a promising way of discriminating between schizophrenic patients and healthy controls. We use 30 patients and 30 healthy subjects and we show the effect of several dissimilarity measures on the classification accuracy of schizophrenia using features extracted by heat diffusion. As a novel approach, we propose an adaptation of random subspace method to select random subsets of bins from the original histograms; and by combining the dissimilarity matrices computed by this operation, we enrich the dissimilarity space and show that we can achieve higher accuracies.

Keywords: heat diffusion, schizophrenia, dissimilarity space, support vector machines, random subspace.

1 Introduction

Recently, it has been popular to use magnetic resonance imaging (MRI) based image analysis to quantify morphological characteristics of different brains [11,1]. The goal of these studies is to distinguish normal subjects with patients affected by a certain disease.

Towards this aim of classification of healthy controls and patients, advanced computer vision and pattern recognition techniques have become popular in the recent years. This process can be summarized in two steps: i) feature (descriptor, distance, etc.) extraction, and ii) application of classification algorithms. The second part of this process is a well-established part of pattern recognition community and there are various techniques for classification ranging from simple classifiers such as decision trees to more advanced ones such as support vector machines [29]; from single classifier to ensembles of classification algorithms [14]; from feature based pattern recognition to dissimilarity based pattern recognition [17]. The list goes on and the research continues in this domain.

---

* Corresponding author.

M. Loog et al. (Eds.): PRIB 2011, LNBI 7036, pp. 306–317, 2011.

Even though there are various classification techniques, there is no single classification algorithm which is the best for all problems and most of the time the algorithm's success depends on the discriminative capability of the features used. Base features change from application to application and one has to find the best representation and the complementary classification algorithm to succeed in the aim towards accurate classification. In this paper, we aim towards the question "Can schizophrenia be detected used MRI images?". Structural and morphological abnormalities have been demonstrated in patients [25,23] and there has been several studies which use different features and pattern recognition techniques [10,27,21,7,30]. Some approaches use the deformation information in the registration of pairwise brains which requires extensive computation time and the solution of complex mathematical problems due to the nonlinearity of the process. The usually preferred approach is to detect volume variations in specific parts of the brain called regions of interest (ROIs) [25,3]. The following works are examples of this methodology which we also used in this study. Gerig *et al.* [10] introduced a ROI-based morphometric analysis by defining spherical harmonics and 3D skeleton as shape descriptors. Timoner *et al.* [27] introduced a new descriptor by encoding the displacement fields and the distance maps for amygdala and hippocampus at the same time. Shape-DNA signature has been proposed in [21]. Liu *et al.* [16] analyzed image features encoding general statistical properties and Law's texture features from the whole brain. In a recently accepted work [7], we used *Heat Kernel Signature* (HKS) [26] using different scales and applied it to classification of schizophrenia. In [30], Yushkevich *et al.* used the Jacobian fields of deformation.

In this work, as the classification technique, we pursue dissimilarity based classification [17] which differs from usual pattern recognition in the sense that objects are represented as pairwise dissimilarities instead of feature vectors per object. This technique has been show to be promising in brain disease research [28]. The dissimilarities are then transferred into a new vector space [17] (the dissimilarity space) where traditional classification algorithms can be applied. This paper is a continuation of our two previous works [7,28] where we use the heat kernel signatures to extract the features and analyse the effect of several dissimilarity measures on the features extracted using HKS and demonstrate the improvement in classification accuracy when we utilize the dissimilarity space.

The novel methodology proposed in this work is the adaptation of Ho's random subspace [12] method to the creation of dissimilarities by the use of four dissimilarity measures on histograms. The method differs from prototype selection [18] because instead of selection prototypes from already available dissimilarities, we attempt at the transpose of this problem where we select (randomly) different bins from the original histograms and then compute the dissimilarities and combine them. As a result we do not reduce the dimension but we use more information and enrich the dissimilarity space. In a similar recent work, Carli *et al.* [6] used the dissimilarities between *local parts* for scene categorization.

The paper is organized as follows: in Section 2, we introduce the heat kernel based shape analysis; in Section 3 we introduce the methods and the application of the methodology. We show our experiments in Section 4 and we conclude in Section 5.

2 Shape Analysis by Heat Diffusion

When we considering a shape M as a compact Riemannian manifold [5], the heat diffusion on shape[1] is defined by the *heat* equation:

$$(\Delta_M + \frac{\partial}{\partial t})u(t, m) = 0; \tag{1}$$

where u is the distribution of heat on the surface, $m \in M$, Δ_M is the *Laplace-Beltrami* operator which, has discrete eigendecomposition of the form $\Delta_M = \lambda_i \phi_i$ in compact spaces. The *heat kernel* can then be decomposed as:

$$h_t(m, m') = \sum_{i=0}^{\infty} e^{-\lambda_i t} \phi_i(m) \phi_i(m'), \tag{2}$$

where λ_i and ϕ_i represent the ith eigenvalue and the ith eigenfunction of the Laplace-Beltrami operator, respectively. The heat kernel $h_t(m, m')$ is the solution of the heat equation with initial point heat source in m at time $t = 0$, and heat value in ending point $m' \in M$ after time t. The heat kernel has been shown to be *isometric invariant, informative*, and *stable* [26].

For volumetric representations, the volume is sampled by a regular Cartesian grid composed of voxels, which then allows us to use the standard Laplacian in R^3 as the Laplace-Beltrami operator. Finite differences are used to evaluate the second derivative in each direction of the volume. The heat kernel on volumes is invariant to volume isometries, in which shortest paths between points inside the shape do not change. Note that in real applications exact volume isometries are limited to the set of rigid transformations [20], however, also non-rigid deformations can be modelled as approximations to volume isometries in practice. It is also worth noting that, for small t the autodiffusion heat kernel $h_t(m, m)$ of a point m with itself is directly related to the *scalar* curvature $s(m)$ [26,20]:

$$h_t(m, m) = (4\pi t)^{-3/2}(1 + \frac{1}{6}s(m)). \tag{3}$$

In practice, Equation 3 states that the heat tends to diffuse slower at points with positive curvature, and viceversa. This gives an intuitive explanation about the geometric properties of $h_t(m, m)$, and suggests the idea of using it to build a shape descriptor [26].

3 Dissimilarities

3.1 Dissimilarity Measures

The computed histograms of the data have been used to calculate dissimilarities between subjects using dissimilarity measures for histograms. There are various

---

[1] The notation is taken from [26,5].

dissimilarity measures that can be applied to measure the dissimilarities between histograms [8,24]. Moreover, histograms can be converted to pdfs and dissimilarity measures between two discrete distributions can be used as well. All in all, we decided to study measures below and the L1 norm (L1).

Given two histograms S and M with n bins, and two discrete probability density functions p and q, we define the number of elements in S and M as $|S|$ and $|M|$ respectively.

Earth Mover's Distance (EMD). Proposed by Rubner *et al.* [22], the basic idea is to use the cost to transform one distribution into another as dissimilarity. It is calculated using linear optimization by defining the problem as a transportation problem. For 1D histograms, it reduces to a simple calculation [8] which was implemented in this study.

$$C_i = \left| \sum_{j=1}^{i} (S_j - M_j) \right| , \; D = \sum_{i=1}^{n} C_i .$$

KullbackÜLeibler (KL) Divergence (KL) . KullbackÜLeibler divergence is defined as

$$D(p,q) = \sum_{i=1}^{n} q_i \log \frac{q_i}{p_i} .$$

This measure is not a distance metric but a relative entropy since $D(p,q) \neq D(q,p)$, i.e., the dissimilarity matrix is not symmetric. There are various ways to symmetrize this dissimilarity. We used the so-called Jensen-Shannon divergence (JS): $D = \frac{1}{2}D(p,r) + \frac{1}{2}D(q,r)$, where r is the average of p and q.

In addition to these dissimilarity measures, we propose to combine different "sub-dissimilarities" computed by random subspace method by using averaging and concatenating. We also test the accuracy of these combinations against the support vector machines on the original feature space. Further details of dissimilarity combination are provided in Section 3.4.

3.2 Dissimilarity Space

Suppose that we have n objects and we have a dissimilarity matrix D of size $n \times n$. And suppose that the dissimilarity between two objects o and $\hat{o}$ are denoted by $D(o, \hat{o})$. There are several ways to transform an $n \times n$ dissimilarity matrix D with elements $D(o, \hat{o})$ into a vector space with objects represented by vectors $X = \{x'_1, \ldots, x'_o, \ldots, x'_{\hat{o}}, \ldots, x'_n\}$ [17]. If the dissimilarities are already proper Euclidean, classical scaling can be used. One can utilize pseudo-Euclidean embedding for arbitrary symmetric dissimilarities. These yield vector spaces in which vector dissimilarities can be defined that produce the given dissimilarities D. When the dissimilarities are non-Euclidean, classification on these pseudo-Euclidean spaces are ill-defined since the corresponding kernels are indefinite.

A more general solution is to work directly in the *dissimilarity space*. It creates an Euclidean vector space by using the given dissimilarities as a set of features. In this case, the dissimilarities in this space do not correspond to the dissimilarities on the original space; which is sometimes an advantage if it is not known that they really are dissimilarities. As this holds in our case we constructed such a dissimilarity space using all available objects by taking X equal to D. In the dissimilarity space basically any traditional classifier can be used. The number of dimensions, however, equals the number of objects, which is 60 in our case. Many classifiers will need dimension reduction techniques or regularization to work properly in this space. Here, we used the linear support vector machine to avoid this.

3.3   Random Subspace Method and Adaptation to Dissimilarity Computation

In classical pattern recognition, there are several classifier combination techniques to get the best of a data set in hand. For the combination to be effective, one has to create diverse classifiers and combine them in a proper way. Most combination methods aim at generating uncorrelated classifiers, and it has been proposed [14] to use different (i) learning algorithms, (ii) hyperparameters, (iii) input features, and (iv) training sets. For example Bagging [4] uses bootstrapping to generate different training sets and takes an average, the random subspace method [12] trains different classifiers with different subsets of a given feature set. In dissimilarity based classification, since most of the time the k-nearest classifier is used, the research has focused on prototype selection [18]. In prototype selection, the aim is to select a subset of the data which explains the whole set of data in a proper way. In this paper, we aim for the transpose of this problem, instead of selecting a set of instances, we aim to select a set of features and combine them as in [12]. This can be done after we have the dissimilarities which was also targeted in [18,19] or before the dissimilarities are actually computed. In this paper, we propose the latter approach and adopt Ho's random subspace method to dissimilarity computation. What we do is we select a random subset of the original set of bins and compute the dissimilarities according to these subsets. Then we combine these subsets to get the final classification. In this way we enrich the dissimilarity space by using more information. This is possible to achieve because we calculate the dissimilarities using histogram bins and is logical because some bins may be more descriptive of the data and computing the dissimilarities using only these bins may increase the classification accuracy. Dividing the feature sets into two and combining Euclidean distances has been investigated recently in [15]. Once these dissimilarities are computed, they are combined using averaging and concatenation. We show the results using both techniques: see below for the combination of dissimilarity matrices.

3.4 Dissimilarity Combination

Combined dissimilarity spaces can be constructed by combining dissimilarity representations. As in normal classifier combination [13], a simple and often effective way is using an (weighted) average of the given dissimilarity measures:

$$D_{combined} = \frac{\sum \alpha_i D_i}{\sum \alpha_i} . \qquad (4)$$

It is related to the sum-rule in the area of combining classifiers. The weights can be optimized for some overall performance criterion, or determined from the properties of the dissimilarity matrix D_i itself, e.g. its maximum or average dissimilarity. In this work, we used equal weights while combining multiple dissimilarity matrices and all the dissimilarity matrices are scaled such that the average dissimilarity is one, i.e.:

$$\frac{D(i,j)}{\frac{1}{m(m-1)} \sum_{i,j} D(i,j)} = 1 \qquad (5)$$

This is done to assure that the results are comparable over the dissimilarities as we deal with dissimilarity data in various ranges and scales. Such scaled dissimilarities are denoted as $\tilde{D}$. In addition, we assume here that the dissimilarities are symmetric. So, every dissimilarity $\tilde{D}(i,j)$ has been transformed by

$$\tilde{D}(i,j) := \frac{\tilde{D}(i,j) + \tilde{D}(j,i)}{2} \qquad (6)$$

After this scaling, we use averaging or concatenation to get the final space for classification.

4 Experiments

4.1 MRI Data Collection

Data collection and processing in MRI-based research faces several methodological issues to minimize biases and distortions. The standard approach is to follow well-established guidelines by international organizations, such as the World Health Organization (WHO), or by respected institutions. The data set used in this work is composed by MRI brain scans of 30 patients affected by schizophrenia and 30 healthy control subjects. All patients received a diagnosis of schizophrenia according to the criteria of the Diagnostic and Statistical Manual of Mental Disorders [2]. After the data collection phase, we employ a ROI-based approach [11], so only a well defined brain subpart has been considered in this study. More specifically, we focus our analysis on the left-Thalamus whose abnormal activity has been already investigated in schizophrenia [9]. ROIs have been manually traced by experts, according to well defined medical protocols.

4.2   Histogram Computation Using HKS

In order to employ a *learning-by-example* approach, we need a collection of samples for both healthy subjects and patients. Source data are MRI scans where shape information can be provided in terms of volumetric data. According to the shape diffusion analysis described in Section 2, for each subject geometric features are extracted at a certain scale according to a time value t, and the autodiffusion value is computed for each voxel m, leading to:

$$H_t(M) = \{h_t(m, m), \forall m \in M\}.$$

Then, such values are accumulated into a histogram $r = hist(H_t(M))$. In this manner, we obtain a shape representation r, encoding the global shape at a certain scale. In our experiments, the number of bins for the histograms have been chosen as 100.

4.3   Experimental Methodology

In our experiments, using the 60 patients, we apply leave-one-out cross-validation to assess the performance of the methods. We first train support vector machines (SVMs) with linear kernels as a base-line for classification accuracies using the original histogram features. The C parameter is chosen using cross validation. We then compute the full dissimilarity space in a transductive way using the dissimilarity measures mentioned in Section 3. We again use the linear kernel on this space to compare the results with the base-line original space accuracy and random subspace accuracy. Random bins are chosen using different number of bins and different number of random subspaces and the dissimilarity space constructed using the combination of these dissimilarities are applied for the final classification. Note that, in this paper we do not aim at combining different modalities or data sets. What we are trying to achieve is to get the best accuracy using the single data set at hand; and with this in mind, we demonstrate the effect of different dissimilarity measures.

4.4   Results

Table 1 reports the accuracies of the SVM with the linear kernel on the original space (LIN), linear SVM on the dissimilarity spaces (DIS, where -AVG shows combining using averaging and -CONC shows combination using concatenation) and random subspace accuracies (RAND-N, where N is the selected number of bins. For this example, we used 20 random subspaces; the results for other numbers are similar. See also Figures 1 and 2). We can see that the dissimilarity space accuracies are always better than the original space accuracies. This also corresponds with our previous work [28]. We can also see that with a suitable number of random bins, one can always achieve a better accuracy than the full dissimilarity space accuracies. We can see that the average operation usually

Table 1. Accuracies using the dissimilarity space and the random subspace. 20 random subspaces are created

					66.67			
LIN								
DIS	76.67		61.67		75.00		68.33	
	L1-AVG	L1-CONC	EMD-AVG	EMD-CONC	KL-AVG	KL-CONC	JS-AVG	JS-CONC
RAND-50	76.67	75.00	70.00	58.33	73.33	68.33	70.00	75.00
RAND-25	78.33	70.00	65.00	58.33	71.67	58.33	71.67	60.00
RAND-20	**86.67**	83.33	65.00	55.00	71.67	70.00	68.33	71.67
RAND-15	80.00	73.33	73.33	68.33	75.00	**88.33**	73.33	63.33
RAND-10	71.67	56.67	65.00	60.00	71.67	66.67	70.00	53.33

creates more accurate results (though the best result is using concatenation on KL). We believe that this is because of the curse of dimensionality. When one uses the concatenation operator, the number of features increase and the classification accuracy may decrease accordingly.

We can see that the random subspace method applied to the creation of the dissimilarity space achieves better accuracies than the full dissimilarity space and the original space. One disadvantage of this method is to choose the proper number of bins and proper number of random subspaces. One can use cross validation to decide these values but we leave the exploration of this topic as a future work. Nevertheless, we want to present the performance of these methods when we apply different number of bins and different number of random subspaces. In Figure 1, we can see the change in accuracy versus the number of subspaces created. The numbers on the plots show the number of bins chosen and the peak accuracy using that number of bins. Although there is no clear correlation between accuracy and the number of subspaces, we can still observe that with a fairly small number of subspaces, one can achieve relatively high accuracies. We can also observe that (as expected), with the increasing number of subspaces, the diversity decreases and the accuracy converges. We can observe that the dissimilarity space is always superior to the original space and (except for KL where the choice of the number of subspaces is critical), the random subspace accuracy is higher than full dissimilarity space accuracy.

In the second set of experiments, we fix the number of random subspaces and see the change in accuracy when we change the number of selected bins. These results are presented in Figure 2. Again, the numbers on the figure show the number of random subspaces and peak accuracy obtained by this selection. We can see a clearer picture in this setup because when we increase the number of bins, we get to a peak point and then the accuracy decreases and in the end levels off as expected. The peak point is usually between 20 and 30 which we suggest to use for a 100 bin histogram representation. This shows us that the selection of number of bins is the critical point in this methodological setup and the proper selection of this parameter yields the best accuracy.

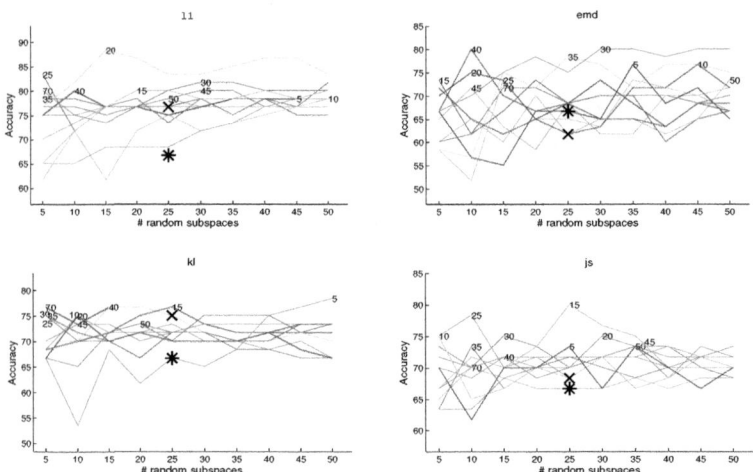

Fig. 1. # of subspaces vs accuracy. The numbers in the plot show the number of histogram bins used. X is the accuracy on the full dissimilarity matrix of the corresponding dissimilarity measure and * is the base-line original space accuracy.

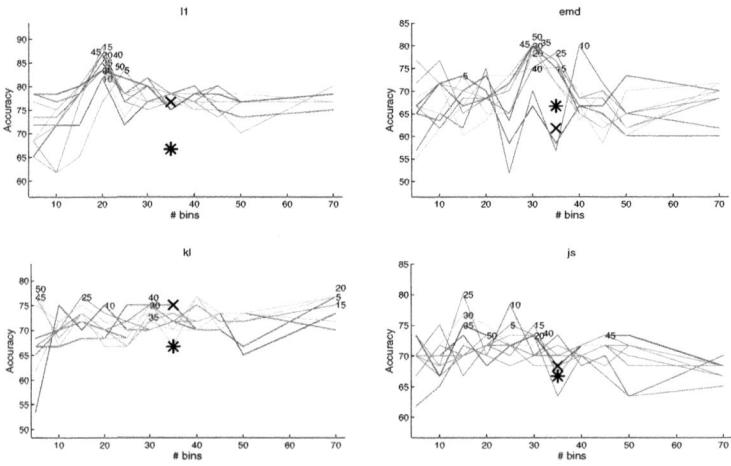

Fig. 2. # of bins vs accuracy. The numbers in the plot show the number of random subspaces used. X is the accuracy on the full dissimilarity matrix of the corresponding dissimilarity measure and * is the base-line original space accuracy.

5   Conclusion

In this paper, we use *Heat Kernel Signatures* to extract histogram based features (see also [7]) from MRI scans of 30 schizophrenic patients and 30 healthy controls to detect schizophrenia. We first create several dissimilarity matrices using histogram based dissimilarity measures in the literature and compare our results with the base-line original space accuracies using the support vector machines with the linear kernel. As the novel part of this study, we propose to adapt the random subspace method [12] to the creation of dissimilarity matrices and show that with the combination of these matrices, we achieve higher detection accuracies. Random subspaces have been used also in the dissimilarity based pattern recognition paradigm [19,18], but the method is applied after the dissimilarity matrices are constructed. The novelty of our method is that we use the random subspaces technique before creating the dissimilarity matrices to combine the useful information of the data.

In our previous work, we used the dissimilarity space and dissimilarity combination [28] using intensity and apparent diffusion coefficient (ADC) based histogram representations. The main aim of that study was to show the effect of dissimilarity space using dissimilarity combination and multiple modalities. In this work, we use only one modality and one ROI; our aim is not to combine data from different sources but to get the best from the data as hand as was also targeted in [18]. For this purpose, we use four different dissimilarity measures and analyze the effect of using the dissimilarity space and the dissimilarity space constructed using random subspaces.

We have seen that the dissimilarity space constructed using the dissimilarity measures mentioned in this work always outperform the base-line original space accuracies. Our novel contribution is the adaptation of Ho's random subspace methodology for selecting a subset of bins used in the computation of the dissimilarity matrices. With this method, we see that we can achieve significantly higher results when proper number of bins and random subspaces are selected. For the combination purpose, we compare combining using averaging and concatenation and we see that averaging usually outperforms concatenation (though the best result is achieved using concatenation) and we believe that this is due to the curse of dimensionality. As a future work, we plan to combine the dissimilarities over random subspaces using *Multiple Kernel Learning* (MKL) paradigm.

Our analysis of number of bins and number of random subspaces show that the important parameter is the number of bins. With relatively small number of random subspaces, one can achieve good results. As expected, when the number of subspaces increase, the individual dissimilarity matrices become less diverse and the combination accuracy converges. Same occurs with the number of bins, when we increase the number of bins, the diversity decreases and the accuracy converges. In this setup, the number of bins become more important because of the peak points. In this study we used a source histogram of 100 bins and we have observed that we get the peak accuracies in the 20-30 bin range. As a future study, we would like to apply this methodology to other data sets and artificial data sets with different number of source bins and draw a theoretical/experimental formula for the selection of the number of bins.

Acknowledgements. We acknowledge financial support from the FET programme within the EU FP7, under the SIMBAD project (contract 213250).

References

1. Agarwal, N., Port, J.D., Bazzocchi, M., Renshaw, P.F.: Update on the use of MR for assessment and diagnosis of psychiatric diseases. Radiology 255(1), 23–41 (2010)
2. American Psychiatric Association: Diagnostic and statistical manual of mental disorders. DSM-IV. Washington DC, 4th edn. (1994)
3. Baiano, M., Perlini, C., Rambaldelli, G., Cerini, R., Dusi, N., Bellani, M., Spezzapria, G., Versace, A., Balestrieri, M., Mucelli, R.P., Tansella, M., Brambilla, P.: Decreased entorhinal cortex volumes in schizophrenia. Schizophrenia Research 102(1-3), 171–180 (2008)
4. Breiman, L.: Bagging predictors. Machine Learning 24(2), 123–140 (1996)
5. Bronstein, A.M., Bronstein, M.M., Ovsjanikov, M., Guibas, L.J.: Shape recognition with spectral distances. IEEE Trans. Pattern Analysis and Machine Intelligence 33(5), 1065–1071 (2011)
6. Carli, A., Castellani, U., Bicego, M., Murino, V.: Dissimilarity-based representation for local parts. In: 2010 2nd International Workshop on Cognitive Information Processing (CIP), pp. 299–303 (June 2010)
7. Castellani, U., Mirtuono, P., Murino, V., Bellani, M., Rambaldelli, G., Tansella, M., Brambilla, P.: A new shape diffusion descriptor for brain classification. In: Fichtinger, G., Martel, A., Peters, T. (eds.) MICCAI 2011, Part II. LNCS, vol. 6892, pp. 426–433. Springer, Heidelberg (2011)
8. Cha, S.-H., Srihari, S.N.: On measuring the distance between histograms. Pattern Recognition 35(6), 1355–1370 (2002)
9. Corradi-DellAcqua, C., Tomelleri, L., Bellani, M., Rambaldelli, G., Cerini, R., Pozzi-Mucelli, R., Balestrieri, M., Tansella, M., Brambilla, P.: Thalamic-insular dysconnectivity in schizophrenia: Evidence from structural equation modeling. Human Brain Mapping (in press, 2011)
10. Gerig, G., Styner, M.A., Shenton, M.E., Lieberman, J.A.: Shape Versus Size: Improved Understanding of the Morphology of Brain Structures. In: Niessen, W.J., Viergever, M.A. (eds.) MICCAI 2001. LNCS, vol. 2208, pp. 24–32. Springer, Heidelberg (2001)
11. Giuliani, N.R., Calhouna, V.D., Pearlson, G.D., Francis, A., Buchanan, R.W.: Voxel-based morphometry versus region of interest: a comparison of two methods for analyzing gray matter differences in schizophrenia. Schizophrenia Research 74(2-3), 135–147 (2005)
12. Ho, T.K.: The random subspace method for constructing decision forests. IEEE Transactions on Pattern Analysis and Machine Intelligence 20(8), 832–844 (1998)
13. Kittler, J., Hatef, M., Duin, R.P.W., Matas, J.: On combining classifiers. IEEE Transactions on Pattern Analysis and Machine Intelligence 20(3), 226–239 (1998)
14. Kuncheva, L.I.: Combining pattern classifiers: methods and algorithms. Wiley Interscience (2004)
15. Lee, W.-J., Duin, R.P.W., Loog, M., Ibba, A.: An experimental study on combining euclidean distances. In: 2010 2nd International Workshop on Cognitive Information Processing (CIP), pp. 304–309 (June 2010)

16. Liu, Y., Teverovskiy, L., Carmichael, O., Kikinis, R., Shenton, M.E., Carter, C.S., Stenger, V.A., Davis, S., Aizenstein, H.J., Becker, J.T., Lopez, O.L., Meltzer, C.C.: Discriminative MR Image Feature Analysis for Automatic Schizophrenia and Alzheimer's Disease Classification. In: Barillot, C., Haynor, D.R., Hellier, P. (eds.) MICCAI 2004. LNCS, vol. 3216, pp. 393–401. Springer, Heidelberg (2004)
17. Pekalska, E., Duin, R.P.W.: The Dissimilarity Representation for Pattern Recognition. Foundations and Applications. World Scientific, Singapore (2005)
18. Pekalska, E., Duin, R.P.W., Paclík, P.: Prototype selection for dissimilarity-based classifiers. Pattern Recognition 39(2), 189–208 (2006)
19. Pękalska, E.z., Skurichina, M., Duin, R.P.W.: Combining fisher linear discriminants for dissimilarity representations. In: Kittler, J., Roli, F. (eds.) MCS 2000. LNCS, vol. 1857, pp. 117–126. Springer, Heidelberg (2000)
20. Raviv, D., Bronstein, A.M., Bronstein, M.M., Kimmel, R.: Volumetric heat kernel signatures. In: Workshop on 3D Object Retrieval (2010)
21. Reuter, M., Wolter, F.E., Shenton, M., Niethammer, M.: Laplace-Beltrami eigenvalues and topological features on eigenfuntions for statistical shape analysis. Computed-Aided Design 41(10), 739–755 (2009)
22. Rubner, Y., Tomasi, C., Guibas, L.J.: The earth mover's distance as a metric for image retrieval. International Journal of Computer Vision 40(2), 99–121 (2000)
23. Rujescu, D., Collier, D.A.: Dissecting the many genetic faces of schizophrenia. Epidemiologia e Psichiatria Sociale 18(2), 91–95 (2009)
24. Serratosa, F., Sanfeliu, A.: Signatures versus histograms: Definitions, distances and algorithms. Pattern Recognition 39(5), 921–934 (2006)
25. Shenton, M.E., Dickey, C.C., Frumin, M., McCarley, R.W.: A review of mri findings in schizophrenia. Schizophrenia Research 49(1-2), 1–52 (2001)
26. Sun, J., Ovsjanikov, M., Guibas, L.: A concise and provably informative multi-scale signature based on heat diffusion. In: Proceedings of the Symposium on Geometry Processing, pp. 1383–1392 (2009)
27. Timoner, S.J., Golland, P., Kikinis, R., Shenton, M.E., Grimson, W.E.L., Wells III, W.M.: Performance issues in shape classification. In: Dohi, T., Kikinis, R. (eds.) MICCAI 2002. LNCS, vol. 2488, pp. 355–362. Springer, Heidelberg (2002)
28. Ulaş, A., Duin, R.P.W., Castellani, U., Loog, M., Mirtuono, P., Bicego, M., Murino, V., Bellani, M., Cerruti, S., Tansella, M., Brambilla, P.: Dissimilarity-based detection of schizophrenia. International Journal of Imaging Systems and Technology 21(2), 179–192 (2011)
29. Vapnik, V.N.: Statistical learning theory. John Wiley and Sons (1998)
30. Yushkevich, P., Dubb, A., Xie, Z., Gur, R., Gur, R., Gee, J.: Regional structural characterization of the brain of schizophrenia patients. Academic Radiology 12(10), 1250–1261 (2005)

Epithelial Area Detection in Cytokeratin Microscopic Images Using MSER Segmentation in an Anisotropic Pyramid

Cristian Smochina[1], Radu Rogojanu[2], Vasile Manta[1], and Walter Kropatsch[3]

[1] Faculty of Automatic Control and Computer Engineering,
"Gheorghe Asachi" Technical University of Iasi, Romania
smochina.cristian@yahoo.com, vmanta@cs.tuiasi.ro
[2] Institute of Pathophysiology and Allergy Research, Medical University, Vienna, Austria
radu@rogojanu.com
[3] Pattern Recognition and Image Processing Group, Institute of Computer Graphics and
Algorithms, Vienna University of Technology, Austria
krw@prip.tuwien.ac.at

Abstract. The objective of semantic segmentation in microscopic images is to extract the cellular, nuclear or tissue components. This problem is challenging due to the large variations of features of these components (size, shape, orientation or texture). In this paper we present an automatic technique to robustly delimit the epithelial area (crypts) in microscopic images taken from colon tissues sections marked with cytokeratin-8. The epithelial area is highlighted using the anisotropic diffusion pyramid and segmented using MSER+. The crypts separation and lumen detection is performed by imposing topological constraints about the epithelial layer distribution within the tissue and the round-like shape of the crypt. The evaluation of the proposed method is made by comparing the results with ground-truth segmentations.

Keywords: Crypt segmentation, Anisotropic diffusion pyramid, MSER, Biomedical imaging, Pathology, Microscopy, Fluorescence microscopy, Image analysis.

1 Introduction

In diagnostic pathology, the pathologists give a diagnostic after a set of biological samples (tissues stained with different markers) are viewed and many specific features of the objects of interest (size, shape, colour or texture) have been analysed. This time consuming and tedious process is an important part in clinical medicine but also in biomedical research and can be enhanced by providing the pathologists/biologists with quantitative data extracted from the images.

To overcome also the possible subjectivity caused by different visual interpretations of different pathologists, image processing techniques are used to allow large scale statistical evaluation in addition to classical eye screening evaluation.

M. Loog et al. (Eds.): PRIB 2011, LNBI 7036, pp. 318–329, 2011.

Image segmentation of microscopic images is a hard task due to several elements of variability noticed in the image dataset like size and arrangement of the crypts inside the tissue sections, intensity levels of both background and cells from either stroma or crypts, as well as a relatively high background compared to the objects of interest.

In this paper we focus on analysing the tissue components like crypts, lumen and stroma, without dealing directly with the small objects like nuclei and cells (Fig. 1a). A rough description and a short overview of the problems to be solved are presented bellow; each step of the proposed method will present more details and justifications:

- some images are slightly underexposed due to weak local biological response.
- some image portions don't contain useful information.
- the crypt appears like a "donut" (or 2D projection of a torus) with many small 'holes' and a central black area inside (lumen).
- the lumen is a black area with different sizes surrounded by the epithelial layer.
- the "donut" has a non-homogeneous organization due to the dark regions smaller than the lumen.
- the pixels within a crypt correspond to three main components: background, dark regions and the noisy pixels, weak biological binding response and strong response (highest values).
- the stroma separates the crypts; situations of touching/very close crypts can appear.
- no relevant information or cells exist between touching/very close crypts.
- the number of crypts may be used in computing distribution statistics.

This paper is organized as follows. The last part of this section points out the goal of this study. The images are enhanced in Section 2 and the epithelial area is highlighted using the anisotropic diffusion pyramid (Section 3.1) and segmented by MSER+ (Section 3.2). In Section 4.1 the false positive results are reanalysed for crypts separation and the lumens are detected in 4.2. The results are evaluated and discussed in Section 5 while the conclusions are elaborated in Section 6.

1.1 State of the Art

The detection of histological structures like gland or crypt has been addressed in many studies. In [1] a threshold is used to identify the gland seeds which are grown to obtain the nuclei chain. In [2] the pixel labelling to different classes is performed using a clustering approach based on the textural properties. In [3] the high level information is preferred against the local one in segmenting the colon glands from hematoxylin-and-eosin stained images. An object-graphs approach is described where the relationship between the primitive objects (nucleus and lumen) is considered. In [4] the prostate cancer malignancy is automatically graded (Gleason system) after the prostate glands are detected.

Many image analysis techniques were successfully used in medical image processing. In [5] the mammographic images are hierarchically decomposed into different resolutions and segmented by analyzing the coarser resolutions while in [6] the multiresolution wavelet analysis is used for texture classification. A Gaussian multiresolution segmentation technique is combined in [7] with the expectation maximization (EM) algorithm to overcome the drawbacks of the EM algorithm.

The maximally stable extremal region is mostly used in situation where a single threshold is difficult or impossible to select. A general analysis of this segmentation approach is described in [8] and an application in medical images can be found in [9].

1.2 Aim of the Study

The basic functional unit of the small intestine is the crypt (crypt of Lieberkühn) [10] and it comprises two main structures of interest: the lumen and the epithelial layer (Fig. 1a). The epithelial layer contains epithelial cells and surrounds the empty area called lumen. The stroma is a heterogeneous compartment between crypts and contains the extra cellular matrix (ECM), fibroblasts, macrophages, vessels etc.

Depending on the marker used on tissues, different components/proteins can be highlighted. To separate the epithelial layer, immunofluorescence staining is performed in paraffin embedded sections with the anti-cytokeratin 8 (CK-8) antibody and a fluorochrome-labelled secondary antibody [11]. The CK-8 is used because it reacts with cytokeratins, proteins found only in the cytoskeleton of epithelial cells. In Fig. 1a, b) the bright area is the binding result between the CK-8 and the epithelial components; the small dark regions are caused by the epithelial nuclei and their lack of cytokeratins.

This work provides specific techniques (Fig. 2) to automatically segment and measure the crypts (epithelial areas) from fluorescence images of colorectal tissue sections. We used 8 bit greyscale images (Fig. 1a) showing cytokeratins from epithelial layer; they were acquired using the automated TissueFAXS^{TM} slide scanner (TissueGnostics GmbH, Austria).

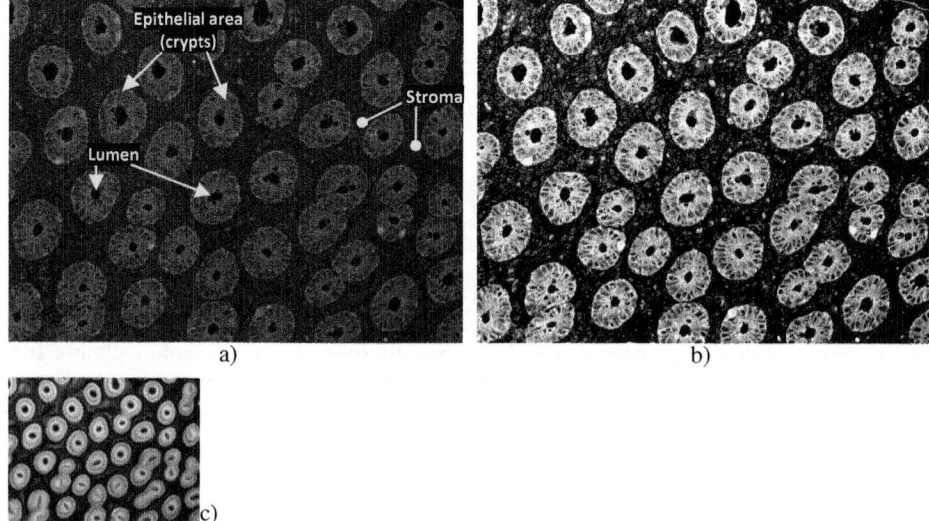

a)   b)   c)

Fig. 1. a) Fluorescence greyscale image showing crypts stained with anti-cytokeratin 8 from colon tissue section. b) The enhanced image by histogram processing; the bright regions indicate the crypts and the inside black regions are the lumens. c) The image from the 4th level of the ADP, up sampled for better visualization.

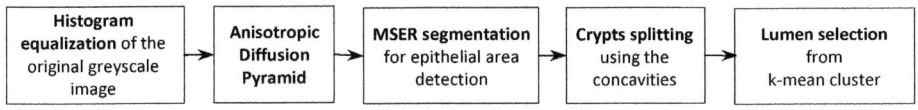

Fig. 2. Overview scheme of the proposed technique

The main motivation for segmenting crypts is to provide the pathologists with quantitative data regarding the epithelial area (crypts boundaries) and epithelium-to-stroma ratios. These ratios may provide important information for the assessment of cancer in colon or other organs [12]. Automatic segmentation is much faster than visual analysis and does not suffer from typical human analysis disadvantages like subjectivity and fatigue of the evaluator. Samples from more patients can be evaluated, thus providing data which is more relevant from a statistical point of view.

2   Image Enhancement

The CK-8 images are slightly underexposed due to tissue variability and local weak binding of the fluorescence marker. The image contrast can be enhanced such that the separation between the crypts and the stroma is better highlighted. Histogram equalization [13] is used because it maximizes the image contrast by transforming the values in an intensity image so that the intensities are better distributed on the image histogram.

Let I denote the input grey scale image. There are many situations in which some image portions don't contain useful information (Fig. 5a), i.e., they have low intensities due to the lack of CK-8 signal. Only the set of pixels with higher intensities than a certain threshold are considered in the histogram enhancement process, since they are the result of binding between the CK-8 and epithelial cells:

$$I_{enh} = \begin{cases} 0, & if \ I \leq c \cdot thr_{Otsu}(I) \\ \text{hist\_enh}(I \mid I > c \cdot thr_{Otsu}(I)), & else \end{cases} \quad (1)$$

where $thr_{Otsu}(\cdot)$ computes the threshold for an image using Otsu's method [14], c is fixed to 0.5 in our experiments and the function $\text{hist\_enh}(\cdot)$ performs the histogram equalization. This technique allows the enhancement of the objects of interest, but keeps the low intensity levels for the background (Fig. 1b).

3   Crypt Outer Borders Detection

The borders detection process must ignore the gaps within crypts. The low level cues (pixels, small areas) don't provide enough information to separate the regions having a particular meaning [15]. A way must be found to keep only the important information and to remove the unnecessary details. In order to detect these regions, the role of *local* information (pixel grey values or gradient) is very important but not sufficient; also *global* information like the region's size and relation with the other region types must be included [15].

In [16] the morphological closing operator is used in a multi-resolution structure; the gaps between nuclei are filled by relating the structure element size to the size of the gaps. This gives good results for images with nuclei, but in our case elements within the stroma near the crypts may be merged by the closing operator. Since an accurate delimitation of the epithelial area is required, the anisotropic filtering is applied to homogenize the crypt regions, fill the inside holes while preserving the (weak) epithelial-stroma edge information.

3.1 Anisotropic Diffusion Pyramid

If we use the conventional spatial filters, the edges will be blurred up to losing completely necessary information for crypt assessment. Anisotropic diffusion is a non-linear filtering method, which improves the image quality significantly while preserving the important boundary information. The image is smoothed by applying a diffusion process whose diffusion tensor acts mainly along the preferred direction [17], [18]. This coherence-enhancing filtering uses the structure tensor of the initial image (in our case the result of the histogram equalization) and a diffusion tensor D. This processes is given by an equation of type:

$$\frac{\partial f}{\partial t} = \nabla \cdot (D \nabla f), \quad D = \begin{pmatrix} a & b \\ c & d \end{pmatrix} \tag{2}$$

where f is the evolving image, t denotes the diffusion time and D is the diffusion tensor.

The preferred local orientations are obtained by computing the eigenvectors of the structure tensor ('second-moment matrix') J_ρ [18]. The local contrast along these directions is given by the corresponding eigenvalues (μ_1 and μ_2). The diffusion tensor D of the anisotropic diffusion uses the same eigenvectors as the structure tensor J_ρ but its eigenvalues λ_1, λ_2 and the entries of D are computed in as follows [18]:

$$\lambda_1 = c_1, \lambda_2 = \begin{cases} c_1 & \text{if } \mu_1 = \mu_2, \\ c_1 + (1 - c_1)e^{\frac{-c_2}{(\mu_1 - \mu_2)^2}} & \text{else} \end{cases}, \quad \begin{bmatrix} a = \lambda_1 \cos^2 \alpha + \lambda_2 \sin^2 \alpha, \\ b = (\lambda_1 - \lambda_2)\sin \alpha \cos \alpha, \\ c = \lambda_1 \sin^2 \alpha + \lambda_2 \cos^2 \alpha \end{bmatrix} \tag{3}$$

where $0 < c_1 < 1$ and $c_2 > 0$.

Let ζ^T denote the anisotropic diffusion process and T the total diffusion time. The anisotropic diffusion pyramid (ADP) [19] Π consists of L levels. The first level is the histogram enhance image $\Pi^1 = I_{enh}$ (width $w = 1392$ and height $h = 1024$) and each level $\ell > 1$ is given by

$$\Pi^l = \zeta^T (\Pi^{l-1}) \downarrow_4, \ell = \overline{2, L} \tag{4}$$

where the $\downarrow_4$ denotes the sub-sampling process which reduce the image size by four times (halving the w and h, a scale pyramid with one octave between scale).

Using the ADP, the edge sharpening properties of anisotropic diffusion can be exploited together with the performance of the multiresolution hierarchical process. The anisotropic diffusion smooths the objects' interior and removes the dark holes. The gaps are continually filled by going to the coarser levels of the pyramid.

A resolution level must be selected from this coarse-to-fine representation such that the crypts area is accurately delineated and all gaps filled. From experimental tests resulted that the 4th level ($w = 174$, $h = 128$) (critical level) is the maximum level of the pyramid which offers the minimum details to extract relevant information (Fig. 1c and Fig. 5b).

3.2 MSER for Epithelial Area Detection

In Fig. 1c and Fig. 5b the bright round areas represent the epithelial regions and the interstitial details were spread out between crypts. Due to the anisotropic diffusion, the pixels intensities in the middle of the "*donut*" decreased because of the small holes. Although the histogram equalization has been applied for image contrast enhancement, the global or adaptive threshold will not offer stable results.

All thresholds are checked and the connected components are extracted judging their change in area in the maximally stable extremal region (MSER) algorithm [20]. The MSERs are regions that are either darker (MSER-) or brighter (MSER+) than their surroundings and are stable according to a stability criterion across a range of thresholds on the intensity function.

A threshold application on the 4th level of ADP gives the binary image B_t. The regions in the binary images became smaller by ranging the threshold form 1 to 256 (for 8 bit images). Let $\Psi(R_i^t)$ defines the stability based on the area change of region R_i^t from the image B_t obtained with threshold t:

$$\Psi(R_i^t) = \frac{\left|R_{i'}^{t-\Delta}\right| - \left|R_{i''}^{t+\Delta}\right|}{\left|R_i^t\right|} \tag{5}$$

where Δ ($\Delta = 15$ in our experiments) is the stability range parameter and $|\cdot|$ gives the region area. The regions $R_{i'}^{t-\Delta}$ and $R_{i''}^{t+\Delta}$ are the ascendant and the descendent regions of R_i^t obtained by decreasing respectively increasing the threshold level and $R_{i'}^{t-\Delta} \supset R_i^t \supset R_{i''}^{t+\Delta}$. A region R_i^t is MSER+ if it has the following properties:

- extremal, i.e., all pixels inside the connected region are strictly brighter than the surrounding ones.
- stable, i.e., its area changes only slightly with the change of the threshold t (across $2*\Delta+1$ thresholds).
- maximally stable, i.e., $\Psi(R_i^t)$ has a local minimum at t.

The advantage of using the MSER is given by the lack of a global or adaptive threshold. All thresholds are tested and the stability of the connected components evaluated. If multiple stable thresholds exist then a set of nested regions is obtained per object. In this case the output of the MSER detector is not a binary image.

The round dark channel within the '*donut*' causes the MSER to produce more than one stable region per crypt, but they are nested. Since we are interested in only one region per crypt, the final binary result is obtained by considering the union of all MSERs.

$$I_{bw} = \bigcup_{t \in T, i_t \in M_t} R_{i_t}^t \qquad (6)$$

where T is the set of thresholds values for which a MSER has been found and M_t contains the MSERs for the threshold t.

This approach has been chosen also for its preference for round regions [8] given by the affine-invariant property [20]. The regions in Fig. 1c and Fig. 5b are homogeneous regions with distinctive boundaries; high detection accuracy especially in these situations has also been concluded in [21].

In case of using the pyramid in image segmentation, the proper reconstruction of the regions found on the top levels must be done by analysing the finer lower levels of the hierarchy where more details are present. In our anisotropic diffusion pyramid this is not required due to the anisotropic filtering which preserves (and even enhance) the boundaries between the epithelial areas and stroma.

The borders of the MSERs+ delimit the epithelial areas and actually represent the outer border of the crypt (the green curves in Fig. 3a). The segmentation produces false positive (FP) results in which two close crypts are detected as a single region (regions A and B in Fig. 3a); the crypts in this situation are rechecked and split in Section 4.1.

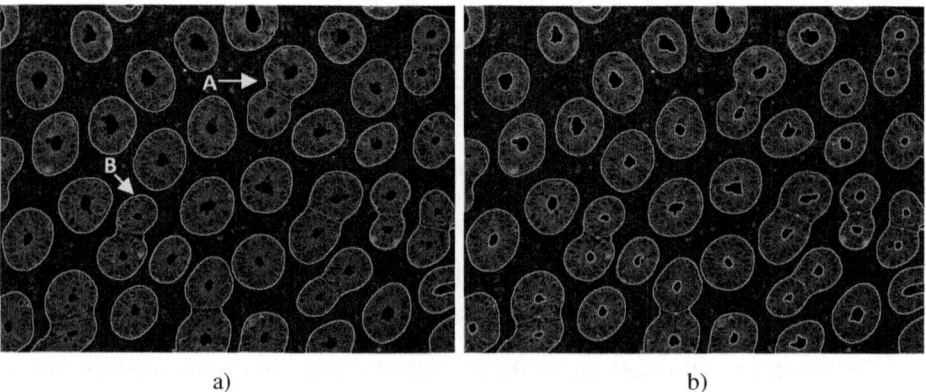

a)   b)

Fig. 3. a) The green curves delimit the epithelial regions and the red lines split the FP results with two close crypts. b) The crypts and the lumens boundaries are depicted in green. In case of two touching crypts the blue line connects the lumens' centroids and the red lines separate the crypts by connecting the two concave points with the middle of the centroids line.

4 Crypts Separation and Lumen Detection

The anisotropic filtration can shrink or even totally fill the lumen; in these cases the MSER- is not able to give good results and MSER+ is unstable. Before performing the lumen detection, the touching crypts segmented as a single region in Section 3.2 are split by considering their geometrical features.

4.1 Crypts Separation

The split of the crypts from a FP result (e.g., regions A, B in Fig. 3a) is made by considering the round-like shape, the size and the concavities near the separation area. The FP results are concave regions and the concavities of these regions give critical information about how the crypts can be separated.

For each region G^i obtained by MSER+, the convex hull G_{convex}^i can be computed. Considering the round shape of the crypt we expect that $G_{convex}^i = G^i$. This is true for the true positive results but for the FP results the difference between the convex hull and the concave region $G_{convex}^i - G^i \neq 0$ offers two regions C_1 and C_2 on both sides of the crypts (depicted with blue in Fig. 4a). These concavities can be approximated with two triangles and their apexes indicate the entrance into the thin separation area between the two crypts. The separation line is determined by the points $p_{C_1} \in C_1$ and $p_{C_2} \in C_2$ such that the distance $d(p_{C_1}, p_{C_1})$ is minimum (pink line in Fig. 4a. This is a rough separation used for detecting the proper lumen per crypt.

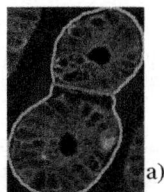 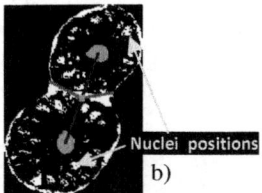

Fig. 4. a) The pink line approximately separates the crypts. b) The cluster with the lowest values contains the lumens (green regions connected by the blue line) and the white regions; the separation line connects the middle of this line with the two concave points.

4.2 Lumen Detection

The lack of cytokeratins in the cell nuclei produces the dark regions considerable smaller than the lumen. Three intensity classes exist within a found region (from Section 3.2) corresponding to the background and the noisy pixels (lowest intensities), weak cytokeratins binding and strong cytokeratins binding. These classes are detected using the k-means clustering [13]. The cluster with the lowest values contains the pixels from the lumen and the small dark holes corresponding to the nuclei (Fig. 4b). The proper region r_L for lumen is selected by considering its size bigger than the nuclei:

$$r_L \mid A(r_L) = \max\{A(r_i, i = \overline{1, N})\} \wedge (A(r_L) > a_{\min}) \tag{7}$$

where $A(\cdot)$ returns de area of a region, N denotes the number of regions from the lowest cluster and a_{min} is the minimum area of a lumen region. The Fig. 4b shows the regions from the lowest cluster (white) and the selected regions as lumen (green).

The set of pixels on which k-means is applied is very important. E.g., in [3], the hematoxylin-and-eosin images are used for glands detection. The pixels are first quantized into a number of clusters and the cluster corresponding to epithelial layer is further processed in the proposed technique. In our case, even though the image was enhanced so that the intensities are better distributed on the image, the k-mean clustering of the entire image doesn't guarantee good results. Instead, the pixels from the small regions found in Section 3.2, which cover one or two crypts, are successfully classified into the proper three clusters.

Even though the exact crypt delineation is not needed (more arguments in Section 5), the separation line can be better approximated by considering the band width of the crypts. By connecting the middle of the centroids line with the concave points found in Section 4.1, a better crypt separation is obtained (Fig. 4b, Fig. 3b). Having better crypt segmentation might enable computing new statistics per crypt once additional markers are used, i.e., percentage of proliferating cells per crypt when staining with a proliferation marker (Ki67).

5   Results

The main issue is to segment the epithelial layer by detecting the area covered by crypts; in the case of two touching/very close crypts the outer boundaries detected in 3.2 delimit the epithelial cells. An exact segmentation of the pixels within this area is not really necessary, but the number of crypts must be accurately found since it may be used in computing statistics about crypts distribution within colon tissue. We tested the proposed segmentation technique on different datasets of images from CK-8 labelled tissue sections; some results are show in Fig. 3b, Fig. 5a and Fig. 6.

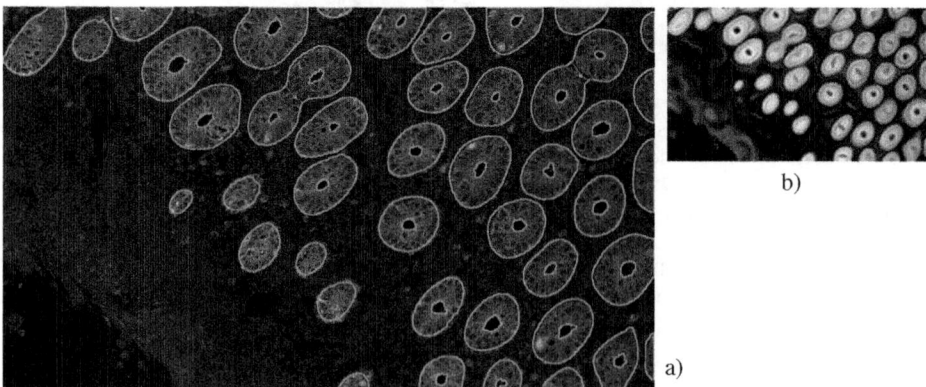

Fig. 5. a) The crypts and the lumens boundaries are depicted in green; the red lines separate the touching crypts. b) The image from the 4th level of the ADP, up sampled for better visualization.

A more rigorous evaluation must be done by comparing the results against the ground truth segmentations. Since a database with reference segmentations for this type of images does not yet exist, a human expert has been asked to manually create a set of ground truth segmentations (GTS). The performance of the algorithm is established by determining how far the obtained segmentation is from the GTS.

The measures precision (P), recall (R) and accuracy (A) [22] are widely used to characterize the agreement between the region boundaries of two segmentations because they are sensitive to over- and under-segmentation. These measures are defined using the true positive (TP), true negative (TN), FP and false negative (FN).

$$P = \frac{TP}{TP + FP} ; R = \frac{TP}{TP + FN} ; A = \frac{TP + TN}{TP + TN + FP + FN} \tag{8}$$

The following results have been obtained by comparing the segmentation of more than 450 crypts: $P = 0.952$, $R = 0.944$ and $A = 0.947$. The results confirmed that the proposed method could efficiently segment the epithelial area with a high accuracy. Since this evaluation is influenced by the accuracy of the crypt boundaries drawn by the expert, a more suitable approach to properly quantify the performance of the algorithm is still a challenge.

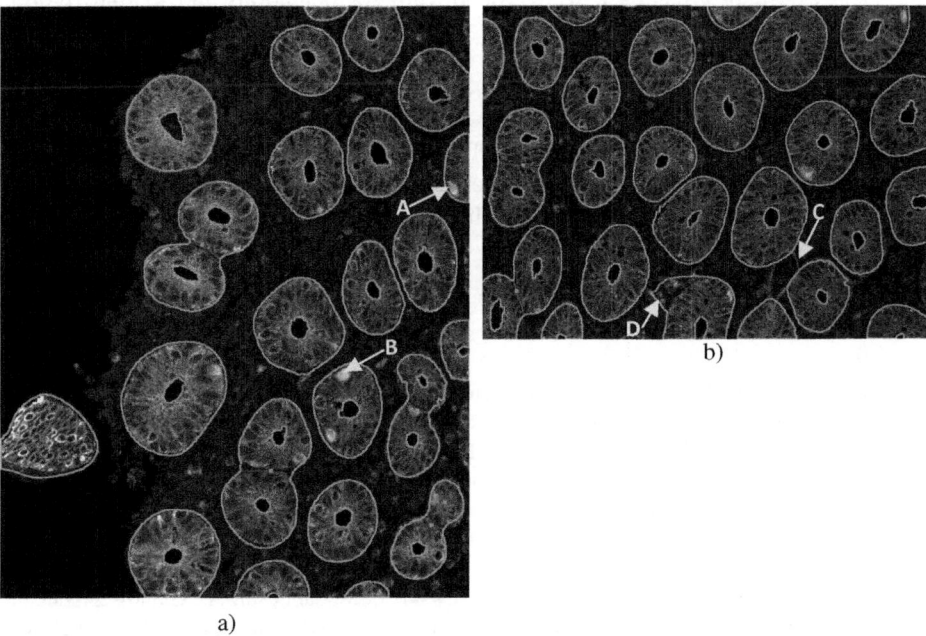

a)   b)

Fig. 6. The crypts and the lumens boundaries are depicted in green; the red lines separate the touching crypts. In a) the bright spots (regions A and B) were correctly included in the crypt region. In b) region C indicates an artefact overlapping the crypts but eliminated due to ADP. Region D is a case FP segmentation.

6 Conclusions

A new automatic technique for robust crypt segmentation based on anisotropic pyramid and stable regions is presented in this paper. The ADP is used to highlight the crypt positions and the MSER algorithm segments the critical level where only the important information persisted. The k-means clustering is used to identify the lumens and the geometrical information is used to split the touching crypts.

The developed segmentation method allows automated assessment of crypts in microscopic images from colorectal tissue sections. Areas of epithelium and stroma are properly identified across a statistically relevant image dataset. The developed technique can be used to assess the areas of epithelium and stroma and their ratios in large tissues sections. The information provided by these measurements allows a better understanding of epithelium-stroma interactions in colon tissue, as well as to confirm the existing qualitative knowledge with new quantitative measurements.

This technique uses a coarser-to-fine approach and can be easily extended on any image with epithelial area from different tissues types (e.g., prostate, breast or lung) but also in any other field in which the objects of interest have the features considered in designing this method. This study will be continued by analysing the topological properties of the graph associated to the tissues components. Considerable effort will be spent to obtain a database with ground-truth segmentations and to find rigorous evaluation criteria of the results.

Acknowledgments. This work was supported in part by the "BRAIN - An Investment in Intelligence" doctoral Scholarship Program within the Technical University of Iasi, Romania, by the Austrian Science Fund under grant no. P20134-N13 and by FFG BRIDGE 818094. The images and necessary biomedical explanations were kindly provided by Dr. Giovanna Bises from the Institute of Pathophysiology and Allergy Research, Medical University of Vienna.

References

1. Wu, H.S., Xu, R., Harpaz, N., Burstein, D., Gil, J.: Segmentation of intestinal gland images with iterative region growing. Journal of Microscopy 220(3), 190–204 (2005)
2. Farjam, R., Soltanian-Zadeh, H., Jafari-Khouzani, K., Zoroofi, R.A.: An image analysis approach for automatic malignancy determination of prostate pathological images. Clinical Cytometry 72B(4), 227–240 (2007)
3. Gunduz-Demir, C., Kandemir, M., Tosun, A.B., Sokmensuer, C.: Automatic segmentation of colon glands using object-graphs. Medical Image Analysis 14(1), 1–12 (2010)
4. Naik, S., Doyle, S., Feldman, M., Tomaszewski, J., Madabhushi, A.: Gland segmentation and computerized Gleason grading of prostate histology by integrating low-, high-level and domain specific information. In: 2nd MICCAI Workshop on Microscopic Image Analysis with Application in Biology, Piscataway, NJ, USA (2007)
5. Chen, C.H., Lee, G.G.: Image Segmentation Using Multiresolution Wavelet Analysis and Expectation-Maximization (EM) Algorithm for Digital Mammography. International Journal of Imaging Systems and Technology 8(5), 491–504 (1997)

6. Roshni, V.S., Raju, G.: Image Segmentation using Multiresolution Texture Gradient and Watershed Algorithm. International Journal of Computer Applications 22(5), 21–28 (2011)
7. Tolba, M.F., Mostafa, M.G., Gharib, T.F., Salem, M.A.: MR-Brain Image Segmentation Using Gaussian Multiresolution Analysis and the EM Algorithm. In: Proc. ICEIS (2), pp. 165–170 (2003)
8. Kimmel, R., Zhang, C., Bronstein, A., Bronstein, M.: Are MSER features really interesting?. IEEE Trans. PAMI (2010) (in press)
9. Donoser, M., Bischof, H.: 3D Segmentation by Maximally Stable Volumes (MSVs). In: 18th International Conference on Pattern Recognition (ICPR 2006), vol. 1, pp. 63–66 (2006)
10. Humphries, A., Wright, N.A.: Colonic Crypt Organization and Tumorigenesis: Human Colonic Crypts. Nature Reviews Cancer 8(6), 415–424 (2008)
11. Moll, R., Divo, M., Langbein, L.: The human keratins: biology and pathology. Histochemistry and Cell Biology 129(6), 705–733 (2008)
12. de Kruijf, E.M., van Nes, J.G.H., van de Velde, C.J.H., Putter, H., Smit Vincent, T.H.B.M., Liefers Gerrit, J., Kuppen, P.J.K., Tollenaar, R.A.E.M., Mesker, W.E.: Tumor-stroma ratio in the primary tumor is a prognostic factor in early breast cancer patients, especially in triple-negative carcinoma patients. Breast Cancer Research and Treatment 125(3), 687–696 (2011)
13. Gonzalez, R.C., Woods, R.E.: Digital Image Processing, 3rd edn. Addison-Wesley, Reading (1992)
14. Otsu, N.: A Threshold Selection Method from Gray-Level Histograms. IEEE Transactions on Systems, Man, and Cybernetics 9(1), 62–66 (1979)
15. Kropatsch, W.G., Haxhimusa, Y., Ion, A.: Multiresolution Image Segmentations in Graph Pyramids. Applied Graph Theory in Computer Vision and Pattern Recognition 52, 3–41 (2007)
16. Smochina, C., Manta, V., Kropatsch, W.G.: Semantic Segmentation of Microscopic Images using a Morphological Hierarchy. In: Real, P., Diaz-Pernil, D., Molina-Abril, H., Berciano, A., Kropatsch, W. (eds.) CAIP 2011, Part I. LNCS, vol. 6854, pp. 102–109. Springer, Heidelberg (2011)
17. Weickert, J.: Anisotropic Diffusion in Image Processing. Teubner, Stuttgart (1998)
18. Scharr, H., Weickert, J.: An anisotropic diffusion algorithm with optimized rotation invariance. In: German Association for Pattern Recognition, Kiel, Germany, pp. 460–467 (September 2000)
19. Acton, S.T., Bovik, A.C., Crawford, M.M.: Anisotropic diffusion pyramids for image segmentation. In: Proceedings of 1st IEEE International Conference on Image Processing (ICIP 1994), vol. 3, pp. 478–482 (1994)
20. Matas, J., Chum, O., Urban, M., Pajdla, T.: Robust wide baseline stereo from maximally stable extremal regions. In: Proc. BMVC, pp. 384–393 (2002)
21. Mikolajczyk, K., Tuytelaars, T., Schmid, C., Zisserman, A., Matas, J., Schaffalitzky, F., Kadir, T., Van Gool, L.: A comparison of affine region detectors. International Journal of Computer Vision 65, 43–72 (2005)
22. Chalana, V., Kim, Y.: A methodology for evaluation of boundary detection algorithms on medical images. IEEE Trans. Med. Imag. 16(5), 642–652 (1997)

Pattern Recognition in High-Content Cytomics Screens for Target Discovery – Case Studies in Endocytosis

Lu Cao[1,*], Kuan Yan[1,*], Leah Winkel[2], Marjo de Graauw[2], and Fons J. Verbeek[1]

[1] Imaging & BioInformatics, Leiden Institute of Advanced Computer Science
[2] Toxicology, Leiden Amsterdam Centre for Drug Research,
Leiden University, Leiden, The Netherlands
{lucao,kyan,fverbeek}@liacs.nl, m.de.graauw@lacdr.leidenuniv.nl

Abstract. Finding patterns in time series of images requires dedicated approaches for the analysis, in the setup of the experiment, the image analysis as well as in the pattern recognition. The large volume of images that are used in the analysis necessitates an automated setup. In this paper, we illustrate the design and implementation of such a system for automated analysis from which phenotype measurements can be extracted for each object in the analysis. Using these measurements, objects are characterized into phenotypic groups through classification while each phenotypic group is analyzed individually. The strategy that is developed for the analysis of time series is illustrated by a case study on EGFR endocytosis. Endocytosis is regarded as a mechanism of attenuating epidermal growth factor receptor (EGFR) signaling and of receptor degradation. Increasingly, evidence becomes available showing that cancer progression is associated with a defect in EGFR endocytosis. Functional genomics technologies combine high-throughput RNA interference with automated fluorescence microscopy imaging and multi-parametric image analysis, thereby enabling detailed insight into complex biological processes, like EGFR endocytosis. The experiments produce over half a million images and analysis is performed by automated procedures. The experimental results show that our analysis setup for high-throughput screens provides scalability and robustness in the temporal analysis of an EGFR endocytosis model.

Keywords: phenotype measurement, image analysis, classification, EGFR endocytosis.

1 Introduction

In this paper we address the problem of deriving a phenotype of a cell in the context of time-lapse cytomics data; in particular we investigate the process of endocytosis and epidermal growth factor receptor (EGFR) signaling.

Enhanced epidermal growth factor receptor (EGFR) signaling triggers breast cancer cells to escape from the primary tumor and spread to the lung, resulting in poor disease prognosis. Moreover, it may result in resistance to anti-cancer therapy. In normal epithelial cells, EGFR signaling is regulated via endocytosis, a process that

---

* Equally contributed.

M. Loog et al. (Eds.): PRIB 2011, LNBI 7036, pp. 330–342, 2011.
© Springer-Verlag Berlin Heidelberg 2011

results in receptor degradation and thereby attenuation of EGFR signaling. However, in cancer cells the endocytosis pathway is often defective, resulting in uncontrolled EGFR signaling. Over the past years, RNA interference combined with fluorescence microcopy-based imaging has become a powerful tool to the better understanding of complex biological processes [18]. Such combined experiment often produces over half a million multi-channel images; manual processing of such data volume is impractical and jeopardizes objective conclusions. Therefore, an automated image and data analysis solution is indispensable. To date, analysis was done with simple extraction of basic phenotypes from EGFR images using tools such as BioApplication[6], ImageXpress[5] and QMPIA[1]. However, these tools are not suitable for a profound study of the dynamics behind EGFR endocytosis which requires more attention. From the existing literature [1, 5, 6, 12, 20, 21, 22] a generic model, defining four major episodes of EGF-induced EGFR endocytosis, can be distilled. Under basic conditions EGFR localizes at the plasma-membrane site, which is in our study defined as "plasma-membrane" (1). Upon binding of EGF to the receptor, EGFR is taken up into small vesicular structures (e.g. early endosomes), which is here defined as "vesicle" (2). Over time EGFR containing vesicles are transported to late endosomes localizing near the nuclear region and form into a larger complex defined here as "cluster" (3). In addition to this EGFR degradation route EGFR can partly also be transported back to the plasma-membrane. Using this dynamic model as the major guideline, the analysis of EGFR-regulation-related gene pathway could be linked to each stage of EGFR endocytosis. Instead of looking at one fixed time point, our current experimental design includes a series of time points at which images are captured. An image potentially contains a ratio in the three characteristic episodes in the EGFR endocytosis process. The image analysis solution should be able to extract basic phenotype measurements as well as to identify the stage of EGFR. In this paper, we illustrate the design and implementation of an automated setup for high-content image and data analysis which can properly capture EGFR dynamics and classify different EGFR phenotypes.

Our workflow for automated analysis solution is depicted in Figure 1. Each high throughput screen (HTS) experiment starts with the design of the experimental scheme, followed by the wet-lab experiment and high throughput microscopy-based imaging. Both experimental schemes and image data are organized and stored in a database. Subsequently, image analysis is used to extract phenotype measurements from these images and classifiers are introduced to recognize each phenotypic stage of EGFR. Finally, comprehensive conclusions are drawn based on comparisons of EGFR expression at each stage and time point.

In this paper, we limit the scope to image analysis and data analysis, some biology will be explained. Accordingly, the organization of this paper is divided into three major sections. In section 2, we introduce the methodology including image acquisition and image analysis; several innovative algorithms will be briefly introduced. After segmentation of the images, EGFR phenotype measurements are obtained. We will illustrate the categorization of phenotypic stages using feature selection and classification. The best combination pair is applied on image data to classify three phenotypic stages and construct a phenotype model. The experimental results are presented in section 3 with two case studies. The first case study tests our solution in identifying dynamic phenotype stages. The second study case examines robustness and scalability of our solution in analyzing a large number of phenotypes.

2   Methodology

Modern techniques in fluorescence microscopy allow visualizing various cell structures so that these can be specifically subject to analysis. Together with a computer-controlled microscope, a high-throughput image acquisition scheme, known as high-throughput screen (HTS), has become feasible. Depending on the biological question at hand, a HTS experiment may produce up to half million. Such a volume of images is beyond the capacity of manual processing and therefore, image processing and machine learning are required to provide an automated analysis solution for HTS experiments. In this section, we will introduce the image acquisition protocol followed by approaches for image analysis and data analysis.

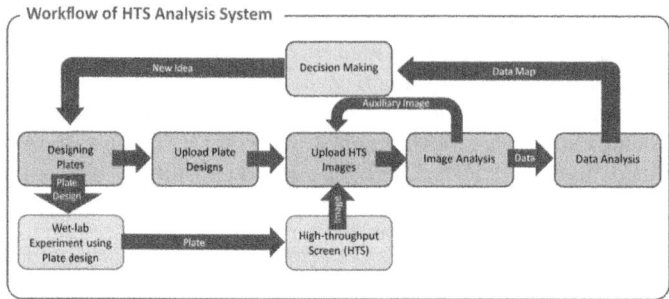

Fig. 1. Workflow of our HTS Analysis System

2.1   Image Acquisition

The workflow for data preparation includes three essential steps: (1) cell culturing, siRNA transfection and EGF exposure, (2) fluorescent staining of proteins of interest and (3) image acquisition. Here we use a design of an EGFR-regulation related siRNA screen to illustrate this workflow. In this design, cells are cultured in 96 well culture plate and transfected using Dharmafect smartpool siRNAs. Subsequently, the transfected cell population was exposed to epidermal growth factor (EGF) for a different duration of time. Cells are fixed at different time points and visualized with a confocal laser microscope (Nikon TE2000). Image acquisition automation was realized with an automated stage and an auto-refocusing lens controller. For each well, images are captured from ten randomly selected locations. For each image three channels are captured: (1) a red channel containing P-ERK expression staining (Cy3), (2) a green channel containing EGFR expression staining (Alexa-488) and (3) a blue channel containing a nuclear staining (Hoechst #33258). Upon completion of the acquisition process all images are uploaded to a database server for image analysis.

2.2   Image Analysis

High-Content Analysis. Basically, the image analysis procedure converts raw microscope images into quantifications characteristic to biological phenomena.

A number of steps are elaborated to achieve this purpose; starting from image acquisition, three steps are distinguished: (1) noise suppression, (2) image segmentation and (3) phenotype measurement. Image segmentation refers to the process of partitioning an image into multiple regions with the goal to simplify and/or change the representation of an image into something that is easier to analyze. For fluorescence microscopy cell imaging we specifically designed a segmentation algorithm: i.e. watershed masked clustering (WMC). The WMC algorithm (cf. Fig. 1d) [23] is an innovative and customized segmentation algorithm that serves different types of cytomics studies like dynamic cell migration analysis [24, 20, 9] and protein signaling modeling [19]. Due to the absence of an indicator for the cell border (cf. §2.1), a border reconstruction and interpolation algorithm is designed to provide artificial representations of the cell borders; i.e. the weighted Voronoi diagram based reconstruction (W-V) algorithm [19]. The W-V algorithm (cf. Fig. 1c) offers the possibility to measure both border-related signal localization [19] and protein expression in terms of continuity and integrity [19]; it does not require a complete cell border or cytoplasmic staining. Both binary mask and artificial cell border are used to derive a number of phenotype measurements for further data analysis.

Phenotype Measurement. In the current experiment and imaging protocol, the phenotype measurements can be categorized into two subgroups: (1) basic measurements of the phenotypes covering shape descriptors and (2) the localization phenotype describing the assessment of the correlation between two information channels. The basic phenotype measurement [2, 11, 24] includes a series of shape parameters listed in Table 2. In addition to the basic phenotype measurement [2, 19, 11, 24], localization measurements can be derived for a specific experimental hypothesis; e.g. the expression ratio between protein channels or shape correlation between objects. The localization phenotypes are quantifications of comparative measurement between information channels such as relative structure-to-nucleus distance or structure-to-border distance [19]. In this paper, we will limit the scope of phenotype measurements to the set employed by the study on EGFR endocytosis. In Table 3 a list of EGFR screen based localization phenotypes is shown. On the basis of the phenotype measurements, objects are classified into phenotypic stages. For the assessment of significance statistical analysis is performed.

2.3   Data Analysis

The aim of the endocytosis study is to quantify the process of EGF-induced EGFR endocytosis in human breast cells and to identify proteins that may regulate this process. The EGFR endocytosis process can roughly be divided into three characteristic episodes: i.e. (1) at the onset EGFR is present at the *plasma-membrane*; (2) subsequently, small *vesicles* containing EGFR will be formed and transported from the plasma-membrane into the cytoplasm; and (3) finally, vesicles are gradually merging near the nuclear region forming larger structures or *clusters*. The characteristic episodes are the read-out for HTS. Based on this model it is believed that EGFR endocytosis regulators may be potential drug targets for EGFR-induced breast cancer. Studying each of the stages (cf. Fig. 3), i.e. plasma-membrane, vesicle and cluster, may provide a deeper understanding of the EGFR endocytosis process.

Table 2. Basic measurements for a phenotype (after segmentation to binary mask)

Feature Name	Description
Size	The size of object, aka as the surface area.
Perimeter	The perimeter of the object
Extension	Derived from 2nd-order invariants of the object [7, 24]
Dispersion	Derived from 2nd-order invariants of the object [7, 24]
Elongation	Derived from 2nd-order invariants of the object [7, 24]
Orientation	Derived from 2nd-order moments of the object [7, 24]
Intensity	Average intensity of all pixels belong to an object
Circularity	Area-to-perimeter ratio; higher compactness suggests a more smooth and less protrusive shape.
Semi-major axis length	Derived from 2nd-order moments of the object [7, 24]
Semi-minor axis length	Derived from 2nd-order moments of the object [7, 24]
Closest object distance	The distance to nearest neighbor of the object, the distance is measured similar to the border distance in Table 3
In nucleus	Boolean describing if the object is included in nucleus mask

Table 3. Localization measurement

Feature Name	Description
Nucleus distance	Distance between structure and nucleus, measured as the average distance between each pixel in an object and the mass center of the corresponding nucleus.
Border distance	Distance between structure and cell membrane, measured as the average distance between each object-pixel and the center of mass of the corresponding cell border (membrane).
Intactness	Overlap between structure expression and cell membrane divided by the total length of cell membrane

Phenotypic Sub Categorization. Here we introduce a profound explanation of the whole procedure employed in the phenotypic sub-categorization including the production of a training set and the procedure for the training of the classifier. The training set is derived from manually delineated outlines of each phenotypic group and subsequent training of a classifier distinguishing three different phenotypes. From two case studies the capability of our solution with respect to identifying characteristic episodes in the process under study stages as well as the scalability in describing different phenotypic groups, is assessed.

Preparation of the Training Set. Ground truth data were obtained by the outlines of the three characteristic episode groups, i.e. cell border/plasma-membrane, vesicle and cluster. These were separately delineated by biologists using our dedicated annotation software (TDR) with a digitizer tablet (WACOM, Cintiq LCD-tablet). From each outline a binary mask is created for each phenotypic stage. In Figure 4(b) the vesicle mask derived from a manually selected vesicle outline is shown. This mask is overlaid with the mask obtained from the WMC algorithm so as to extract the intersection set of two masks as shown in Figure 4(d). Finally, the phenotype measurements are computed with this mask. In similar fashion the ground truth datasets for the plasma-membrane and cluster groups are prepared.

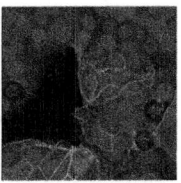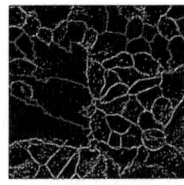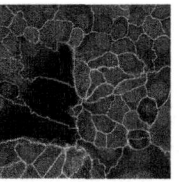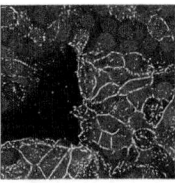

Fig. 2. (a) Original image: PERK (red), EGFR (green) and nucleus (blue), (b) Component definition: artificial cell border (red) and binary mask of protein expression (green), (c) cell border reconstruction : artificial cell border (W-V), (d) image segmentation: binary mask of EGFR channel by WMC

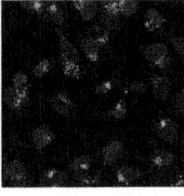

Fig. 3. Sample images of the 3 phenotypic groups with (a) Plasma-membrane, (b) Vesicle, (c) Cluster

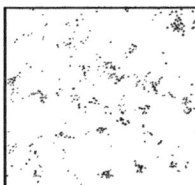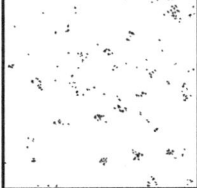

Fig. 4. Ground truth data production. (a) Original image, (b) manual mask, (c) WMC mask, (d) overlay of the mask.

The training dataset includes three characteristic episode groups with 2254 objects and 14 features. Given the huge differences in the feature ranges, it is necessary to normalize the dataset. Normalization is accomplished by shifting the mean of the dataset to the origin and scaling the total of variances of all features to 1. In this way the magnitude effect is successfully removed and the recognition accuracy can be significantly improved [17]. The normalized dataset is used for training of the EGFR classifier.

Feature Selection. First, it is crucial to make a selection of the probabilistic distance criterion for the discriminability estimation. For this we have chosen the Mahalanobis distance [14] since it takes the correlations among the variables into consideration and, in addition, it is scale-invariant. Other distance criteria, such as the Euclidean or Manhattan distance, are, more or less, related to the assumption that all features are independent and have an equal variance. We cannot be certain that all features in our dataset are independent and therefore the Mahalanobis distance is preferred.

Second, we have selected three representative search algorithms including parametric and non-parametric search algorithms; i.e. the branch and bound procedure [10], best individual N features and sequential backward selection [8]. Branch and bound is a top-down procedure, beginning with the set of variables and constructing a tree by deleting variables successively; i.e. an optimum searching procedure requiring the evaluation of partially constructed or approximate solutions without involving exhaustive search. Best individual N features procedure is the simplest suboptimal method for choosing the best N features by individually assigning a discrimination power estimate to each of the features in the original set. In some cases, especially if the features from original set are uncorrelated, this method results in a well-defined feature sets. Sequential backward selection is another suboptimal search algorithm. Variables are deleted one at a time until the required number of measurements remains [4]. The advantage of backward selection is its capability for global control during the feature selection.

Third, we choose three classifiers covering both linear and non-linear categories; i.e. the linear classifier (LDC), the quadratic classifier (QDC) and k-nearest neighbor classifier (KNNC). A linear classifier makes a classification decision based on the value of a linear combination of the characteristics [16]. If the data are strongly non-Gaussian, they can perform quite poorly relative to nonlinear classifiers [3]. A quadratic classifier, which is generalization of the linear classifier; it separates measurements of classes by a quadric surface. Finally, the k-nearest neighbor classifier classifies an object by a majority vote of its neighbors, with the object being assigned to the class most common amongst its k nearest neighbors. The k-nearest neighbor rule achieves a consistent high performance, without a priori assumptions about the distributions from which the training examples are drawn. Moreover, it is robust with respect to noisy training data and still effective if the training dataset is large. By permutation we obtained 9 pairs of combinations. The result of the error estimation is shown in Figure 5. An interesting characteristic can be observed in these plots. The weighted error of the quadratic classifier jumps abruptly when the number of features exceeds a certain threshold (10 for individual feature selection, 12 for branch & bound, and 5 for backward feature selection). This is caused (1) by including a feature with which it is hard to distinguish three phenotypic groups and (2) by the fact that the distribution of the three classes might be more properly classified by the linear and k-nearest neighbor classifier rather than quadratic classifier.

Feature Extraction. Feature extraction is another category to manage multi-dimensional features by reducing dimensionality of features trough combining. For the final result, we also tested the performance of the feature extraction combined with the three classifiers selected. As our starting point is a labeled training dataset, a supervised feature extraction method is most suitable. The Fisher mapping [4] was chosen as extraction method. Fisher mapping finds a mapping of the labeled dataset onto an N-dimensional linear subspace such that it maximizes the between-scatter over the within-scatter. It should be taken into account that the number of dimensions to map is less than the number of classes in the dataset. We have three phenotype classes and consequently the labeled dataset can only be mapped onto a 1D or 2D linear subspace. The result of the performance estimation is shown in Figure 6(a). In addition, in Figure 6(b,c,d), the scatter plots of mapped data with corresponding classifiers are shown.

Comparison of the Results. Each weighted classification error curve (cf. Fig. 5 and 6(a)) represents a combination of a feature selection/extraction method and a classifier algorithm. For each combination, we selected the lowest point value representing the best feature selection/extraction performance of the combination and, subsequently, compared the weighted error and standard deviation of each lowest point. The combination of branch and bound feature selection with k-nearest neighbor classifier has the lowest minimal value and relatively small standard deviation, as can be concluded from Table 4.

Table 4. Minimal value of Mean Weighted Errors and its Standard Deviation

	Individual		B&B		Backward		Fisher	
	min	σ	min	σ	min	σ	min	σ
LDC	0.0586	0.0093	0.0562	0.0098	0.0534	0.0109	0.0555	0.0105
QDC	0.0609	0.0119	0.0626	0.0117	0.0815	0.0113	0.0589	0.0125
KNNC	0.0502	0.0092	0.0450	0.0091	0.0535	0.009	0.0587	0.0124

The three selected features, derived from branch and bound feature selection with the best performance, are *closest object dist*, *object intensity* and *area*. The "*closest object dist*" is a distance measurement between an object and its nearest neighbor. It defines the local numerical density of an object. The cluster and vesicle categories usually have a much lower *closest object dist* since they tend to appear in clusters. The amount of fluorescence therefore directly relates to the amount of EGFR and can be measured as intensity at a certain spot. We suppose that plasma-membrane, vesicle or cluster are all composed of EGFR and the expression of EGFR is more evenly distributed in the plasma-membrane and gradually increases concentration in vesicle and cluster. Intensity represents the amount of EGFR and is significant. Size is undoubtedly the major feature for describing three characteristic episode groups. The results confirm our expectations. We have chosen the combination of branch and bound feature selection with k-nearest neighbor classifier as the best classifier for the case studies.

Statistical Analysis. We provide two case studies in order to sustain the performance of our solution. The first case study is aimed at a better understanding of EGFR endocytosis across time series. The EGFR endocytosis procedure is as follows: in the absence of EGF, EGFR localizes at the cell membrane (e.g. cell border localization). Upon EGF exposure, a portion of the plasma membrane containing EGFR is invaginated and pinched off forming a membrane-bounded vesicle. Some vesicles will be accumulated in clusters in the peri-nuclear region. As for the experimental design, the cells in separate wells are treated with EGF for a variable amount of time. In this way each well represents a fixed time point. After fixation, cells are stained and visualized. The images that have a clear representation of phenotype stage are carefully selected by a specialist. The result of image and data analysis based on selected images provides a notable capability of our solution on identifying the dynamics in the characteristic episodes. The source images include a total of 13 time points with 2 pairs of images each.

The second case study is on identification of mediators of EGFR endocytosis. The results demonstrate that our automated high-content analysis solution can properly describe different phenotypic groups and is capable to manage large quantities of phenotypes. Per culture plate ten images are acquired per well; i.e. 96×10 images are used in the image and data analysis. In order to evaluate the phenotype difference between wells, we calculate the number of each phenotypic group (vesicle, plasma-membrane, and cluster) per nucleus in each well. The plasma-membrane, representing the composed EGFR evenly distributed on the cell membrane, is always continuously linked between cells. The quantification is accomplished by calculating the pixels of plasma-membrane per nucleus.

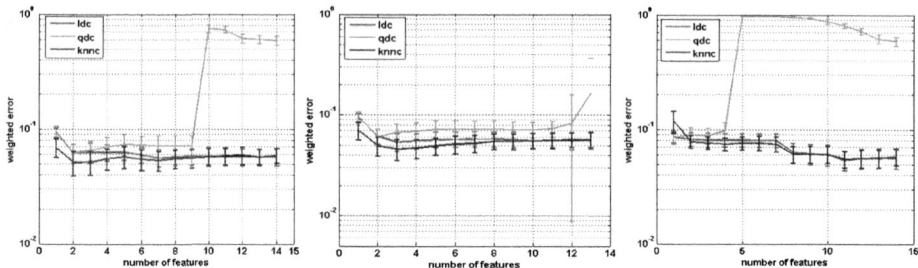

Fig. 5. Weighted classification error curves, with (a) Individual feature selection, (b) Branch and bound feature selection and (c) Backward feature selection

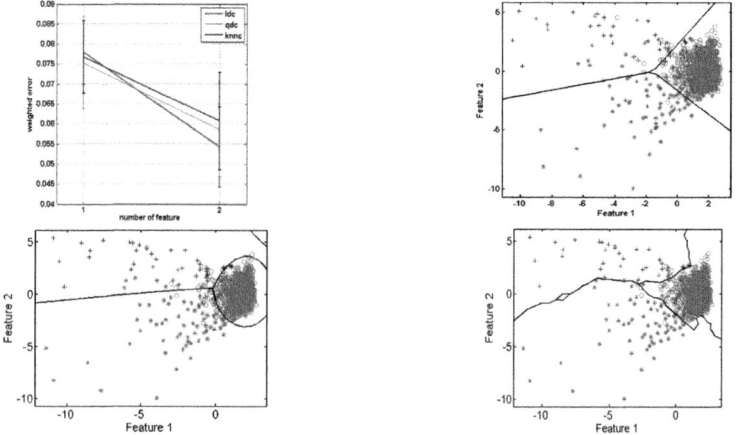

Fig. 6. Results of feature extraction: (a) Weighted classification error curve of Fisher feature extraction, (b) Fisher feature extraction with Linear Discriminant Classifier, (c) Fisher feature extraction with Quadratic Discriminant Classifier, (d) Fisher feature extraction with K-Nearest Neighbor Classifier

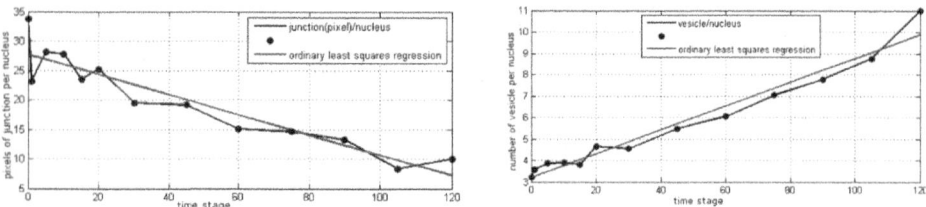

Fig. 7. Average number of plasma-membrane (a) and vesicle (b) per nucleus

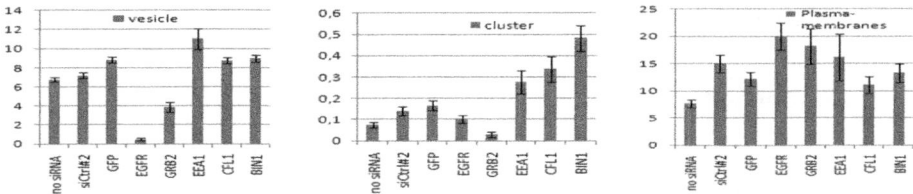

Fig. 8. (a) Number of vesicles per nucleus, (b) Number of clusters per nucleus (c) Plasma-membranes (pixel) per nucleus

An analysis with both the Jarque-Bera [9] and Lillie test [13] established that over 80% of our measurement data of the composition vesicle/plasma-membrane/cluster is not normally distributed. We, therefore, use the Kolmogorov-Smrinov test [15] with siCtrl#2 as control sample to identify significant changes in EGFR endocytosis.

3   Experimental Results

3.1   Dynamic Phenotype Stage

The results shown in Figure 7a illustrate that the amount of EGFR localized at the plasma-membrane (e.g. number of plasma-membranes, expressed as pixel/nucleus) decreases over time. This fits with the EGFR endocytosis process during which EGF exposure causes a gradual EGFR re-distribution from the plasma-membrane into vesicles. Meanwhile, the number of vesicles per nucleus increases caused by the formation vesicles as illustrated in Figure 7b. These graphs indicate the trend of the endocytosis process and are representative to illustrate phenotype stage dynamics.

3.2   Phenotype Classification

We validated our automated high throughput image analysis using siRNA-mediated knock-down of several known EGFR endocytosis regulators (e.g. siGrb2, siEEA1, siCFL) To this end images were selected from WT cells (not treated with siRNA), control siRNA treated cells (siCtrl#2 and siGFP), siEGFR treated cells and three target siRNAs. In Figure 8a-c the comparison of selected results with three phenotypic groups is shown. In Figure 9 some sample images are depicted to check the correctness of our solution for phenotype description. Our analysis shows that cells treated with siCtrl#2 resemble non-treated WT cells, while siGFP differs significantly; indicating that siCtrl#2 is the best control for further analysis. As expected, siEGFR showed decreased levels of all three classifiers since treatment of cells with siEGFR results in > 90% knock-down of EGFR. In addition, siGrb2, siEEA1 and siCFL behaved as expected. These results demonstrate that the automated high throughput analysis could be used for large scale siRNA screening.

A comprehensive overview of the results of a complete experiment is shown in the heatmaps depicted in Figure 10. The data are derived from a siRNA screen of more than 200 potential regulators of EGFR endocytosis. The y-axis represents different siRNA targets (regulators) and the x-axis represents the features plus the number of different phenotypic groups.

4   Conclusions

This paper provides an efficient solution to analyze the high-throughput image data sets on the level of protein location. The experimental results of both case studies show that our automated analysis procedure can be involved in the identification of the characteristic episodes in the EGFR process and provides a set of robust and precise phenotypic descriptions. From the case studies it is illustrated that our solution is suitable for a robust analysis of different phenotypes in a siRNA based HTS. Furthermore, the whole process, from image segmentation, phenotypic quantification to classification, is part of a successfully automated procedure. Our solution can be easily extended to cope with studies utilizing fluorescence microscopy.

Fig. 9. Sample images, (a) no siRNA no EGF, (b) EGFR, (c) GRB2, (d) BIN1 (> response)

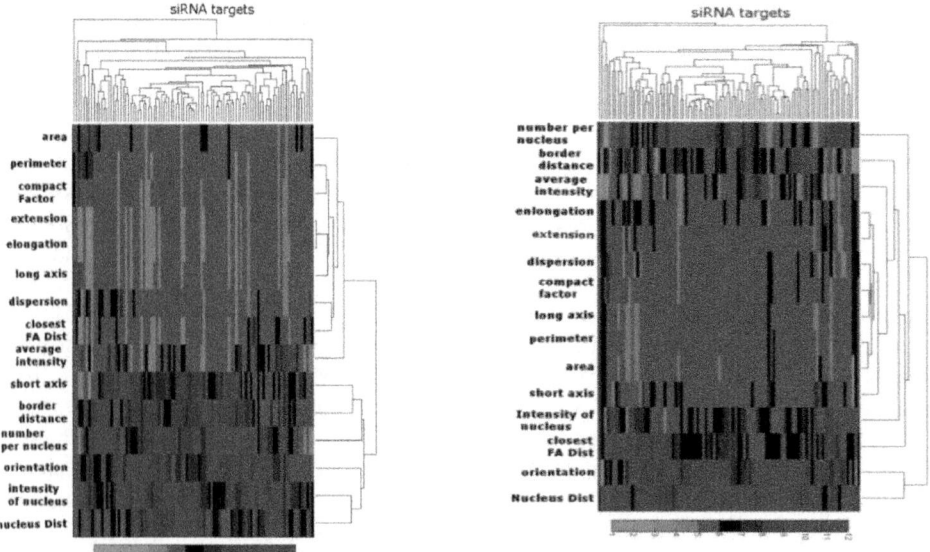

Fig. 10. (a) Vesicle p-value heat map (b) Plasma-membrane p-value heat map

References

1. Collinet, C., Stöter, M., Bradshaw, C.R., Samusik, N., Rink, J.C., Kenski, D., et al.: Systems survey of endocytosis by multiparametric image analysis. Nature 464, 24–249 (2010)
2. Damiano, L., Le Dévédec, S., Di Stefano, P., Repetto, D., Lalai, R., Truong, H., Xiong, J.L., Danen, E.H., Yan, K., Verbeek, F.J., Attanasio, F., Buccione, R., van de Water, B., Defilippi, P.: p140Cap suppresses the invasive properties of highly metastatic MTLn3-EGFR cells via paired cortactin phosphorylation. Oncogene 30 (2011)
3. Devroye L, Lugosi G.: A probabilistic theory of pattern recognition (1999)
4. Fukunaga, K.: Introduction to Statistical Pattern Recognition, 2nd edn. Academic Press (1990)
5. Galvez, T., Teruel, M.N., Heo, W.D., Jones, J.T., Kim, M.L., Liou, J., et al.: siRNA screen of the human signaling proteome identifies the PtdIns(3,4,5)P3-mTOR signaling pathway as a primary regulator of transferrin uptake. Genome Biology 8(7), 142 (2007)
6. Ghosh, R.N., DeBiasio, R., Hudson, C.C., Ramer, E.R., Cowan, C.L., Oakley, R.: Quantitative cell-based high-content screening for vasopressin receptor agonists using transfluor technology. Journal of biomolecular screening: the official journal of the Society for Biomolecular Screening 10(5), 47–84 (2005)
7. Hu, M.K.: Visual pattern recognition by moment invariants. Information Theory 8(2), 179–187 (1962)
8. Jain, A.K., Duin, P.W.: Statistical pattern recognition: a review. IEEE Transactions on Pattern Analysis and Machine Intelligence 22(1), 4–37 (2000)
9. Jarque, C.M., Bera, A.K.: Efficient tests for normality, homoscedasticity and serial independence of regression residuals. Economics Letters 6(3), 255–259 (1980)
10. Land, A.H., Doig, A.G.: An Automatic Method of Solving Discrete Programming Problems. Econometrica 28(3), 497–520 (1960)
11. Le Dévédec, S., Yan, K., de Bont, H., Ghotra, V., Truong, H., Danen, E., Verbeek, F.J., van de Water, B.: Systems microscopy approaches to understand cancer cell migration and metastasis. Science 67(19), 3219–3240 (2010)
12. Li, H., Ung, C.Y., Ma, X.H., Li, B.W., Low, B.C., Cao, Z.W., et al.: Simulation of crosstalk between small GTPase RhoA and EGFR-ERK signaling pathway via MEKK1. Bioinformatics (Oxford, England) 25(3), 358–364 (2009)
13. Lilliefors, H.W.: On the Kolmogorov-Smirnov test for the exponential distribution with mean unknown. Journal of the American Statistical Association 64, 38–389 (1969)
14. Mahalanobis, P.C.: On the generalised distance in statistics. Proceedings of the National Institute of Sciences of India 2, 49–55 (1936)
15. Massey, F.J.: The Kolmogorov-Smirnov Test for Goodness of Fit. Journal of the American Statistical Association 46(253), 68–78 (1951)
16. Mitchell, T.: Generative and Discriminative Classifiers: Naive Bayes and Logistic Regression (2005)
17. Okun, O.: Feature Normalization and Selection for Protein Fold Recognition. In: Proc. of the 11th Finnish Artificial Intelligence Conference, pp. 207–221 (2004)
18. Pelkmans, L., Fava, E., Grabner, H., Hannus, M., Habermann, B., Krausz, E., et al.: Genome-wide analysis of human kinases in clathrin- and caveolae/raft-mediated endocytosis. Nature 436(7047), 78–86 (2005)
19. Qin, Y., Stokman, G., Yan, K., Ramaiahgari, S., Verbeek, F.J., de Graauw, M., van de Water, B., Price, L.: Cyclic AMP signalling protects proximal tubular epithelial cells from cisplatin-induced apoptosis via activation of Epac. British Journal of Pharmacology (in Press, 2011)

20. Roepstorff K, Grøvdal L, Grandal M, Lerdrup M, Deurs B van.: Endocytic downregulation of ErbB receptors: mechanisms and relevance in cancer. Histochemistry and cell biology. 129(5):563–78(2008)
21. Tarcic, G., Boguslavsky, S.K., Wakim, J., Kiuchi, T., Liu, A., Reinitz, F., et al.: An unbiased screen identifies DEP-1 tumor suppressor as a phosphatase controlling EGFR endocytosis. Current Biology 19(21), 88–98 (2009)
22. Ung, C.Y., Li, H., Ma, X.H., Jia, J., Li, B.W., Low, B.C., et al.: Simulation of the regulation of EGFR endocytosis and EGFR-ERK signaling by endophilin-mediated RhoA-EGFR crosstalk. FEBS Letters 582(15), 2283–2290 (2008)
23. Yan K, Verbeek FJ.: Watershed Masked Clustering Segmentation in High-throughput Image Analysis (submitted, 2011)
24. Yan, K., Le Dévédec, S., van de Water, B., Verbeek, F.J.: Cell Tracking and Data Analysis of in Vitro Tumour Cells from Time-lapse Image Sequences. In: VISSAPP, vol. 1, pp. 281–286 (2009)

Author Index

GPSR Compliance

The European Union's (EU) General Product Safety Regulation (GPSR) is a set of rules that requires consumer products to be safe and our obligations to ensure this.

If you have any concerns about our products, you can contact us on ProductSafety@springernature.com

In case Publisher is established outside the EU, the EU authorized representative is:

Springer Nature Customer Service Center GmbH
Europaplatz 3
69115 Heidelberg, Germany

Batch number: 09478804

Printed by Printforce, the Netherlands